AN INTRODUCTION TO
BIOLOGICAL
EVOLUTION

Kenneth V. Kardong
Washington State University

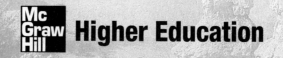
McGraw Hill **Higher Education**

Boston   Burr Ridge, IL   Dubuque, IA   Madison, WI   New York   San Francisco   St. Louis
Bangkok   Bogotá   Caracas   Kuala Lumpur   Lisbon   London   Madrid   Mexico City
Milan   Montreal   New Delhi   Santiago   Seoul   Singapore   Sydney   Taipei   Toronto

Higher Education

AN INTRODUCTION TO BIOLOGICAL EVOLUTION

Published by McGraw-Hill, a business unit of The McGraw-Hill Companies, Inc., 1221 Avenue of the Americas, New York, NY 10020.
Copyright © 2005 by The McGraw-Hill Companies, Inc. All rights reserved. No part of this publication may be reproduced or distributed in any form or by any means, or stored in a database or retrieval system, without the prior written consent of The McGraw-Hill Companies, Inc., including, but not limited to, in any network or other electronic storage or transmission, or broadcast for distance learning.

Some ancillaries, including electronic and print components, may not be available to customers outside the United States.

✪ This book is printed on recycled, acid-free paper containing 10% postconsumer waste.

1 2 3 4 5 6 7 8 9 0 QPD/QPD 0 9 8 7 6 5 4

ISBN 0-07-238579-0

Publisher: *Margaret J. Kemp*
Senior developmental editor: *Deborah Allen*
Executive marketing manager: *Lisa L. Gottschalk*
Lead project manager: *Joyce M. Berendes*
Lead production supervisor: *Sandy Ludovissy*
Lead media project manager: *Judi David*
Senior media technology producer: *Jeffry Schmitt*
Senior designer: *David W. Hash*
Cover designer: *Rokusek Design*
Cover images: Charles Darwin (1809-82), ©*Getty Images;* Tyrannosaurus, ©*Corbis Vol. 5;* Zebra group, ©*Corbis Vol. 86*
Senior photo research coordinator: *Lori Hancock*
Photo research: *Pam Carley/Sound Reach*
Compositor: *Electronic Publishing Services Inc., NYC*
Typeface: *10/12 Times Roman*
Printer: *Quebecor World Dubuque, IA*

The credits section for this book begins on page 315 and is considered an extension of the copyright page.

Library of Congress Cataloging-in-Publication Data

Kardong, Kenneth V.
    An introduction to biological evolution / Kenneth V. Kardong. — 1st ed.
        p. cm.
    Includes index.
    ISBN 0-07-238579-0
    1. Evolution (Biology)—Textbooks. I. Title.

QH366.2.K355   2005
576.8—dc22   2004040211
                                              CIP

www.mhhe.com

dedication

To Jason and Debbie,
to Kyle and Darcie,
and to their good friends and company

brief contents

contents

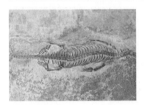

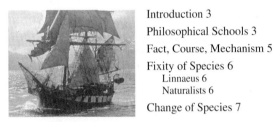

Chapter 10 Co-Evolution 154

Chapter 11 Life History Strategies 172

Chapter 12 Life in Groups 182

Chapter 13 Extinctions 202

Chapter 14 Human Evolution: The Early Years 228

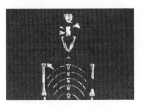

Chapter 15 Human Evolution: Building Modern Humans 248

Chapter 16 Evolutionary Biology: Today and Beyond 270

preface

I wrote this book for a general college audience, as an introduction to the principles and significance of Darwinian evolution. As the book unfolds, the student will meet intellectual challenges; visit the richness of evolutionary science today; explore evolution's insights in biodiversity, mass extinctions, and modern diseases; and examine his or her current and precarious place in nature. This textbook is intended to be the centerpiece of a course on evolution or an evolution-based course, around which instructors can build a course that suits their interests and that focuses on the students. To this end, this textbook departs in three fundamental ways from most other textbooks on evolution.

GENERAL COLLEGE AUDIENCE

First, this book is intended for a general college audience, especially for students early in their studies. Over the last two decades, I have taught such a course for biology and for nonbiology majors. It is surprising to me that even many biology majors, with a substantial number of life science courses behind them, are still inept at discussing or using basic evolutionary concepts. Their words are right, but they lack a true understanding of evolutionary principles. With some small effort, these students can spout dates, names, and jargon, and even some genetic laws they have memorized. But their basic grasp of scientific reasoning and ability to think critically lag substantially.

States are evaluating their K–12 school curricula, and times are changing. I hope for the better. But for the moment, students still largely learn about evolution from the pulpit and from pop culture. The misunderstandings they bring to the subject of evolution are the same fundamental misunderstandings they harbor about science in general. We can parade the facts before them, but we must also develop the intellectual equipment students use to think critically about a subject. For many students, the process of evolution is still mysterious, even threatening. Most students who enter college have heard at least of evolution's offense to religious beliefs, but not of its service in unifying all of the modern life sciences. This book is intended for just such an audience.

INTELLECTUAL SIGNIFICANCE

Second, this book departs from others in that it examines the intellectual significance of Darwinian evolution. Darwin's ideas on biological evolution touch every aspect of modern life, from genetic engineering to human medicine. This book explores the history of Darwin's thought; the evidence for his ideas, and the implications these have for the appearance and loss of life on Earth; and the significance to our own future as a species.

To begin, students need to be unburdened of the major misconceptions about evolution. At least in my course, they all want to be Lamarckians. Most students think that need drives change, progressively on to better and better organisms. Certainly, Lamarck's ideas initially seem sensible. Bolstered by our society's technological bias, many students enter a university under the influence of Western culture's boast of ever-evolving and improving innovations to meet our ever-escalating needs. Students expect biological evolution to be driven by the same engines of progress. Nothing, of course, could be farther from the truth. Biological evolution is not driven by needs. Therefore, a modern student arrives on the doorstep of such a course already harboring mistaken expectations that must be addressed head-on before any of this can be understood. Understanding what Darwin had to say requires much more than just examining the formal evidence for evolution. In a sense, each student must make an intellectual journey, and step outside the pervasive grip of the industrial and technological culture of our day.

ORGANISMS, SURVIVAL, FITNESS

Third, I have also chosen to depart from the standard treatment of evolution in other textbooks on evolution, wherein the arguments are reductionist, molecular, and overwhelmingly genetic in flavor. Certainly genetics plays a large and indispensable role in evolutionary thought, and genetics is covered here within these pages as well. But the much-deserved pride geneticists take in the success of their field has led to a hubris, tempting some to believe that evolution is little more than a genetic shell game. As one such scientist rhapsodized, "it is only in genetics that there can be said to be a 'theoretical biology.'" Such claims, taken alone, are much too narrow to provide a full understanding of Darwinian evolution. This is especially true now that the successes of evolutionary morphologists and ecologists have taken the science from laboratories into the wild, and documented the workings of natural selection in nature. Organisms in populations are the center of the evolutionary action, so it is here, on organisms, that I center this book.

A BOOK TO BE READ

I have tried hard to write an inviting book that will be read—by students. The World Wide Web offers wonderful resources that can be accessed through powerful search engines. This is a comfortable medium for today's students, who have grown up mostly in a virtual world of information. But printed books remain important and even unique vehicles of information and ideas—they are convenient; they are reviewed and edited; they are organized; they can be thought provoking; they are coherent with a common voice. With a textbook in hand, the instructor of a course knows what information and what slant on the subject will be presented to students. The instructor can then build the course she or he wants, expanding or amending or even constructively disagreeing. But to serve this central and supportive role, the textbook must be read.

Many current textbooks are excellent encyclopedias of information. But too many are overbearing and stately—often mannered in presentation. They go unread by students. And that is a shame. I therefore experiment here with a writing style that I hope will engage students and encourage them to read, to pause and consider, then to read further.

ORGANIZATION AND RATIONALE

I do not pit evolution against fundamentalism, red in tooth and claw. Instead, the book is focused on the science of Darwinian evolution. I open by placing Darwin's ideas in historical context, but for the purpose of identifying particular scientific obstacles he faced—time and heredity. The remainder of the book brings evolution into a modern context, explores its biological implications, and concludes with a timely consideration of the significance of evolution to human society today. Topics and chapters can be reordered to suit particular courses. Reasonably, everyone using this book will follow his or her own strategy, and the book supports this. Let me share my thinking on one approach.

CHAPTERS AND TOPICS

Chapter 1 (Evolution of Evolution)
Chapter 2 (Time)
Chapter 3 (Heredity)

Darwin's great achievement was not in proposing that organisms evolve. In fact, evolution was discussed by ancient Greek philosophers before the time of Christ, and thus was old news in Darwin's day. Most of the evidence was available to nineteenth-century contemporaries, but Darwin was the first to see around the distractions, misinformation, and scientific confusion and put it all together. His insight was profound, even disturbing, and is still the subject of serious debate, and is still being analyzed for its implications to ourselves, to our politics, and to our basic ethical systems. To understand this larger significance of evolutionary principles, most students must make much the same intellectual journey as did Darwin. What did Darwin say? And what were the scientific objections of his contemporaries? With all the contentious religious objections receiving attention, then and now, we tend to forget that Darwin faced serious scientific challenges to his views. As a unit, these first three chapters set up the issues, and then show how each is resolved today.

Chapter 4 (Emergence of Life)
Chapter 5 (Diversity of Life)

What can be done about students' background in biology? At best, it is uneven. In some states, all students now have at least a brush with biology on their way through high school. And we should credit television with bringing nature programs to a general public. These chapters cannot, of course, substitute for a proper background course in biology. But I have tried to pack into these two chapters enough to refresh the memories of students with prior courses in biology. For those without such backgrounds, I have tried to be sure that all biological principles they meet elsewhere in the book are at least mentioned here, in these chapters, in a cohesive context. Chapter 4 notes the major, early transitions of life, and introduces students to the prokaryotes; here the step to organic evolution occurs, heterotroph to autotroph, the appearance of a eukaryotic cell, and the basic coding structure of life itself is described—DNA, RNA. Chapter 5 places before us the further cast of characters in the evolutionary story, the eukaryotes—protists, plants, fungi, and animals.

Chapter 6 (Evidence of Evolution)

Now we are ready to critically examine the formal, scientific evidence favoring a natural explanation for evolutionary change, piece by piece. This chapter especially offers the chance to engage students in critical thinking. Formally, the chapter provides the student a chance to join in hypothesis testing, which is somewhat of the form "If . . . , then . . . , thus." A hypothesis is stated, its implications or predictions are identified, and then these are tested. If the predictions are accurate, then the hypothesis is supported, and thus a tentative conclusion is drawn.

Chapter 7 (Selection)
Chapter 8 (Variation: Spice of Life)
Chapter 9 (Speciation)
Chapter 10 (Co-Evolution)
Chapter 11 (Life History Strategies)
Chapter 12 (Life in Groups)

If their curiosity is tweaked, students will want to know more about how the process of evolution works. Chapter 7 explores the workings of natural selection. Evolution runs on variation. The sources of this variation—genetic shuffling and mutations—receive attention in chapter 8. Chapter 9 examines how new species arise, and chapter 10 reflects on how species interact in co-evolutionary relationships. We are reminded in chapter 11 that life histories, not just adults, evolve. Levels of selection and life in groups raise a controversial issue in chapter 12—individual versus group benefits. Here we also meet issues of microevolution and macroevolution, because they can help us understand how evolution sometimes advances in macro steps to produce rapid changes.

Chapter 13 (Extinctions)

In chapter 13 we explore the other side of evolution—extinctions wherein old species become extinct in a boom (mass extinctions) or a whimper (background extinctions). Here the story of "hot-" or "cold-blooded" dinosaurs is also discussed.

Chapter 14 (Human Evolution: The Early Years)
Chapter 15 (Human Evolution: Building Modern Humans)

By this point in the book, we have examined the workings of natural selection and the way in which complex associations arise. We are ready for our own evolution, discussed in these two chapters.

Chapter 16 (Evolutionary Biology: Today and Beyond)

Evolution happens. It happens today. It is an ongoing process that cannot be ignored, especially as humans take over much of the responsibilities for building a better genetic and social future. This concluding chapter considers evolution's implications in modern society, including its implications for molecular genetics with its possibilities and pitfalls, and how peoples and plagues continue the evolutionary race.

FEATURES

- **Study aids include:**

    Key terms set in bold where they first appear, and defined in the glossary

    End-of-chapter overviews that summarize and discuss the chapter's most significant points

    End-of-chapter selected references

    End-of-chapter Web links

    "Consider This" boxed essays that draw attention to the personal or scientific significance of one of the chapter's topics

    Glossary

- **Appendixes:**

    Primer on Cell Division

    Techniques of Taxonomy

    Molecular Clocks

- **Website:**

    Color images, matched to each chapter, further enhance the book's contextual discussion

    Interactive labs

    Case studies

    Guide to writing lab reports and scientific papers

    Practice quizzing

    Guide to electronic research

ACKNOWLEDGMENTS

From Charles Darwin to the present, evolutionary biology has grown into a major scientific discipline that unites all of the life sciences and holds a view of life that touches us all. Bringing this to a wider audience is a passion I shared with a team of publishing professionals and special reviewers, without whom this first edition would not have been feasible. I am indebted to the core publishing group who believed in this project and contributed their talent and time. I am especially grateful to several in particular: As always, it was a delight to work with Sue Dillon, who brought her special copyediting talents and sympathetic support to early versions of a scruffy manuscript. At McGraw-Hill, Marge Kemp (publisher) from the beginning saw merit and possibilities in this unique book, and helped nurture it through all stages of production. It is hard to overstate the remarkable good judgment and dedicated effort Kathy Loewenberg (developmental editor) brought to this project; she could be "red in tooth and claw" or encouraging, depending on what was needed to keep our team on track. Additionally, my thanks go to many others who helped at various stages, including Joyce Berendes (project manager), Francine Banwarth (proofreader), Wendy Nelson (copy editor), David Hash (design manager), John Rokusek (cover design), Sandy Ludovissy (production supervisor), Judi David (media producer), and Lori Hancock/Pam Carley (photo research). I thank as well the McGraw-Hill field staff who link the summary effort of all who helped in this book to faculty and students who use it. In turn, these field reps return your comments of what you do and do not like, and thereby aid in the improvement of this textbook, making it a shared work in progress. My thanks also to Donna Nemmers for her dependable good humor.

I am most grateful to the wonderful team of patient and thoughtful reviewers who offered considered comments on earlier drafts of this text. Besides being scientists in their own right, they are exceptional instructors dedicated to improving science education in general, and the teaching of evolution in particular. In large and small ways, their insightful comments and helpful suggestions have been incorporated into this text. I hope that they find the finished product much improved, and I sincerely appreciate their efforts to bring a better book into the reader's hands. Throughout, Tamara L. Smith was colleague, friend, and critical advisor, and for all of this I am always grateful.

Finally, I thank my colleagues, students, and family for their enduring encouragement throughout the long process that brought this book into its first edition.

REVIEWERS

Christine A. Andrews
University of Chicago
Robin M. Andrews
Virginia Polytechnic Institute and State University
Randall Breitwisch
University of Dayton
Sherryl Broverman
Duke University
Paul J. Bybee
Utah Valley State College
Arthur H. Harris
University of Texas, El Paso
Anne E. Houde
Lake Forest College
Scott L. Kight
Montclair State University
Charles J. Kunnert
Concordia University
Brian Morton
Barnard College
Richard S. Peigler
University of the Incarnate Word
Christopher G. Peterson
Loyola University
Cathy Schaeff
American University

AN INTRODUCTION TO
BIOLOGICAL
EVOLUTION

1

"It is interesting to contemplate a tangled bank, clothed with many plants
of many kinds, with birds singing on the bushes, with various insects flitting
about, and with worms crawling through the damp earth, and to reflect that
these elaborately constructed forms, so different from each other,
and dependent on each other in so complex a manner,
have all been produced by laws acting around us."

Charles Darwin, *The Origin of Species*

Evolution of Evolution

INTRODUCTION

The name Charles Darwin (figure 1.1) and the concept of evolution are almost synonymous. Yet most persons are surprised to learn that Charles Darwin was not the first to propose that populations of organisms change through time—that is, they evolve. In fact, the idea that organisms evolve is very old, dating back to respected schools of ancient Greek philosophical thought in the sixth century before Christ.

PHILOSOPHICAL SCHOOLS

Anaximander (610–547 B.C.E.) developed ideas about the course of change from fishlike and scaly forms to land forms. Empedocles (490–430 B.C.E.) saw original creatures coming together in oddly assembled ways: men with heads of cattle, animals with branches like trees. Most perished, he thought—but a few, those that came together in practical ways, survived. That was more than 2,000 years ago.

Although more poetic than scientific, such philosophical ideas had been knocking around for a long time before Darwin, so when he came forth with his own ideas on evolution, they should have shocked no one. Yet in 1859, when Darwin published his ideas, they caused instant controversy—and still do.

Why should ideas over a century old still excite controversy? The usual reason given is that evolutionary facts contradict religious beliefs, and this feeds the ongoing social debate. Yet many scientific facts contradict religious beliefs without raising eyebrows. The Bible says that Jonah was swallowed by a "great fish" and survived to tell of it, but physiologists point out that this is at odds with the scientific facts of basic digestive processes. Jonah's life was witness to temptations of the flesh and, being flesh, he would have been digested like any other meal—three days and three nights would have been plenty of time to turn Jonah from prophet to protein. Or consider Joshua in the heat of battle against the Canaanites. The Bible describes Joshua commanding sun and moon to stand still, lengthening the day, and giving the Israelites light to complete their victory. Yet such a biblical claim contradicts basic physical laws of planetary motion. If the Earth came suddenly to a wrenching halt, bringing the sun to a stand-

3

FIGURE 1.1   Charles Darwin (1809–1882)
Darwin at about 30 years old, and three years back from his voyage aboard HMS *Beagle*. Although *The Origin of Species* was still just a few notebooks in length and several decades away from publication, Darwin had several accomplishments behind him, including his account *The Voyage of the* Beagle, a collection of scientific observations. At this time, he was also engaged to his cousin, Emma Wedgwood, with whom he would live a happy married life.

still overhead, our planet's continents would be thrown about like unstrapped, back-seat passengers when brakes are suddenly applied in a speeding car. These contradictions of fact between science and religion raise no significant social controversies, and we have no "Society for the Stationary Sun Research" to redress the supposed insult to religious faith. Differences over facts do not explain the heated debates provoked by evolution.

What did Darwin say that was so disturbing? At face value, his proposal was very simple. He proposed the *conditions* for—and the *mechanism* of—evolutionary change. Darwin proposed three conditions: First, members of any species, if left unchecked, increase naturally in number. All species possess a high reproductive potential. Even slow-breeding elephants, Darwin pointed out, could increase from a pair to millions in a few hundred years. But we are not up to our rooftops in elephants because as numbers increase, resources are consumed at an accelerating rate and become scarce. This brings about the second condition—competition among contending organisms for declining resources. This leads to condition three, the survival of the few. In

this competition for increasingly scarce resources, Darwin termed the mechanism that determines which individuals survive and which do not **natural selection** —nature's way of weeding out the less fit. Those with superior adaptations would, on average, fare better in this competition and survive, passing on their successful adaptations to their offspring (table 1.1). Evolutionary modification resulted, then, from the preservation—by natural selection—of individuals bearing favorable characteristics. Darwin envisioned that new species derive from common ancestors by modifications favored in the natural process of selection.

This vision seems tame indeed, hardly the stuff of scientific revolutions or explosive social controversy. But it is a profound insight. What the Darwinian revolution brought was a new way of looking at the biological part of our world. Darwin saw species—including humans—as descending from common ancestors by a process as natural as the forces that guide celestial motion. Many religious persons viewed humans as the special products of Creation. The Darwinian view involved no divine act or even divine purpose—humans evolved as parts of the animal kingdom, physical products of natural processes like any other organism populating the planet. Many social philosophers believed in progress and perfection in nature. The Darwinian view replaced these beliefs with concepts of chance and adaptation.

As we look out into a thick forest of trees and flowering plants and the animals harbored there, the complexity is not chaotic, but understandable. Just as the motion of planets abides by the natural laws of physics, the evolution of species abides by the action

Table 1.1	Darwin's Basic Evolutionary Theory

Conditions

1. Intrinsic increase in the number of individuals within a species.
2. Competition for limited resources
3. Survival of the few

Mechanism: Natural Selection

Those individuals with more favorable features would, on average, fare better than competitors and survive, passing on to their offspring those advantageous characteristics.

of laws of natural selection. Evolution is controversial because it opens a whole new perspective on ourselves, on our culture, and on our future. It is an alternative explanation of the biological world and an alternative view of why we arose as part of that world. That is why many fundamentalist preachers even today fume over the mention of Darwin's name, but they are not alone. Marxists and other political activists—anyone with a dogma to peddle—are likely to take offense at the Darwinian revolution. Today, even in many grade schools, laws of planetary motion are taught to children but Darwin is still too hot to handle.

The Greek philosopher Heraclitus (fifth century B.C.E.) asked, "If we dip our foot into a river and then do so again a moment later, do we dip the second time into the same or a different river?" Water flows by, the river changes. Or does it? Change in the Darwinian world is more profound than this—it holds a significance beyond the quaint or esoteric. It is also quite disturbing to comfortable beliefs because it challenges all that we generally hold about purpose in life, the significance of what we do, and the provincial perspective we cherish about ourselves. If we think deeply about, understand, and follow the significance of the Darwinian revolution, we are changed forever. Regardless of whether it is loathed or loved, the Darwinian view is the foundation of all modern life sciences from DNA to human diseases.

Certainly, evolution happens in other realms besides biology, without raising controversy. For instance, language evolves. A few decades ago, if the exhilaration of a warm summer day and its freedom led you to tell a friend that you felt quite "gay," you would be suggesting a mood. Today if you told friends that you were gay, you would be proclaiming a lifestyle. The wiener wurst, a Vienna sausage, and the frankfurter sausage, popular in Frankfurt, Germany, both inspired a popular snack served on Coney Island early in the twentieth century. Known as Coney buns, the sausage was served in a bread bun. Vendors gained a reputation of worrying little about the meat content of the sausage, and patrons, betraying their suspicions about what cheap meats were substituted, dubbed the sausages "hot dogs." Similarly, when I want to use a projector before my class to show 35-mm slides, I dial up the services of Instructional Support Services. Not long ago, I would have called the Instructional Media Services, and before that the Audio Visual Center, and before that the Bureau of Visual Teaching. The same people arrive with the same equipment, but over the years their name has changed. The language evolves. I do not know how much envy I would evoke if I were to tell colleagues that last weekend I engaged in a whole evening of social intercourse. An innocent expression a few generations ago, but today it suggests an athletic ability beyond my means.

FACT, COURSE, MECHANISM

Organisms change too, and controversies about their change might occur at one of three levels. For example, scientists might differ over the *fact* of evolution. A simple first question: Do organisms change through time? From many lines of evidence ranging from gene changes to the fossil record, we can conclude that, in fact, evolution has occurred and still occurs even today. But this does not mean that all controversies over evolution are today comfortably and amicably settled. We might secondly differ about its *course*. Anthropologists who study human evolution usually agree on the fact of evolution—humans did evolve—but they often differ violently about the course that evolution took. Some see "Lucy" (*Australopithecus afarensis*) as a direct ancestor on the path to modern humans; others consider this early hominid species divergent—a side branch to the main course of hominid evolution.

Finally, we might agree on the fact and course of evolution, but differ about the *mechanism* that produced the changes. Verbal scuffles over evolution often become steamy until combatants realize that they are arguing over different parts of the evolutionary issue. Historically, too, each of these questions—*fact, course, mechanism*—had to be settled to bring us to our current understanding of the evolutionary process.

Darwin's unique contribution to evolutionary thought was his answer to the third question. He proposed the mechanism of change. To understand how profound this was and to appreciate the preceding attempts to clarify the fact of evolution and its course, we should review the ideas of those who immediately preceded Darwin. Even in eighteenth- and nineteenth-century scientific circles, opinion was strongly divided on the issue of the "transmutation" of species, as evolution was then called. Did species change and, if so, how? Two scientists, Carl von Linné and J. B. de Lamarck, illustrate the general conflicting views on evolution preceding Darwin. These two scientists were not alone, and many contemporary scientists fall into one camp or the other. But these two are perhaps best known—their views illustrate the divided intellectual climate in the early nineteenth century. Carl von Linné thought species were fixed (no evolution); J. B. de

Lamarck thought that species changed (evolution). Let's start with Carl von Linné and the view that species are, and always have been, fixed and unchangeable.

FIXITY OF SPECIES

Linnaeus

Foremost among the scientists who felt that species were fixed and unchangeable was Carl von Linné (1707–1778) (figure 1.2). This Swedish biologist followed the custom of the day and Latinized his name to Carolus Linnaeus, the name by which he is usually recognized. Linnaeus devised a system for naming plants and animals that is still the basis for modern taxonomy. Philosophically, he also argued that species are unchanging—created originally as we find them today. For several thousand years, Western thought kept company with the biblical view that each species resulted from a single and special act of divine Creation as described in Genesis, and thereafter each species remained unchanged.

Although most scientists sought to avoid strictly religious explanations, the biblical view of Creation was an overbearing presence in Western intellectual circles, was conveniently at hand, and meshed comfortably with the philosophical arguments put forth by

FIGURE 1.2   Carolus Linnaeus (1707–1778)
This Swedish biologist devised a system still used today for naming organisms. He also firmly abided by and promoted the view that species do not change.

scientists such as Linnaeus who argued that species were immutable (unchanging). But the secular idea of immutable species was appealing for more than its comfortable compatibility with the biblical Genesis. At the time, evidence for evolution was not easily assembled. The available evidence was ambiguous; it could be interpreted both ways, for and against evolution.

Naturalists

Today we understand that the adaptations of animals—trunks of elephants, long necks of giraffes, wings of birds—are the natural products of evolutionary change; diversity of species is the consequence of adaptive change. To scientists of an earlier time, though, adaptation was evidence of the care exercised by the Creator. The diversity of plant and animal species was proof of His almighty power. Animated by this conviction, many sought to learn about the Creator by turning to the study of what He had created. One of the earliest to do so was the Reverend John Ray (1627–1705), who summed up his beliefs, along with a natural history, in a book entitled *The Wisdom of God Manifested in the Works of the Creation* (1691). William Paley (1743–1805), Archdeacon of Carlisle, also articulated the common belief of his day in his book, *Natural Theology, or Evidences of the Existence and Attributes of the Deity Collected from the Appearances of Nature* (1802). Louis Agassiz (1807–1873), curator of the Museum of Comparative Zoology at Harvard University, found much public support for his project of building and stocking a museum to exhibit the remarkable creatures of this world—manifestations of the divine mind that produced them. For most scientists, philosophers, and laypeople there was no change and, thus, no evolution in the biological world of species. Even in secular circles of the mid nineteenth century, the intellectual obstacles to the idea of evolution were formidable.

Linnaeus, like many scientists before and immediately after him, insisted on the constancy and fixity of species and the sharp discontinuities between them. No intermediates existed between species. Species were literally created at the beginning of the world—a single pair of each species, multiplying but remaining unchanged thereafter. Ironically, the clarity of Linnaeus's conceptual opposition to species change helped define the issue of *species evolution*, or *transmutation of species* as it was then called. However, naturalists, among them Darwin, noticed apparent intermediate species. Perhaps species were not fixed forever. One of the first to build on this possibility was Lamarck. As we are about to see, however, his personal contribution to evolutionary theory was a mixed blessing.

Consider This—

"Intelligent Design"... But Even Lice?

William Paley cast his gaze across nature and saw exquisitely fashioned features of plants and animals—features that served them well. The eye is a "contrivance" for perceiving light; the webbed feet of swans paddle them elegantly across the water; mole hair sheds clinging soil. Paley reasoned, to wit, as because there is a well-crafted watch, there is a watchmaker. So too, he thought, as there are well-crafted contrivances in nature, there is a contrivance maker, and that maker is God. John Ray did Paley one better. He tackled the tricky question of why the divine Creator made obnoxious creatures. To paraphrase Ray, consider lice: They harbor and breed in clothes, "an effect of divine providence, designed to deter men and women from sluttishness and sordidness, and to provoke them to cleanliness and neatness."

The idea that well-crafted design demands a designer survives today, almost 200 years later, in the argument of "intelligent design." This surviving idea of Paley and the naturalists similarly suggests that complex contrivances of organisms are so well and thoughtfully assembled, so intelligently constructed, that a designer must be behind it. Unfortunately, like beauty, "intelligent" design resides in the eye of the beholder. What is intelligent depends on the criteria applied—what is intelligent to one person is just an obvious biological adaptation to another. As with Paley before, intelligent design reduces to a poetic indulgence, but not a useful scientific insight.

CHANGE OF SPECIES

J. B. de Lamarck

Among those taking the side of evolution, few were as uneven in their reputation as Jean-Baptiste de Lamarck (1744–1829) (figure 1.3a). He lived most of his life on the border of poverty, although he held the equivalent of a professorship at the Jardin du Roi in Paris (later the Muséum National d'Histoire Naturelle) (figure 1.3b). Abrupt speech, inclination to argument, and strong views did little to endear him to colleagues. Yet Lamarck's *Philosophie Zoologique*—mostly dismissed when published in 1809 as the amusing ruminations of a "poet"—eventually came to establish the theory of evolutionary descent as a respectable scientific generalization.

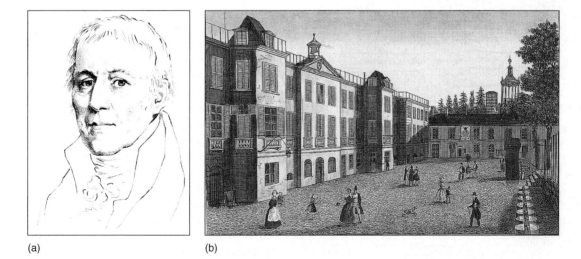

(a)   (b)

FIGURE 1.3   J-B. de Lamarck (1744–1829) (a) J-B. de Lamarck worked most of his life at the Muséum d'Histoire Naturelle (b). His academic position gave him a chance to promote the idea that species change.

Lamarck's ideas spoke to the three issues of evolution: *fact, course,* and *mechanism.* As to the fact of evolution, Lamarck argued that species change through time. Curiously, he thought that the simplest forms of life arose by spontaneous generation, springing ready-made in muck from inanimate matter, but thereafter evolved onward and upward into higher forms. As to the course of evolution, Lamarck proposed a progressive change, along an ascending ladder or scale from the lowest on one end to the most complex and "perfect" (meaning humans) on the other. As to the mechanism of evolution, he proposed that "need" itself produced changes inherited through subsequent generations. When environments or behavior changed, an animal developed new needs to meet the new demands placed upon it by the new environment. These new needs altered the animal's metabolism and changed its internal physiology, triggering the appearance of a new body part to address the new need. Continued use of a body part tended to develop that part; disuse led to its withering. Environments changed, a need arose, metabolism adjusted, new organs were created. Once acquired, these new characteristics were passed on to offspring. This, in summary, was Lamarck's view. It has been called evolution by means of the **inheritance of acquired characteristics**. Characters were "acquired" to meet new needs and then "inherited" by future generations.

Need, Use

For Lamarck, *need* is a technical term denoting the motor that runs evolutionary change. Therefore, when you invoke the idea of "need" or "use" to explain biological change, you are employing a technical term loaded with historical meaning and implying that you keep intellectual company with Lamarck. What's wrong with "need" as a biological explanation of change? Lots.

Central to Lamarck's thinking was an inadvertent confusion between physiology and evolution. Any person who begins and stays with a regular weight-lifting program can expect to see his or her strength increase and muscles enlarge. With added weight, use (need) increases, and big muscles appear. But this is a physiological response, limited to the exercising individual. The new, big muscles are not passed on genetically to offspring. Charles Atlas, Arnold Schwarzenegger, and other bodybuilders do not pass on their newly acquired muscles to wimpy children. If their children seek large muscles, they too must start from scratch with their own training program. *Somatic characteristics*—body characteristics, such as big muscles, acquired through use—cannot be in-

herited. Inherited traits reside within genes carried in germ cells (eggs and sperm), not in somatic cells.

Unlike such physiological responses, evolutionary responses involve changes in an organism that are inherited and thus pass from one generation to the next. We know today that such characteristics are genetically based. They arise by gene mutation, not from somatic alteration due to exercise or metabolic need.

Acquired Characteristics

Lamarck's proposed mechanism—inheritance of acquired characteristics—confused immediate physiological response with long-term evolutionary change. His mechanism failed. Yet most people today still inadvertently think in Lamarckian terms—in terms of parts arising to meet needs. Recently, an actor/moderator of a television nature program on giraffes spoke what was probably on the minds of viewers when he said that the long neck originated to help giraffes meet the "need" of reaching treetop vegetation. Environmental demands do not reach into the genetic material and directly produce inherited improvements to address the new needs or new opportunities. That route for modification does not exist in any organism's physiology.

The other side of the Lamarckian coin is disuse—the loss of a part following the loss of a need. Some fish and salamanders live in deep caves that daylight cannot reach. These species lack eyes. Even if they return to the light, they do not form eyes. The eyes are lost. It is tempting to attribute this loss of eyes to disuse in a dark environment. That, of course, would be invoking a Lamarckian mechanism—use produces a needed part, disuse eliminates it. But because somatic traits acquired or lost during a single lifetime are not inherited, they cannot have long-term evolutionary consequences.

Because it comes easily, a Lamarckian explanation is difficult to purge from our own reasoning. We fall automatically and too comfortably into the convenient habit of thinking of parts arising to meet "needs," one linked to the other. For Darwin, and for students new to the study of evolution, the theory that new needs produce new parts is an intellectual obstacle to clear reasoning, an obstacle that, unfortunately, Lamarck helped popularize and current culture perpetuates.

Upward to Perfection

The proposed course of evolution championed by Lamarck was also in error. The concept of the "ladder, or scale, of nature" (*scala naturae*) goes back to Aristotle, who proposed it as a ranking system. Since then, it has expanded, adding the notion that the ranking, in fact, represents an evolutionary direction: Life starts with the lowest and moves to the highest organisms, progressively upward toward perfection. Evolutionists, like Lamarck, viewed life metaphorically as ascending a ladder, one rung at a time, up toward the complex and perfected. After a spontaneous origin, organisms progressed up this metaphorical ladder of nature through the course of many generations. Unfortunately, remnants of this idea still linger on in modern society. You may hear it expressed something like this: "If humans evolved from apes, then why are apes still around?" Translation: If humans are perfected, then why do less-perfected animals persist? Humans certainly are equipped to meet environmental demands they face, but no more so than any other organism. Moles and mosquitoes, bats and birds, earthworms and anteaters, water lilies and Venus flytraps—each species achieves an equal match of parts-to-performance-to-environmental-demands.

The concept of a ladder of progress was misleading because it suggested that animal evolution was driven internally in a particular direction from the early, soft-bodied forms up toward humans, the perfected forms. As water runs naturally downhill, descent of animals runs naturally to the perfected. Simple animals were seen, not as adapted in their own right, but rather as improving steps along the way to a better future. Animal species were a series of progressive improvements. Sponges were improvements over amoebas, worms over sponges, fishes over worms, amphibians over fishes, reptiles over amphibians, and so on, scaling up to humans.

The idea of perfection entrenched in Western culture is built upon our experience with continued technological improvements. We bring it like unnoticed excess intellectual baggage into biology, where it clutters our interpretation of evolutionary change. When we use terms *lower* and *higher,* we risk resurrecting this discredited idea of perfection. Lower animals and higher animals are not poorly designed and better designed animals, respectively. *Lower* and *higher* refer only to order of evolutionary appearance—"lower" animals arose first; "higher" animals arose after them. Nothing more is implied or claimed. To keep this clear and to avoid any suggestion of increasing perfection, many scientists replace the terms *lower* and *higher* with the terms **primitive** (**ancestral** or **basal**) and **derived** to emphasize only the evolutionary sequence of appearance—early and later, respectively.

Darwin's mechanism of evolutionary change was external. Organisms live or die based on how well they meet current environmental demands. Descent with modification was the result. In contrast, for those adopting the ladder-of-nature perspective, the driving force of evolutionary change was internal. To Lamarck and other evolutionists of his day, nature got better. Animals improved as they evolved "up" the evolutionary ladder. Thus, Lamarck's historical contribution to evolutionary concepts was a mixed blessing. On the one hand, he defended the mistaken view that acquired characters are inherited, and thereby confused physiological response with evolutionary adaptation. Further, by incorporating the notion of a ladder of nature into his theories, he adopted the flawed view that unspecified internal engines drive evolutionary change outward. On the other hand, Lamarck vigorously defended the view that plants and animals, indeed, evolve. By so doing, he eased the route to Darwin. For many years, textbooks have been harsh in their treatment of Lamarck, mostly to ensure that modern students do not repeat his mistakes. But we must allow him his deserved place in the history of evolutionary ideas. By arguing for the view that species change, Lamarck helped blunt the sharp antievolutionary dissent of contemporaries such as Linnaeus. He gave respectability to the idea of evolution and helped prepare the intellectual environment for those who would solve the species question.

THE MECHANISM OF EVOLUTION: NATURAL SELECTION

Such was the divided European intellectual climate of the early nineteenth century. This intellectual climate is represented here by Linnaeus and Lamarck, fixity versus evolution, but many others joined sides with their own opinions on the species question. What may also be surprising is that the proposed mechanism of evolution—natural selection—was unveiled publicly by two persons jointly in 1858, although it was conceived independently by each. One was Charles Darwin, as we have seen; the other was Alfred R. Wallace. Both were part of the respected naturalist tradition in Victorian England that encouraged physicians, clergymen, and persons of leisure to devote time observing plants and animals in the countryside. Such interests were not seen as harmless pursuits that passed idle time. Observation of nature was deemed respectable because it brought personal contact with the Creator's handiwork.

Despite the reason for the tradition, the result was to encourage thoughtful attention to the natural world.

A.R. Wallace

Alfred Russel Wallace (1823–1913) (figure 1.4) was 14 years younger than Darwin. Although he lived the life of a naturalist, Wallace lacked the comfortable economic circumstances of most gentlemen of his day. Therefore, he turned to a trade for a livelihood. Initially, he surveyed land for railroads in his native England; later, following his interest in nature, he collected biological specimens in foreign lands to sell to museums back home. His search for rare plants and animals in exotic places took him to the Amazonian jungles and later to the Malay Archipelago in southeastern Asia now known as Indonesia. We know from his diaries that he was impressed by the great variety and number of species he encountered in his travels. In early 1858, Wallace fell ill while on one of the Spice Islands (the Moluccas) between New Guinea and Borneo. During a fitful night of fever, he recalled a book by the Reverend Thomas Malthus entitled *An Essay on the Principle of Population, As It Affects the Future Improvement of Society.* Writing of human populations, Malthus observed that unchecked breeding causes populations to grow geometrically, whereas the food supply grows more slowly. The simple, if cruel, result is that the number of people increases faster than the amount of food—there is not enough to go around—so some people survive, but most die. It flashed into Wallace's mind that the same principle applied to all species. In his own words:

> It occurred to me to ask the question, why do some die and some live? And the answer was clearly; that on the whole the best fitted lived. From the effects of disease the most healthy escaped; from enemies, the strongest, the swiftest, or the most cunning; from famine, the best hunters or those with the best digestion; and so on.
>
> Then I at once saw, that the ever present variability of all living things would furnish the material from which, by the mere weeding out of those less adapted to the actual conditions, the fittest alone would continue the race.
>
> There suddenly flashed upon me the idea of the survival of the fittest.
>
> The more I thought over it, the more I became convinced that I had at length found the long-sought-for law of nature that solved the problem of the Origin of Species.

Wallace began writing that same evening and within two days had sketched out his idea in a paper. Knowing that Darwin was interested in the subject, but unaware of how far Darwin's own thinking had progressed, he mailed the manuscript to Darwin for his opinion. The post was slow—the journey took four months. Wallace had not only hit upon the same central mechanism of evolution as Darwin, but even unknowingly coined some of the same phrases and terminology. Darwin was stunned. But what to do?

Charles Darwin

Unlike Wallace, Charles Darwin (1809–1882) was born into economic security. His father was a successful physician; his mother was part of the successful Wedgwood (pottery) fortune. Darwin tried medicine at Edinburgh, but those days predated anesthetics, and he was squeamish with the human pain that accompanied raw operations. Fearing creeping idleness, his father redirected him to Cambridge and a career in the church, but Darwin proved uninterested. At formal education, he seemed a mediocre student. At Cambridge, however, Reverend John S. Henslow, a professor of botany, encouraged Darwin's long-standing interest in natural history. Darwin took geology excursions, col-

FIGURE 1.4   Alfred Russel Wallace (1823–1913) Wallace in his thirties. (By courtesy of the National Portrait Gallery, London.)

lected biological specimens, and upon graduation joined as a *de facto* naturalist the government ship HMS *Beagle* (figure 1.5a) over the objections of his father, who wanted Charles to pursue a more conventional career in the ministry.

Darwin spent nearly five years on the ship and explored the coastal lands it visited (figure 1.5b). The experience intellectually transformed him. Darwin's belief in the special creation of species, with which he began the voyage, was shaken by the vast array of species and adaptations the voyage introduced to him. The issue would come especially into focus on the Galápagos Islands, off the west coast of South America. Each island contained its own assortment of species, some found only on that particular island. Local experts could tell at sight from which of the several islands a particular tortoise came. The same was true of many of the bird and plant species Darwin collected.

What all of this implied came gradually to Darwin after his return.

Darwin arrived back in England in October 1836 and set to work sorting his collection, obviously impressed by the diversity he had seen but still wedded to misconceptions about the Galápagos collection in particular. He had, for instance, thought the Galápagos tortoise was introduced from other areas by mariners stashing reptile livestock on islands they might visit later. He apparently dismissed reports of island differences between the tortoises as merely changes that attended the animal's recent introduction to new and dissimilar habitats. However, in March 1837, six months after his return and almost a year and a half after departing the Galápagos, Darwin met in London with John Gould, a respected specialist in ornithology. Gould insisted that the mockingbirds Darwin had collected on three different

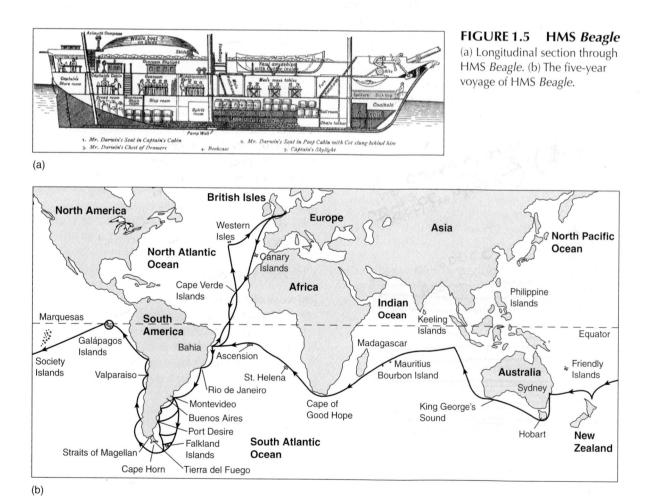

FIGURE 1.5   HMS *Beagle*
(a) Longitudinal section through HMS *Beagle*. (b) The five-year voyage of HMS *Beagle*.

(a)

(b)

Galápagos islands were actually three distinct species. In fact, Gould emphasized that all the birds were endemic to the Galápagos—distinct species, not just varieties—although clearly each was related to species on the South American mainland. It seems to have suddenly dawned on Darwin that not only birds, but also plant and tortoise varieties were distinct as well. Geographically isolated on the Galápagos, these organisms were derivatives of ancestral stocks but now distinct island species of their own.

Darwin noted the contrast of the Galápagos with the Cape Verde Islands off the coast of Africa. Like the Galápagos, the Cape Verde Islands were of similar soil (volcanic) and climate (tropical, equatorial). If animals were created for one island archipelago, they might be found in the other. But this was not the case. The plant and animal communities were distinct. Those of the Cape Verde Islands were clearly related to Africa; those of the Galápagos to South America. In Darwin's words from *The Origin of Species*:

> there is a considerable degree of resemblance in the volcanic nature of the soil, in the climate, height, and size of the islands, between the Galapagos and Cape Verde Archipelagoes: but what an entire and absolute difference in their inhabitants! The inhabitants of the Cape Verde Islands are related to those of Africa, like those of the Galapagos to [South] America.

For Darwin, the issue came down to this,

> Why should the species which are supposed to have been created in the Galápagos Archipelago, and nowhere else, bear so plainly the stamp of affinity to those created in [South] America?

Was each of these Galápagos species of tortoise or bird or plant an act of special creation? Although distinct, each species was also clearly related to those on the other islands and to those on the nearby mainland of South America. To account for these species, Darwin had a serious choice to make: Either they were products of a special creation, one act of creation for each species, or they were the natural results of evolutionary adaptation to the different islands. If acts of special divine Creation, then each of the many hundreds of the island species represented a distinct and separate act of creation. But that seemed odd. All were similar to each other—the tortoises to other tortoises, the birds to other birds, and the plants to other plants on the various islands, almost as if the Creator was running out of new ideas. On the other hand,

if they were the natural results of common evolutionary descent, then similarity and diversity might be expected. For example, the tortoise species on different islands illustrated adaptation to distinct environments. The similarities they shared were explained by descent to the present traced from a common ancestor from the past. The first animals or plants washed or blown to these oceanic islands would constitute the common stock from which similar, but eventually distinct, species evolved. Darwin sided with a natural evolution.

The pieces began to fall into place. Darwin now became convinced that, in fact, evolution had occurred here, and he had a good idea of the course of this evolution within these islands. But he needed a mechanism by which such evolutionary descent might proceed, and at first he had none to suggest. While in the midst of writing up his results of other studies from the *Beagle*, he read for amusement the essay on population by Malthus, the same essay Wallace would come upon years later. The significance struck Darwin immediately. If animals, like humans, outstripped food resources, then competition for scarce resources would result. Those with favorable adaptations would fare best, and new species incorporating these favored adaptations would arise. "Here then I had at last got a theory by which to work" wrote Darwin. In a moment of insight, he had solved the species question.

That was 1838, and you would think the excitement would have set him to work on papers and lecturing. Nothing of the sort happened. Four years, in fact, elapsed until he wrote a first draft, and that was only 35 pages, handwritten in pencil; two years later, he expanded it to two hundred and some odd pages, in ink, but placed it quietly into a drawer along with a sum of money and a sealed letter to his wife instructing her to publish it if he met an untimely death. A few close friends knew what he had discovered, but most did not—including his wife, with whom he otherwise enjoyed a close and loving marriage.

This was Victorian England. Science and religion fit hand-in-glove. Darwin's delay testifies to how profoundly he understood the larger significance of his insight. He wanted more time, to gather more evidence to write the volumes he thought necessary to make a compelling case. Then, in June 1858, 20 years after he had first come upon the mechanism of evolution, the manuscript from Wallace arrived. Darwin was dumbfounded. Wallace even spoke of "natural selection," the same terminology. Mutual friends intervened, and much to the credit of both Wallace and

Darwin, a joint paper was read in the absence of both before the Linnaean Society in London the next month, July 1858. Wallace was, as Darwin described him, "generous and noble." Wallace, in "deep admiration," later dedicated his book on the Malay Archipelago to Darwin as a token of "personal esteem and friendship." Oddly, this first paper made no stir. But Darwin's hand was now forced.

Critics and Controversy

Darwin still intended a thick discourse on the subject but agreed to a shorter version running to "only" 500 pages. This was *The Origin of Species*, published at the end of the next year, 1859. By then word was out, and the entire first edition of 1,250 copies was sold completely on the day it was published. Largely because Darwin produced *The Origin of Species*, the expanded statement of the case for evolution by means of natural selection, and because of a continued series of related works, he is remembered more than Wallace for formulating the basic concept. He brought a scientific consistency and cohesiveness to the concept of evolution, and that is why it bears his name, Darwinism. Science and religion had been tightly coupled, especially in England. For centuries, a ready answer was at hand for the question of life's origin—a divine explanation, as described in Genesis. Darwinism countered with a natural explanation. Controversy was immediate and still lingers even today in some circles. Darwin, a thoughtful and kind person, could not sleep if he answered a critic harshly. And for all of his adult life his health was fragile. He retired from the fray, leaving to others the task of public defense of the ideas of evolution. Sides quickly formed. Speaking before the English Parliament, the future prime minister Benjamin Disraeli safely chose his friends: "The question is this—Is man an ape or an angel? My lord, I am on the side of the angels."

The media's reaction included much puffing and posturing. Ignorance and misinformation ruled in public discourse. But in scientific circles, not all was misguided and hysterical. Two serious criticisms of his theories stuck, and Darwin knew it. One was the question of variation; the other the question of time. As to time, there seemed not enough. If the evolutionary events Darwin envisioned were to unfold, then the Earth must be very old to allow time for life to diversify. In the early part of the nineteenth century, Dr. John Lightfoot, vice chancellor at Cambridge University, calculated from his biblical studies of who begot whom that humans were created in 4004 B.C., October the 23rd, at 9:00 in the morning, presumably Greenwich mean time. Many took this date, if not literally, then at least as being indicative of the recent origin of humans, leaving no time for evolution from apes or angels. Lord Kelvin, author of the second law of thermodynamics and a preeminent scientist of his day, made a more scientific effort to date the Earth. Using temperatures he measured in deep mine shafts, he reasoned that the Earth would cool from its primitive molten state to present temperatures at a constant rate. He extrapolated backward to calculate that the Earth was probably no more than 100 million years old. Later, he revised this estimate downward to possibly as little as 20 million years. Unfortunately, no one then knew of the natural radioactivity in the Earth's crust. This deep radioactivity produces new heat to keep the surface hot and deceptively makes it seem close in temperature, and thus in age, to its molten temperature at first formation. Modern scientists estimate that the true age of the Earth is actually several billion years; unfortunately for Darwin, this was not known until long after his death.

As to the question of variation, critics also pointed to inheritance of variation as a weak point in Darwin's theory of evolution. The basis of heredity was unknown in his day. One popular view held that inheritance was a matter of blending—like mixing two paints, offspring received a blend of characteristics from both parents. This mistaken view was taken seriously by many and created a pair of scientific problems for Darwin. Where did variation come from? How was it passed from generation to generation? If natural selection favored individuals with superior characteristics, what ensured that these superior characteristics were not blended and diluted out of existence in the offspring? If favored characters were blended, they would effectively be lost from view, and natural selection would not work. Darwin could see this criticism coming and devoted much space in *The Origin of Species* to discussing *sources* of variation.

Today we know the answers to this paradox. Mutations in genes and chromosomes produce new variations; genes carry characteristics unaltered and without dilution from generation to generation. But this twentieth-century mechanism of inheritance was unknown and unavailable to Darwin and Wallace in the nineteenth century. Given the social and scientific obstacles of the time, their insights are remarkable.

OVERVIEW

It was probably no coincidence that the intellectual breakthroughs of both Darwin and Wallace were fostered by voyages of separation from the conventional scientific climate of their day. Certainly, the study of nature was encouraged in England, but that culture also supplied a ready-made interpretation of the diversity and order they observed in nature. The biblical story of creation in Genesis was conveniently at hand and taken by some literally to supply explanations for the presence of species. But there were scientific obstacles as well. The confusion between physiological and evolutionary adaptation (Lamarck), the notion of a ladder of nature, the idea of fixity of species (Linnaeus and others), the alleged short history of the Earth (Kelvin), the mistaken views of variation and heredity (blending inheritance)—all of these differed from predictions of evolutionary events or confused the picture. It is thus testimony to their intellectual insight that Darwin and Wallace could see through such obstacles when others were stymied.

Perhaps the greatest historical irony, at least for Darwin, was that the mechanism of heredity was actually discovered during his lifetime, but he did not know it. This is a strange quirk of fate we shall examine in chapter 3. As to evolutionary time, there was plenty, but Darwin did not know that fact either. In chapter 2 we will turn to this story of time because it is probably one of the most misunderstood and least appreciated aspects of evolutionary events.

Selected References

Darwin, C. R. (ed.). 1841. *The zoology of the voyage of H. M. S. Beagle, under the command of Captain FitzRoy R. N., during the years 1832–1836. Part III: Birds.* London: Smith, Elder, & Co.

Darwin, C. R. 1845. *Journal of researches into the natural history and geology of the countries visited during the voyage of H. M. S. Beagle round the world, under the command of Captain FitzRoy, R. N.* 2nd ed. London: John Murray.

Darwin, C. R. 1859. *On the origin of species by means of natural selection, or the preservation of favoured races in the struggle for life.* London: John Murray.

Darwin, C. R. 1868. *The variation of animals and plants under domestication.* 2 vols. London: John Murray.

Darwin, C. R. 1887. *The life and letters of Charles Darwin, including an autobiographical chapter.* Edited by F. Darwin. 3 vols. London: John Murray.

Darwin, C. R. 1903. *More letters of Charles Darwin: A record of his work in a series of hitherto unpublished letters.* Edited by F. Darwin and A. C. Seward. 2 vols. New York: D. Appleton.

Darwin, C. R. 1933. *Charles Darwin's diary of the voyage of H. M. S. Beagle.* Edited by N. Barlow. Cambridge: Cambridge University Press.

Darwin, C. R. 1958 [1876]. *Autobiography: With original omissions restored.* Edited with appendix and notes by his granddaughter, N. Barlow. London: Collins.

Darwin, C. R. 1963 [1836]. Darwin's ornithological notes. Edited with introduction, notes, and appendix by N. Barlow. *Bull. British Mus. (Nat. Hist.) Historical Series*, 2, no. 7.

Darwin, C. R. 1980 [1836–37]. The red notebook of Charles Darwin. Edited with introduction and notes by S. Herbert. *Bull. British Mus. (Nat. Hist.) Historical Series*, 7 (Ithaca, N. Y.: Cornell University Press).

Darwin, F. 1888. Charles Robert Darwin (1809–1882). *Dict. of Nat. Biog.*, 14:72–84.

Darwin, F. 1909. *Introduction to the foundations of the origin of species: Two essays written in 1842 and 1844, by Charles Darwin.* Edited by F. Darwin. Cambridge: Cambridge University Press.

Darwin, F. (ed). 1903. *More letters of Charles Darwin: A record of his work in a series of hitherto unpublished letters.* Edited by F. Darwin and A. C. Seward. 2 vols. New York: D. Appleton.

FitzRoy, R. 1839. *Narrative of the surveying voyages of His Majesty's ships Adventure and Beagle, between the years 1826 and 1836. Describing their examination of the southern shores of South America and the Beagle's circumnavigation of the globe. Vol. 2. Proceedings of the second expedition, 1831–1836, under the command of Captain FitzRoy, R. N., with appendix.* London: Henry Colburn.

Ghiselin, M. T. 1969. *The triumph of the Darwinian method*. Berkeley and Los Angeles: University of California Press.

Gould, J. 1837. Remarks on a group of ground finches from Mr. Darwin's collection, with characters of the new species. *Proc. Zool. Soc. (London)*, 5:4–7.

Grant, P. R. 1983. The role of interspecific competition in the adaptive radiation of Darwin's finches. In *Patterns of evolution in Galápagos organisms*, edited by R. I. Bowman, M. Berson, and A. E. Leviton. Speciation Publication 1. San Francisco: AAAS, Pacific Division, pp. 187–99.

Lack, D. 1945. The Galápagos finches *(Geospizinae)*: A study in variation. *Occ. Paprs. Calif. Acad. Sci.*, no. 31.

Linnaeus, C. 1781. *Politica Naturae* [Amoen. Academicae]. Translated by F. J. Brand. London, pp. 131–32.

Mayr, E. 1991. *One long argument*. Cambridge, Mass.: Harvard University Press.

Ray, J. 1691. *The wisdom of God manifested in the works of the creation*. London: W. Innys and R. Manby.

Ruse, M. 1979. *The Darwinism revolution: Science red in tooth and claw*. Chicago and London: University of Chicago Press.

Ruse, M. 1982. *Darwinism defended: A guide to the evolution controversies*. Reading, Mass.: Addison-Wesley.

Ruse, M. 1986. *Taking Darwin seriously: A naturalistic approach to philosophy*. Oxford: Blackwell Press.

Ruse, M. 1988. *But is it science: The philosophical question in the creation/evolution controversy*. Buffalo, N.Y.: Prometheus.

Ruse, M. 2001. *Can a Darwinian be a Christian?* Cambridge: Cambridge University Press.

Simpson, G. G. 1944. *Tempo and mode of evolution*. New York: Columbia University Press.

Sulloway, F. J. 1982. Darwin's conversion: The *Beagle* voyage and its aftermath. *J. Hist. Biol.* 15(3): 325–96.

Sulloway, F. J. 1984. Darwin and the Galápagos. *Biol. J. Linnean Soc.* 21:29–59.

Wallace, A. R. 1855. On the law which has regulated the introduction of new species. *Ann. Mag. Nat. Hist.*, 2nd ser., 16:184–196.

Wilson, E. O. 1975. *Sociobiology: The new synthesis*. Cambridge, Mass.: Harvard University Press.

Learning Online

Testing and Study Resources

Visit the textbook's website at **www.mhhe.com/ evolution** (click on the book's title) to take advantage of practice quizzing, study/writing tips, timely news articles, and additional URLs for research on the topics in this chapter.

2

"... paleontologists have discovered several superb examples of intermediary forms and sequences, more than enough to convince any fair-minded skeptic about the reality of life's physical genealogy."

Stephen Jay Gould, evolutionary biologist

Time

INTRODUCTION

Many contemporaries worried that Darwin was out of time—that is, out of evolutionary time. So did Darwin. By estimates of his day, not enough evolutionary time was available for natural selection to operate and produce the rich diversity of organisms that populated the world. Darwin became even more concerned about this in his later life, especially as Lord Kelvin revised his estimated age of the Earth from 100 million downward to 20 million years and declared that safe surface temperatures—temperatures that did not roast living organisms—would have been reached only a few million years ago. Many who shared Darwin's views and accepted Kelvin's estimates simply saw natural selection as operating at a faster speed than previously thought. Darwin must have felt that the dates would be proven wrong (as, in fact, they were years later), but he saw Kelvin's calculations as a major, immediate challenge to his view of natural selection as a stately and gradual process.

With the discovery of radioactivity early in the twentieth century, the central premise of Kelvin's argument collapsed. Radioactive decay, not old residual heat left over from the Earth's formation, produced the high temperatures Kelvin measured in deep mine shafts. His formulas were elegant, but the data he entered were wrong.

The actual age of the Earth is a little over 4.6 billion years (figure 2.1). The oldest rocks, lifeless, on Earth are in Australia and date to 4.2 billion years ago. Old continents from even earlier times have since disappeared, remelted back into the Earth's molten mantle. But meteorites that survived the fiery passage through the atmosphere have been dated to 4.6 billion years ago. Current views on the formation of the solar system hold that meteors and everything else in it—such as planets, asteroids, comets—formed at the same time. Dating a part of the solar system dates it all, and therefore 4.6 billion years (a value taken from meteorites) is considered the best estimate of the age of the Earth.

Such an age should humble and give us pause. We may speak of 4.6 billion years glibly, but it is an almost incomprehensible span of time. We cannot gauge such vastness of time by our own brief lives. Billions are staggering numbers. Yet just such depth to time is the key to understanding the diversity and complexity of life on

FIGURE 2.1   Geologic Time The gathering of cosmic gases under gravity's pull created Earth some 4.6 billion years ago. Yet life became neither abundant nor complicated until the Cambrian period, or slightly earlier, when the first vertebrates appeared. **Source:** *After* U.S. *Geological Survey publication*, Geologic Time.

Earth. Time's vastness has made time evolution's partner. Time has allowed for evolutionary trial and error and experimentation. Deep time has converted the impossible into the possible, and the possible into the probable. Looking deep into a starry sky, we can be told that the light reaching our eyes left the stars billions of years ago. We can nod, recognize the claim, but not quite understand how enormous are the time and the distance. The numbers are beyond the reference of our own short experience.

Even the whole of human history is brief against this evolutionary time scale. Compress, for example, the history of life on Earth into a 24-hour period, with the first microscopic organisms arising at midnight to begin the day. On our 24-hour evolutionary clock, three-quarters of the evolutionary day is gone by 6:00 P.M. in the evening, before life in the sea becomes abundant. The first amphibians arise at 8:00 P.M., the first reptiles at about 9:00 P.M., and dinosaurs at 10:30 P.M. All of human recorded history would not appear until one-quarter of a second before the end of this compressed evolutionary day.

DATING FOSSILS

If the age of the Earth is measured in billions of years, how do we date life's evolutionary chronology? Obviously no one was there to record events directly as they marched by.

The story of life on Earth is a story of evolving life—which forms came first, which next, and which later. Yet, most life ever to exist on Earth is now extinct. Therefore, much of our understanding of evolutionary history comes from the surviving bits of these preceding organisms, the fossils that remain to tell their story. Most dating techniques center upon the aging of these fossils, putting them in sequence. The techniques for dating fossils vary, and preferably several are used to cross-check a fossil's age.

Stratigraphy

One fossil-dating technique is **stratigraphy**, a method that places fossils in relative sequence to each other

(figure 2.2). It occurred to the Italian geologist Giovanni Arduino as early as 1760 that rocks could be arranged from oldest (deepest) to youngest (surface). By the time the British geologist Charles Lyell published his great classic, the three-volume work entitled *Principles of Geology* (1830–1833), a system for assigning relative dates to rock layers was well established. The principle is simple: Sheets of similar rock, termed *strata*, layered one upon another, are built up in chronological order. Like the construction of a tower, the oldest rocks are at the bottom, with later rocks in ascending sequence to the top, where the most recent rocks reside. Each layer of rock is spoken of as a *time horizon* because it contains the remains of organisms from one thick slice in time. Any fossils contained in the separate layers might reasonably be ordered,

FIGURE 2.2   Stratigraphy Sediment settling out of water collects at the bottom of lakes. As more sediment collects, the deeper layers are compacted by the ones above until they harden and become rock. Animal remains become embedded in these various layers. Deeper rock forms first and is older than rock near the surface. Logically, fossils in deeper rock are older than those above, and their position within these rock layers gives them a chronological age relative to older (deeper) or younger (surface) fossils.

then, from oldest-to-recent, bottom-to-top. Although this method yields no absolute age, it does produce a chronological sequence. We can determine which fossil species arose first and which later, relative to other fossils in the same geological exposure.

An overall chronology, greater than any single exposure taken separately, can be built up by placing separate rock exposures into overlapping register with each other (figure 2.3). Where different sets of strata have been studied, those sharing the same strata can be matched. The overall chronology is the range of the time horizons spanned collectively by the matched exposures.

Index Fossils

By matching rock strata in one location to comparable rocks in another exposed location, it is possible to build

up an overlapping chronological sequence longer than that represented at any single site by itself. The actual correlation of rock strata between two distantly located sites is done by comparison of mineral content and structure. This method also takes advantage of index fossils (figure 2.4)—distinctive marker fossils. An index fossil is an organism, usually a hard-shelled invertebrate animal, that previous work has shown to occur only within one specific time horizon of rock. The presence of an index fossil in a rock layer confirms that the stratigraphic layer belongs to the same time horizon as strata elsewhere that contain the same fossil.

Radiometric Dating

The relative position in time of one species to another is useful, but to actually place an age on a fossil requires

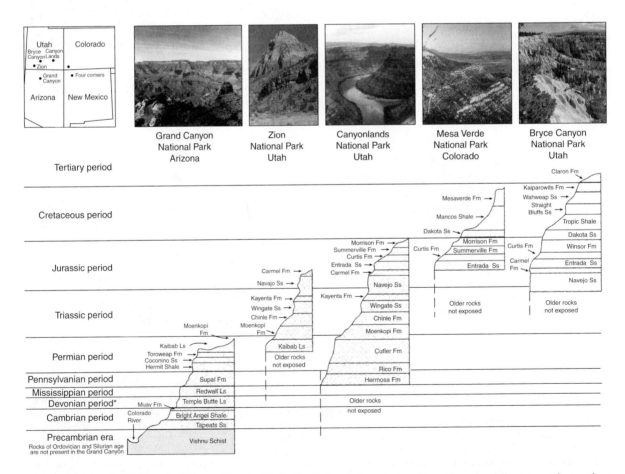

FIGURE 2.3 Building a Chronology of Fossils Each exposure of rocks can be of a different age from other exposures. To build up an overall sequence of fossils, various exposures can be matched where they share similar sedimentary layers (layers of the same ages). From five sites in the southwest United States, overlapping time intervals allow paleontologists to build a chronology of fossils greater than that at any single site.

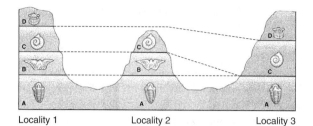

Locality 1   Locality 2   Locality 3

FIGURE 2.4   Index Fossils After careful study at many well-dated sites, paleontologists can confirm that certain fossils occur only at restricted time horizons (in specific rock layers). These distinctive index fossils are diagnostic fossil species used to date rocks in new exposures. In this example, the absence of index fossils confirms that layer B does not exist at the third location. Perhaps rock-forming processes never reached the area during this time period, or the layer was eroded away before layer C formed. After Longwell and Flint.

a different technique. This is **radiometric dating**, a technique that takes advantage of the natural transformation over time of an unstable elemental isotope to a more stable form or product (figure 2.5). Such radioactive decay of the isotope of an element into another isotope occurs at a constant rate, expressed as the characteristic **half-life** of an isotope—the length of time that must pass before half the atoms in the original sample transform into the product. Common examples include uranium-235 to lead-207 (half-life of 713 million years) and potassium-40 to argon-40 (half-life of 1.3 billion years). When rocks form, these radioactive isotopes are often incorporated. By comparing product-to-original (argon-from-potassium, for example, in rock) and knowing the rate at which this transformation occurs, the age of the rock—and hence the age of fossils it holds—can be calculated. For instance, if our sample of rock held relatively lots of argon and not much potassium, then the rock would be quite old and our estimated age quite high. Most of the potassium would have decayed to its product, argon. Conversely, if there were little argon compared to potassium, then less time would have passed and our calculated age would be young.

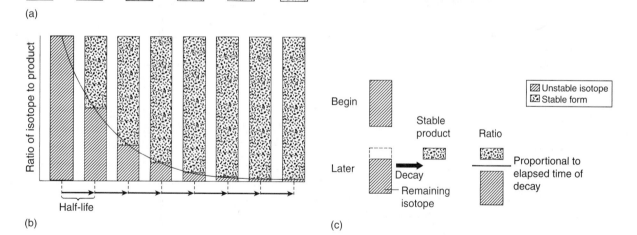

(a)

(b)

Half-life

(c)

FIGURE 2.5   Radiometric Dating (a) Sand flows regularly from one state (upper portion) to another (lower portion) in an hourglass. The more sand in the bottom, the more time has passed. By comparing the amount of sand in the bottom with that remaining in the top and by knowing the rate of flow, we can calculate the amount of time that has elapsed since the flow in an hourglass was initiated. Similarly, knowing the rate of transformation and the ratios of product to original isotope, we can calculate the time that has passed for the radioactive material in rock to be transformed into its more stable product. (b) **Half-life.** It is convenient to visualize the rate of radioactive decay in terms of half-life, the amount of time it takes an unstable isotope to lose half its original material. Shown in this graph are successive half-lives. The amount remaining in each interval is half the amount present during the preceding interval. (c) A radioactive material undergoes decay, or loss of mass, at a regular rate that is unaffected by most external influences, such as heat and pressure. When new rock is formed, traces of radioactive materials are captured within the new rock and held along with the product into which it is transformed over the subsequent course of time. By measuring the ratio of product to remaining isotope, paleontologists can date the rock and thus date the fossils it contains.

Because of the sometimes capricious uptake of isotopes when rocks form, not all rocks can be dated radiometrically. But when available and cross-checked, the technique yields absolute ages for the rocks and therefore for the fossils they contain.

GEOLOGICAL AGES

The span of Earth history, from 4.6 billion years ago to the present, is divided into four unequal eons from the earliest time: Hadean (molten rocks), Archean (ancient rocks), Proterozoic (early life), and Phanerozoic (visible life) (figure 2.6). Poorly known and delineated in the early study of geology, the first three eons—Hadean, Archean, Proterozoic—were sometimes lumped together into terms of convenience, the *Precambrian* (before the Cambrian) or *Cryptozoic* (hidden life). Although such collective terms are still useful, the particular eons are better known today and each embraces a distinctive time in Earth history. The earliest eon is the *Hadean*, when most water existed in gaseous form and the Earth was still largely molten, leaving no rock record. The oldest dated rocks, at 4.2 billion years ago, mark the beginning of the *Archean*, and its conclusion is by convention taken as 2.5 billion years ago.

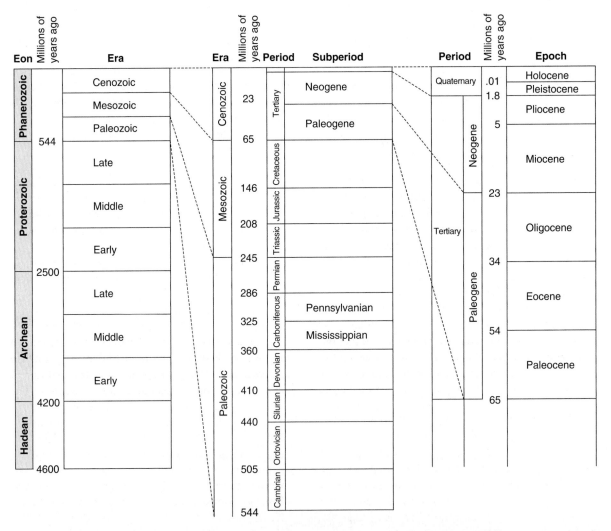

FIGURE 2.6   Geological Time Intervals The Earth's history, from its beginnings 4.6 billion years ago, is divided into four major eons of unequal length—Hadean, Archean, Proterozoic, and Phanerozoic. Each eon is divided into periods, and periods into epochs. Only epochs of the Cenozoic are listed in this figure.

Fossils of the middle Archean include impressions of microorgansims and stromatolites, layered mats of trapped bacteria. These are prokaryotic microorganisms, simple cells that lack a nucleus. We shall meet them and define them formally in chapter 4. For now, let's just note their early debut in Earth history. Through the early Archean, the Earth and its moon received heavy meteorite bombardment. Around each impact the crust melted and was perhaps punctured, allowing enormous outpourings of lava that flooded the surrounding surface. The heavy meteorite bombardment pitted much of the early Earth's crust, leaving it cratered. That is how we would find our Earth today, cratered and scarred, were it not for extensive geological remixing. In the subsequent several billion years since the early Archean, crustal movements—plate tectonics—have churned and reworked the Earth's pockmarked surface. These geological processes continue today, remelting old continents and building up new ones. In the process, early rocks and early-cratered continents are obliterated. That is why surviving traces of these early rocks are so rare. However, similar geological processes on the moon stopped very early in its history, due to the moon's smaller size, thereby preserving a glimpse of the ancient, cratered Archean landscape. Scientists survey our moon to study this early and relatively undisturbed companion record of geological events preserved there, but mostly eradicated on Earth.

From Archean into the start of the *Proterozoic* (2.5 billion years ago), the fossil record changes little. Stromatolites and other prokaryotic microfossils are still present. Early in the Proterozoic, eukaryotes appear as microfossils—microscopic impressions in rocks 2 billion years old. Unlike prokaryotes before them, eukaryotes have a nucleus, specialized cellular equipment, and the ability to reproduce sexually, rather than just by dividing (see chapter 4). This was also a time when the world's one large continental landmass broke up into small continents. By the late part of the Proterozoic, these smaller continents were scattered around the equatorial belt. At about 700 to 800 million years ago, the Earth descended into the most severe ice age ever experienced before or since, and remained in its frozen grip for almost 200 million years. Ice accumulated to a kilometer thick from the poles all the way to the equator, and average worldwide temperature dipped to $-50\,°C$, much colder than the average temperature today. Evidence of this glaciation comes from equatorial rocks of this age, which were clearly formed from compacted deposits of dirt and debris left behind

when the ice finally melted. What precipitated this ice age is currently a guess. Apparently, it started with the fragmentation of the single large continent into several smaller continents, all clustered around the equator, and carbon dioxide (CO_2) levels plummeted. The heavy equatorial rain fell on land, washing CO_2 from the air, reducing its greenhouse effect; temperatures dropped, and ice accumulated first at the poles. Accumulating polar ice reflected the sun's heat back into space, further plunging temperatures and extending the growing glacial sheets to the equator. Oceans would have frozen all the way to the bottom if it were not for deep heat seeping up from the Earth's core. Many microorganisms perished; some found refuge around deep-sea thermal vents, others around active volcanoes, and some along the edges of still-open ocean. Earth was a snowball and would have remained so had it not been for continued volcanic activity. CO_2 levels, which had declined, now started to rise. Like great chimneys above the ice, the tallest volcanoes spewed their plumes of ash into the atmosphere. Carbon dioxide now accumulated, a greenhouse effect grew, and the glaciers began to melt, first at the equator and then toward the poles.

As the worldwide ice age ended, life rebounded. At 544 million years ago (or, as we now know, slightly earlier), complicated, multicellular organisms made a sudden appearance. This explosion of multicellular organisms marks the start of the fourth eon, the *Phanerozoic*.

The Phanerozoic divides into three eras: *Paleozoic* (old animal life), *Mesozoic* (middle animal life), and *Cenozoic* (recent animal life). Invertebrates predominated during the Paleozoic era, as they still do today. But among the vertebrates, fishes were the most conspicuous and diverse, so the Paleozoic might be termed the "Age of Fishes." The term **biological radiation** applies to an evolving group that spreads into different environments and exhibits diversity of structure. The first vertebrates to live on land, the amphibians, appeared in the Paleozoic, and by late in this era an extensive radiation of their own was well underway. One later group derived from the amphibians, the reptiles, underwent an extraordinary radiation in the Mesozoic that took them into nearly every conceivable environment. So extensive was this radiation that the Mesozoic is often termed the "Age of Reptiles." The following era, Cenozoic, is often called the "Age of Mammals." Until the Cenozoic, mammal species were small in size and few in number. The vast extinctions of dinosaurs and many allied groups of reptiles that mark the end of the Mesozoic opened evolutionary opportunities for mammals. As a result, mammals enjoyed a period of their own expansive radiation into the ensuing Cenozoic, taking up dominant positions in most terrestrial ecosystems. However, this radiation must be kept in perspective. Although the numbers and types of mammals increased dramatically during this period, mammals still remained relatively rare compared to fishes and birds. If the Cenozoic were to be named for the vertebrate group with the most species, it would properly be named the Age of Fishes, or secondly, the Age of Birds, or thirdly still, the Age of Reptiles. Despite the previous Mesozoic extinctions that depleted their ranks, reptiles today still outnumber mammals in terms of numbers of species. Because we are mammals and it is our taxonomic class that is on the rise, the Cenozoic is known to most, not unexpectedly, as the "Age of Mammals."

Eras divide into periods, names for which originated in Europe. The Paleozoic era includes the *Cambrian, Ordovician,* and *Silurian* periods, named by British geologists working in Wales—*Cambria* was the Roman name for Wales, and the *Ordovices* and *Silures* were names for Celtic tribes that existed there before the Roman conquest. *Devonian* was named for rocks near Devonshire, also on British soil. The *Carboniferous* (coal-bearing) period similarly celebrates the British coal beds upon which so much of Great Britain's participation in the Industrial Revolution depended. In North America, coal-bearing rocks of this age match with the Lower and Upper Carboniferous, although American geologists sometimes prefer the terms *Mississippian* and *Pennsylvanian*, respectively, after Carboniferous rocks in the Mississippi Valley and the state of Pennsylvania (figure 2.6). The *Permian,* although named by a Scotsman, is based on rocks in the province of Perm in western Siberia. The Mesozoic era divides into three periods: The *Triassic* takes its name from rocks in Germany; the *Jurassic* from the Jura Mountains between France and Switzerland, and the *Cretaceous* from the Latin word for chalk *(creta),* which refers to the white chalk cliffs along the English Channel.

It was once thought that geologic eras could be divided into four parts—Primary, Secondary, Tertiary, and Quaternary—oldest to youngest, respectively. This proved untenable for the eras, but two names, *Tertiary* and *Quaternary*, survive in U.S. usage as the two periods of the Cenozoic. Internationally, however, these terms are mostly replaced with *Paleogene* and *Neogene.*

Periods divide into epochs, usually named after a characteristic geographic site of that age. Sometimes, boundaries between epochs are marked by changes in characteristic fauna. For example, in North America the late part of the Pliocene epoch is recognized by the presence of particular species of fossil deer, voles, and gophers. The early part of the succeeding Pleistocene is recognized by the appearance of mammoths. The boundary or transitional time between these epochs is defined by a fauna that includes extinct species of jackrabbits and muskrats, but not mammoths. Most names of epochs are not in general use and will not be referred to in this book.

The story of life on Earth is a narrative spoken partially from the grave, because of all animal species ever to evolve, roughly 99.9% are today extinct. In this story of life on Earth, most of the casts of characters are dead. What survive are their remnants—the fossils and the sketchy story these fossils tell of the structure and early history of life.

FOSSILS AND FOSSILIZATION

When we think of fossil vertebrates, we probably think of bones and teeth, the hard parts of a body that more readily resist the destructive processes following death and burial. Most fossil vertebrates are known from their skeletons and dentition. In fact, some extinct species of mammals are named on the basis of only a few distinct teeth—all that remains of the whole animal—because the mineral composing bones and teeth, a calcium phosphate compound, is usually preserved indefinitely with little change in structure or composition. If the bones lie in earth with seeping groundwater, then over time other minerals such as calcite or silica may also soak into the tiny spaces of bone to add further minerals and harden it.

But fossils are more than bones and teeth. Occasionally products of vertebrates will fossilize, such as eggs (figure 2.7). If the tiny bones of the young are preserved inside the eggs, then we can identify the animals and the general group to which they belong. This tells us more than just the structure of this species; it also tells us something about its general reproductive biology (figure 2.8). The recent discovery in Montana of fossilized clumps of eggs of duckbill dinosaurs testifies to the reproductive style of this species, but there was accompanying circumstantial evidence to imply even more. The clumps, or clutches, of eggs were in the same vicinity, about two adult di-

FIGURE 2.7   Fossil Eggs This clutch of dinosaur eggs from about 70 million years ago is thought to be from a *Segnosaur*, an enigmatic carnivorous (or omnivorous) species about which we know little. These eggs, found in China, were laid together in pits or holes in the ground that may have been lined with plant material, which did not fossilize. Each egg is about 6 cm in diameter. Photograph kindly supplied by Lowell Carhart, Carhart Chinese Antiques.

nosaur body lengths apart, suggestive evidence that this was a breeding colony. Analysis of the rock sediments in which the eggs were found suggests that the colony was on an island in the middle of a run-off stream from the nearby Rocky Mountains. The presence of bones of duckbills of different ages implies further that the young stayed around the nests until grown. If true, then perhaps the parents even gathered food and brought it back to nourish the newly hatched young. At least for this species of duckbill dinosaur, the picture to emerge is one, not of a dispassionate reptile that laid its eggs and departed, but instead of a reptile that provided sophisticated parental care and probably supportive social behavior. Food gathering, protection of the young, learning from parents, pair bonding, and more are at least plausibly implied here by the fossils that remain.

Still more comes to us from fossils. If a full animal skeleton is discovered, the location of its stomach in life can be determined. Although the soft tissue of the stomach decayed away soon after the animal's death, microscopic analysis of that stomach region still might reveal the types of hard foods it ate shortly before death. The dung of animals is sometimes fossilized. Although we might not know which animal dropped it, we can gain some notion about what types of foods were eaten.

Soft parts usually decay quickly after death and so are seldom directly fossilized. A dramatic exception to this was the discovery in Alaska and separately in Siberia of woolly mammoths frozen whole. When these were thawed, they yielded actual hair, muscle, viscera, and digested food. But these are exceptional cases. Rarely are the people who study fossils—paleontologists—so lucky.

Although soft parts usually do not fossilize, occasionally they leave an impression in the earth in which they are buried. The impressions of feathers in the rock around the skeleton of *Archaeopteryx* make a clear statement that this animal was a bird (figure 2.9).

The behavior of vertebrates, or at least how they ambled across ancient landscapes, is implied in the spacing of fossilized footprints they left. Such footprints are unlabeled, of course—we do not know who made them. But their size and shape, together with knowledge of known animal assemblages of the time, give us a good idea of who stamped them out. From sequential spacing of dinosaur tracks, the speed of the

FIGURE 2.8
Fossil Ichthyosaur Small skeletons are seen within the adult's body and next to it. This may be a fossilized birth, with one young already born (outside), one in the birth canal, and several more still in the uterus. Such special preservations suggest the reproductive pattern and live-birth process in this species.

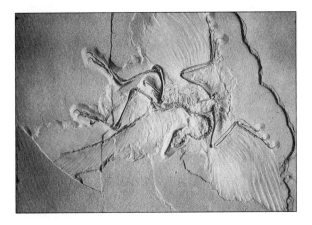

FIGURE 2.9 *Archaeopteryx* The original feathers have long since disintegrated, but their impressions left in the surrounding rock confirm that the associated bones are those of a bird.

Recovery and Restoration

Paleontologist and artist combine talents to visually recreate the extinct animal as it might have looked in life. Remnants of long-dead vertebrate animals provide source material from which their basic anatomy is reassembled. After so long a time in the ground, even mineral-impregnated bone becomes brittle. If the original silt sediments around buried bone have hardened to stone, then this compacted bony matrix must be chipped or cut away to expose the fossilized bone within (figure 2.10a). With picks and chisels, the upper surface and sides of the bone are partially exposed, wrapped in protective plaster, and allowed to harden. The remainder of the bone is then exposed and the plaster wrap is extended to completely encase the bone (figure 2.10b). Now supported by the plaster, the brittle bones are shipped to laboratories where the plaster and any additional clinging rock are removed. Tiny needles once used to pick away the rock might be replaced now with dentist's picks. Or a stream of fine sand from a pencil-sized nozzle may be used to sandblast or carve away the rock, finally freeing the fossil.

Confidence in a restored version of a fossil rests largely upon the direct fossil evidence and upon the knowledge of modern, living counterparts to supply the likely biology of the fossil (figure 2.11a). Size and body proportions are readily determined from the skeleton. Muscle scars on adjacent bones help determine how muscles were stretched across limbs and body. When muscle mass is added to the skeleton, it

animal can be calculated at the time the tracks were made. Similarly, 3-million-year-old volcanic ash, since hardened to stone, holds the footprints of ancestral humans. Discovered in present-day Tanzania by Mary Leakey, the footprints are those of a large individual, a smaller individual, and a still smaller individual walking in the steps of the first. A family? Male, female, and young offspring? Could be. But at the least, these human footprints confirm a view independently deciphered from skeletons—namely, that our ancestors of 3 million years ago walked upright on two hindlegs.

Consider This—

Lewis and Clark—In Search of Mastodons

Thomas Jefferson, when vice president, reported before a scientific society, and subsequently published a paper in 1797, on *Megalonyx*, a fossil ground sloth whose bones he had discovered in Virginia. He also knew of large bones of mastodons and other great fossil animals from the eastern United States. When president, he set up the Lewis and Clark expedition to find a passage to the Northwest. One of the expedition's other goals was to see if mastodons or any other animal known only from fossils still lived in the vastness of the western continent. In 1806, the expedition found a giant leg bone near Billings, Montana, that was certainly from a dinosaur.

Unfortunately, no living mastodons were found; we now know they disappeared from North America at least 8,000 years earlier.

Jefferson was irked into including this mastodon search into the Lewis and Clark mission because of an earlier comment by a European scientist. Prior to the American Revolution, the French naturalist George Louis LeClerc de Buffon proposed that, compared to his rich European environment, the North American environment was impoverished—unable to support any animals robust in character. His patriotic pride stung, Jefferson countered, using the mastodon as an example of such an animal that had once thrived in the New World.

(a)

(b)

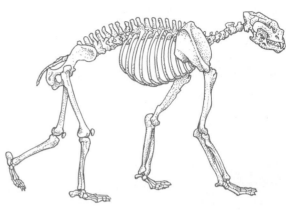

(a)

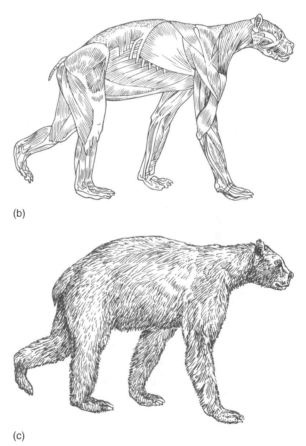

(b)

(c)

FIGURE 2.10   Fossil Dig in Wyoming (a) Partially exposed dinosaur bones. The work crew prepares the site and notes the location of each excavated part. (b) This *Triceratops* femur is wrapped in a plastic jacket to prevent disintegration or damage during transport back to the museum. Photos courtesy of Dr. David Taylor, Executive Director, NW Museum of Natural History, Portland, Oregon.

gives us a general idea of body shape (figure 2.11b). The general feeding type—herbivore or carnivore—is implied by the type of teeth. The general lifestyle— aquatic, terrestrial, aerial—is determined by the presence of specialized features, claws, hooves, wings, fins, and so forth. The type of rock from which the fossil was recovered might further testify to its lifestyle—marine or terrestrial, swamp or dry land. Comparison to related and similarly structured living vertebrates can help confirm walking posture and general environmental requirements.

Ears, a proboscis (trunk), nose, hair, and other soft parts that have no internal bones must be guessed. Living relatives help. For instance, all living rodents have

FIGURE 2.11   Restoration of a Fossil (a) This skeleton of the extinct short-faced bear, *Arctodus simus*, is positioned in a likely posture in life. (b) Scars on the bones from muscular attachments and knowledge of general muscle anatomy from living bears allow paleontologists to restore muscles and create the basic body shape. (c) Hair added to the surface completes the picture and gives us an idea of what this bear might have looked like in its Alaskan habitat 20,000 years ago.

vibrissae (long hairs on the snout), so these can be included in restorations of extinct rodents. Except for some burrowing or armored forms, most mammals have a coat of fur, so it is fair to cover a restored mammal with hair (figure 2.11c). All living birds have feathers and reptiles have scales, so these can logically be added to restored bird or reptile fossils, although the length or size must be guessed. Surface color or patterning, stripes, or spots almost never are preserved in an extinct animal. Certainly, living animals are colored with patterns that camouflage their appearance, or emphasize courtship and territorial behaviors. Reasonably the same strategy held among extinct animals. A restoration should be faithful to this, although the specific color and pattern chosen must be produced from the artist's imagination. A flashy Hollywood movie showing dinosaurs at battle may satisfy our curiosity for what these great animals might have looked like in life, but we should remember that in any such restorations, much interpretation stands between the actual bones and the computer-animated reconstruction.

New fossil finds—especially of more complete skeletons—improve the evidence upon which to build a view of extinct vertebrates. But often new insights into old bones arise from an inspired reassessment of the assumptions upon which earlier restorations were based. Such inspiration encourages our new view of dinosaurs. They were likely warm-blooded beasts enjoying a lifestyle less like that of reptilian lizards and turtles and more like that of mammals and birds. Certainly new fossil discoveries get us thinking, but the major change in the way dinosaurs are restored today by artists and paleontologists reflects new courage in interpreting them as active and predominant land vertebrates of the Mesozoic.

Reconstruction of human fossils has followed fashion as well as new discoveries. When first unearthed in the mid nineteenth century, Neandertal bones were thought to be those of a robust individual human killed during the Napoleonic wars a few decades earlier. In the early part of the twentieth century, this view gave way to the stoop-shouldered, beetle-browed, and dim-witted image. Neandertals were thought to be a breed apart from modern *Homo sapiens*, and restoration reflected this demoted image. In the mid twentieth century, Neandertals were back again as a human subspecies, *Homo sapiens neanderthalensis*. Artists produced restorations making Neandertals look like modern humans. Currently, Neandertals are again their own species, *Homo neanderthalensis*—the consequence of a recent reevaluation.

The point before us is not to scoff at those who err or follow fashion, but to recognize that any restoration of

a fossil is several steps of interpretation away from the direct evidence of the bones themselves. Efforts to reconstruct the history of life on Earth improves with new fossil discoveries, but also with improved knowledge of basic animal biology. Our understanding and image of restored fossils today will likely give way to a better understanding tomorrow. The better we understand the function and physiology of animals, the better our assumptions when restoring life to the bones of dead fossils. But it is worth the risks and pitfalls to recreate the creatures of the past, because in so doing we recover the unfolding story of life on Earth they have to tell.

From Animal to Fossil

The chance is extremely remote that an animal, upon death, will eventually fossilize (figure 2.12). Too many carrion eaters await within the food chain. Disease or age or hunger may weaken an animal, but a harsh winter or a successful predator is often the immediate instrument of death. Carnivores consume the flesh, and the bones are then broken and picked over by marauding scavengers that follow. On a smaller scale, insect larvae and bacteria then feed on the diminishing remains. By stages, the deceased animal is broken down and its chemical components recycled into the food chain. Even in a small forest, hundreds of animals die each year; yet as any hiker or hunter can attest, it is rare to come upon an animal that has been dead for any length of time. The scavengers and decomposers are quick to work. Even rodents, whose customary food is seed or foliage, will gnaw at bones of dead animals for the calcium. To escape this onslaught, something unusual must intervene before all traces of the dead animal are literally eaten up.

One route to fossil production is via burial by volcanic ash. In August 79 C.E., Mt. Vesuvius, in Italy, erupted spewing tons of ash on surprised citizens in Pompeii and nearby Herculaneum, burying them under many feet of volcanic debris. Grass and vines gradually grew over the smothered cities and their names were mostly forgotten. Not until 1738 (Herculaneum) and 1748 (Pompeii) were they rediscovered, and the rich cultural history of the Roman Empire represented in these cities was recovered from the protective ash, which had cocooned them for seventeen centuries.

In northeastern Nebraska (U.S.A.), 10 million years ago, volcanic spouts disgorged ash across the landscape interring the local fauna and flora. The quick burial in volcanic ash produced remarkable lifelike skeletons of rhinos, horses, camels, birds, and smaller animals, some of which contained stomach contents

FIGURE 2.12   Making Fossils   The remains of extinct animals that persist have escaped the appetites of scavengers, decomposers, and later tectonic shifting of the Earth's crustal plates in which they reside. Most surviving fossils are of dead animals that quickly became covered by water and escaped the notice of marauding scavengers. As more and more silt is deposited over time, the fossil becomes even more deeply buried in soil compacted into hardened rock. For the fossil held in the rock to be exposed, the Earth must open either by fracture or by the knifing action of a river.

and unborn young. The volcanic ash was a kind of "time capsule" of this ancient world. The ash even caught many animals engaged in their natural social behavior. An entire herd of rhinos was preserved in its nocturnal defensive formation—males at the perimeter facing out, females and young in the center.

Animals living in water or near the shore are more likely to be covered by mud or sand when they die than are upland animals. Upland animals die on ground exposed to scavengers and decay. Thus, most fossil-bearing rocks are sedimentary—formed in water. Even if an animal is successfully buried, its bones are still in peril. Under pressure and heat, the silt turns to rock; shifting and churning and settling of the rock layers can pulverize the fossils within. The longer a fossil lies buried, the greater the chance these tectonic events will obliterate it. Thus, the older the rock, the less likely it will harbor fossils. Finally, to be known, the fossil must be discovered. Theoretically, one could dig straight down anywhere through the Earth's crust and eventually hit fossil-bearing rocks. Excavations for roads or buildings occasionally unearth fossils in the process. But such a freelance approach to fossil discovery is too chancy and expensive. Instead, paleontologists visit natural exposures where the sheets of surface rock have fractured and slipped apart or been cut through by rivers. The edges of rock layers are now exposed, perhaps for the first time in millions of years. In these strata the search begins for fossils that have survived.

Missing Fossils

The complaint by some laypeople that unsatisfactory gaps occur in the recovered fossil record is based on a misunderstanding of the process of fossilization. The onslaught of scavengers when an animal dies and the destructive geological processes while bones are buried make it unlikely that any bits or pieces of ancient animals will survive to be discovered millions of years later by paleontologists. Even recent extinctions leave only faint traces.

Until late into the nineteenth century, the Great Plains of North America teemed with bison from horizon to horizon. They lived and died in great numbers on these lands for most of the previous 10,000 years, producing many millions of carcasses in that time—and, therefore, potentially many millions of fossils. Yet today, when driving through these open lands, the traveler does not meet great heaps of bison bones stacked high from horizon to horizon. Scavengers and decomposers have done their work.

Early in the nineteenth century, eastern North America supported 3 to 5 billion passenger pigeons. Reports from back then describe passing migratory flocks of passenger pigeons that darkened the skies for days. Wholesale shooting and netting of these communal birds led to over-exploitation for their favored meat, reduced their numbers below recovery, and the last passenger pigeon died in 1914. Yet despite the thousands of birds that lived and died each year until 1914, their bones cannot be easily found in the recent soils of eastern North America.

OVERVIEW

Fossilization is a chancy process. It is remarkable that the recovered fossil record is as complete as it is today. Further, the economics of finding old dinosaurs is a practical problem as well. More money was spent to produce the artificial dinosaurs in the movie *Jurassic Park* than to recover some of the real fossils in the first place. Nevertheless, the fossil record becomes better known all the time as the work of paleontologists steadily continues. And certainly today the mistaken nineteenth-century views of Earth's age have been corrected. The Earth has been a hospitable planet for life's emergence and evolution for most of its 4.6 billion years. This vastness of time has been evolution's partner. Although unresolved during his lifetime, Darwin's assumption that the Earth was very old has been vindicated. But you will recall that besides the issue of the age of the Earth, Darwin's proposed principles of evolution also faced another scientific challenge—sorting out the method by which favored characteristics are inherited. Although Darwin, like most of the nineteenth-century scientific community, did not know it, the correct mechanism of heredity strangely enough was discovered during Darwin's lifetime. As we shall see in chapter 3, the discovery and subsequent neglect of the principles of heredity represent one of the most puzzling episodes in the history of scientific discovery.

Selected References

Alexander, R. McN. 1976. Estimates of speeds of dinosaurs. *Nature* 261: 129–30.

Alexander, R. McN. 1989. *Dynamics of dinosaurs and other extinct giants.* New York: Columbia University Press, p. 167.

Appleby, R. 1979. Fossil aches and pains. *New Scientist*, 16 August, pp. 516–17.

Gould, S. J. 1983. False premise, good science. *Nat. Hist.* 92(10): 20–26.

Hoffman, P. F., and D. P. Schrag. 2000. Snowball Earth. *Sci. Amer.* 282(1): 68–75.

Lanham, U. 1973. *The bone hunters*. New York: Columbia University Press.

Mayr, E. 1963. *Animal species and evolution*. Cambridge, Mass: Belknap Press.

Mayr, E. 1982. *The growth of biological thought*. Cambridge, Mass: Belknap Press.

Rudwick, M. J. S. 1985. *The meaning of fossils*. Chicago: University of Chicago Press.

Simpson, G. G. 1934. *Attending marvels: A Patagonian journal*. New York: Macmillan.

Simpson, G. G. 1983. *Fossils and the history of life*. New York: Scientific American Books.

Learning Online

Testing and Study Resources

Visit the textbook's website at **www.mhhe.com/ evolution** (click on the book's title) to take advantage of practice quizzing, study/writing tips, timely news articles, and additional URLs for research on the topics in this chapter.

3

"Mendel must have done a good deal of observation and experiment before the formal work, in order to tease out these [seven characters] and convince himself that seven qualities or characters was just what he could get away with. There we glimpse the great iceberg of the mind in that secret, hidden face of Mendel's on which the [research] paper and the achievement float."

Jacob Bronowski, physicist and mathematician

Heredity

INTRODUCTION

When Darwin finally settled in to write *The Origin of Species*, he faced one of the major problems that had contributed to his writing delay—the problem of heredity. If, as Darwin proposed, natural selection preserves individuals with favorable traits, then how are these favored traits passed undiminished to offspring of the next and to succeeding generations? And in fact, how do new traits arise in the first place? No scientific consensus existed, no established body of laws or concepts on heredity were available to come to his aid. In fact, much of what biologists of his day believed was not just wrong, it was misleading. In Darwin's own words in *The Origin of Species,*

> The laws governing inheritance are for the most part unknown. No one can say why the same peculiarity in different individuals of the same species, or in different species, is sometimes inherited and sometimes not so.

Darwin could see this challenge brewing, so he filled *Origin* with examples of ideas and possibilities for sources of variation and methods to preserve variation—more than three chapters' worth. Let's look at these two problems systematically. First we will examine the nature of inheritance, and then we will turn to the sources of variation.

INHERITANCE BY INTUITION

Early Intuition

Attempts to control heredity go back a long way in human history. And many of these attempts were successful, at least by the standards of their day. Unheralded people far back in human history learned to improve domestic animals and crops by inbreeding and crossbreeding the best among the parent stock. The process of fertilization was appreciated in much of the ancient world, too. For instance, male and female flowers of the date palm are found on different trees. "Bring males to the female," wrote Theophrastus (371–287 B.C.E) of the date palm "for the male makes them ripen and persist." What was not appreciated was how characters were passed to each new generation—the mechanism of heredity. In the seventeenth century, Anton van Leeuwenhoek, a Dutch drape maker by day but glass grinder in his spare time, became one of the most accomplished lens makers in Europe. With this skill, he fashioned a microscope, crude by today's standards, but a remarkable tool giving entry into the tiny world in a drop of pond water. Bee wings, blood, and muscles received his examination. In fresh semen from various animals, including human, van Leeuwenhoek saw sperm swimming about. He called the sperm "animalcules"—literally, "little animals." Other scientists quickly followed and turned new microscopes upon animalcules. A vivid imagination led some to claim that they could see within each human sperm a "homunculus," or tiny human—a future human being in miniature. They deduced that once implanted in the womb of the female, the sperm's homunculus was simply nurtured there and grew to term. Any resemblance the child bore to the mother was attributed to maternal influences during incubation in the womb. The scientists who gave a central hereditary role to the sperm were termed "spermists."

Very quickly, a rival school arose, termed the "ovists." They followed the discoveries of another Dutchman, Régnier de Graaf. In 1672, de Graaf described what he thought was the human egg, or ovum, in a human ovary. As it turned out, what he saw was not the actual ovum, but the coat of cells surrounding it, the ovarian follicle. (Von Baer was actually the first to describe the true mammalian ovum, in 1827.) Nevertheless, de Graaf's discovery inspired the counterclaim that the ovum actually contained the future human in miniature, and the sperm merely stimulated its growth. Both the ovists and the spermists reasoned a step further. If the homunculus, whether embedded in the ovum or in the sperm, bore the characters of the adult-to-be, then each homunculus must also contain, telescoped within it, the preformed homunculus of the generation to follow thereafter; and in turn within it the preformed homunculus of the next generation after that—and so on, generation upon generation nested within and within and within.

Today, the high-resolution electron microscope gives us a direct, visual inspection of eggs and sperm, and an answer: No tiny, preformed miniatures are stacked up awaiting their turn. But the views of spermists and ovists alike fell into eclipse early in the nineteenth century, long before the advent of the electron microscope. This was largely because the implied mechanism of inheritance was incompatible with the practical experience of master gardeners of ornamental flowers. The theories of ovists and spermists implied that one sex carried the characteristics passed to offspring. Scientists differed in their view as to whether the ovum or the sperm was the carrier, but they agreed that one sex bore the larger hereditary influence on offspring. However, the experience of ornamental gardeners showed that both sexes contributed to a new plant variety. In artificial breeding of plants (crossing), it seemed to matter little which plant supplied pollen (which contain the male sex cells) and which supplied egg cells (female sex cells). Regardless, offspring could be produced that carried the desired characteristics.

Blending Inheritance

The pattern of heredity was baffling. Often a trait would seemingly disappear in offspring only to suddenly reappear in the third or fourth generation. In other crosses, the character seemed to merge with related characters. By Darwin's time, the concept of a blending inheritance held great sway in scientific circles and seemed a promising concept to address the mechanism of heredity.

The concept of blending inheritance simply proposed that when male and female sex cells, the **gametes** (*gamos*, marriage) combine, the characters they carry blend. Like mixing two colors of paint, the mix is a blend of the two pigments. Even in its day, the concept of blending inheritance was unsatisfactory, at least for many types of crosses. If distinct characters blended, then how could they sometimes return again as distinct characters after several generations?

Nevertheless, this concept was no small obstacle for Darwin. Natural selection acted upon character variation. But if characters blended like paints, how could favored varieties be preserved and avoid being diluted into oblivion? The chapters on variation in *The Origin of Species* were packed with the evidence. Certainly, variation existed in domestic breeds and in nature, he argued. Look at the various races of dogs bred under domestication—greyhound, bloodhound, terrier, spaniel, bulldog; or look at the domestic breeds of pigeons—English carrier, short-faced, tumbler, runt, turbit, Jacobin. And in nature, look at the many varieties of plants, such as the oak tree family with its many varieties. Variety exists. No one doubted that, but the mechanism of its inheritance was a difficult question.

MENDELIAN INHERITANCE

Darwin's life was touched again by dramatic irony. As a quirk of fate, the mechanism of heredity, and thus the answer to some of his critics, was discovered during Darwin's lifetime, but he did not know of it. The mechanism of inheritance that eluded Darwin was unfolding as a quiet drama in a remote monastery garden in what is today the Czech Republic. This is the story of an Austrian monk, Gregor Mendel.

Gregor Mendel

Gregor Mendel (1822–1884) (figure 3.1) attended the University of Vienna for two years, but a nervous student he unsuccessfully tested for a teaching certificate. He "lacks insight and the requisite clarity of knowledge" wrote one examiner, and failed him. Mendel thus returned to the monastery of Brno, today part of the Czech Republic. The year was 1853. Several years later, he began his experiments with peas. By his own account, these experiments lasted eight years. They were to lead to a new insight into the mechanism of inheritance.

Specifically, Mendel made two unique proposals. First, inherited characters did not blend, but were transmitted in discrete parcels he called Elemente, and today we call **genes**. Second, each gene (elemente) had two possible expressions in the same individual—two **alleles** as they are now called—one dominant, one recessive. To begin, Mendel picked seven characteristics of peas and crossed plants with these characteristics. The adults used to begin the experimental crosses represented the **parental generation**, noted as "**P**". The progeny they produced represented the F_1 **generation** (L. *filius*, son), noted as F_1. If he was right about inheritance, then when this first generation was crossed, their progeny, the F_2 **generation**, should express predictable ratios of dominant to recessive traits, specifically a 3:1 ratio, dominant to recessive. For example, the trait *seed color* was expressed as yellow (dominant) or green (recessive). In actual numbers, Mendel's second generation exhibited yellow (6,022):green (2,001), or 3.01:1, statistically close to the 3:1 ratio he predicted (figure 3.2).

Mendel knew nothing of DNA, genes, or chromosomes. No one in the nineteenth century would know of these fixtures of twentieth-century genetics. But still, he got the answer right. Was Mendel only lucky? Almost certainly not. His paper on pea inheritance was written with a confidence beyond the data, his experimental design was modern, and his thinking carried a

FIGURE 3.1
Gregor Johann Mendel (1822–1884)   Mendel discovered a mechanism of inheritance while conducting experiments on garden peas at a monastery in Brunn, Austria (now Brno, Czech Republic).

polish that reveals a clear idea of the mechanism of heredity he sought to demonstrate. Yet when he published his results, no one noticed. Too much statistical mathematics on a little garden pea.

Mendel wrote to several prominent botanists of his day. They apparently saw his work as no more than one tiny set of experiments, on a rather common plant, by an unestablished amateur. More indifference. He publicly read his paper on pea plants before a scientific organization, the Brunn Natural Science Society; no one noticed. Eventually, an abstract was prepared of his work and published in an enormous German encyclopedia of plant breeding. There, in this German encyclopedia, Mendel's results lay buried. The results—which would put blending inheritance to rest, which could give Darwin the answer to his critics, which could place the theory of natural selection on a solid footing—went unnoticed.

In fact, Darwin's quest for insight into the mechanism of heredity led him independently on a course surprisingly close and parallel to Mendel's, but with snapdragons instead of garden peas. In basic setup, Darwin's approach was identical. He crossed two true-breeding parent plants (the **P** generation), one carrying spiral and the other radial arrangement of flowers on stems. In the first-generation hybrids (F_1), the radial

Character	Dominant vs. recessive trait	F₂ generation		Ratio
		Dominant form	Recessive form	
Flower color	 Purple × White	705	224	3.15:1
Seed color	 Yellow × Green	6022	2001	3.01:1
Seed shape	 Round × Wrinkled	5474	1850	2.96:1
Pod color	 Green × Yellow	428	152	2.82:1
Pod shape	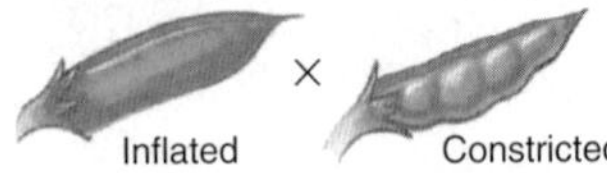 Inflated × Constricted	882	299	2.95:1
Flower position	 Axial × Terminal	651	207	3.14:1
Plant height	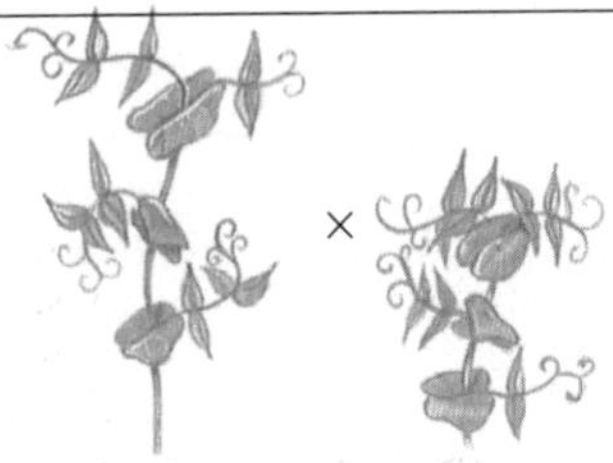 Tall × Dwarf	787	277	2.84:1

FIGURE 3.2   Mendel's Experimental Results   The table summarizes results of Mendel's experiments following seven characters during crosses of the garden pea. For example, Mendel observed that in the F_1 generation, the character (or trait) *seed color* occurred 6,022 yellow (dominant) to 2,001 green (recessive)—a ratio of 3.01:1, very close to his predicted ratio of 3:1.

flower trait was not expressed; only spiral-flowered plants appeared. Next, he crossed the spiral offspring to get the second generation (F_2). In actual numbers, the second generation showed 88 to 37, spiral to radial, or 2.37:1. That was close to a confirming 3:1 ratio, but it meant nothing to Darwin. He attributed the reappearance of the spiral trait in the second generation to latency—"invigoration" during the preceding F_1 generation rather than to dominance and recessiveness, as Mendel correctly recognized. Here, Darwin had come within a hair's breadth of making the second major biological discovery of the nineteenth century: particulate inheritance. But he missed it.

There is a special historical footnote here too. Following Darwin's death, a copy of the great German encyclopedia on plant breeding with Mendel's abstract in it was discovered in Darwin's laboratory. On the page facing Mendel's abstract, the article was marked with notations in Darwin's own handwriting. He had read it. But the abstract of Mendel's paper bore no marks, no marginal notes. Darwin was a page away from the paper that would clarify his own snapdragon research, provide him with a mechanism of heredity, and give his evolutionary theory of natural selection the support he sought. Darwin overlooked it.

In 1900, independently, several scientists who were determined to settle the issue of inheritance combed the literature. Mendel's paper was rediscovered and its larger significance was finally appreciated. Mendel's work at last took its deserved place in modern biology, but, too late for Mendel personally—he had died 16 years earlier (1884); and too late for Darwin—he had died 18 years earlier (1882).

Mendel's Experiments

Modern genetics begins with the pioneering work of Gregor Mendel. To understand his contribution, we must look at his methods. First, the pea plant. The *stamen* is the male part of the plant; the *carpel* (pistil), the female. In some plant species, stamens and carpels reside in different flowers or even in different individual plants. But in the garden pea, both reside together in the same flower (figure 3.3). Pollen grains, holding sperm, are produced in the male part and travel to and settle on the tip of the female carpel. Upon arrival, the pollen grains sprout pollen tubes that grow down the carpel to its base, where they reach and discharge sperm that fertilize the eggs.

In the pea, petals of each flower protectively embrace their anthers and stigma and do not open until after fertilization. Thus, whereas most plants **cross-fertilize** between stamen and carpels on different individual plants, the pea **self-fertilizes** between stamen and carpels of the same plant—in fact, of the same flower. Mendel took advantage of this. With tweezers, he carefully parted the petals of each flower, removed the still immature stamens, and dusted the female carpel with male pollen of his choosing (figures 3.3 and 3.4). When finished, the petals snapped back into place to prevent entrance of errant wind-borne pollen from another plant. Thus, Mendel knew exactly which plants, and with what characters, he had crossed. The ovary now grew into a pod and the ovules within grew into peas. Mendel collected the ripened peas, planted them, and recorded the characteristics of the offspring that sprouted from them—the first generation (F_1) of his artificial crosses.

We can illustrate the genetics involved by looking at the inheritance of a single character—flower color (figure 3.5). Mendel crossed a plant with purple flowers with one with white flowers. The first generation of offspring of this cross (F_1) all bore purple flowers. However, when he then crossed the F_1 offspring with each other, the next generation (F_2) produced some individual plants with white flowers as well as plants with purple flowers. The white character had not blended or disappeared, but persisted.

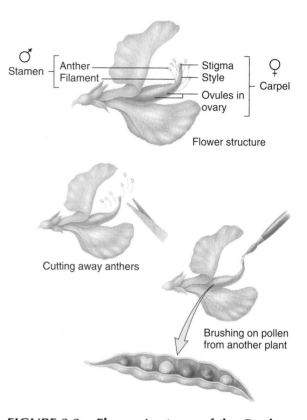

FIGURE 3.3   Flower Anatomy of the Garden Pea   Pollen, containing sperm, is produced in the male anther; the ovules contain eggs in the female carpel. To prevent self-pollination, Mendel cut away the male anthers and dusted the female carpel with pollen from another plant of his choosing to control the cross. Seeds ripened in the pod, their characteristics were scored, and then these seeds were planted to grow and flower. The characteristics of these full plants were then scored.

Mendel reasoned that flower color was contributed by a pair of particles (alleles), each with two discrete expressions—one for purple flowers and one for white flowers. Each parent contributed one of the two particles at fertilization to the offspring. He called the trait predominating in the F_1 generation *dominant* (here, purple flowers), and the character that reappeared in the F_2 generation he termed *recessive* (here, white flowers). Mendel further reasoned that even though the recessive trait did not visibly appear until the F_2 generation, it must have been present in pea plants of the preceding F_1 generation.

Let's examine the genetics underlying the observed appearance of the plants (figure 3.5). The parent plants are true breeding—one parent plant is purple and genetically carries only dominant alleles *(PP);* the other is white and carries only recessive alleles *(pp).*

The **F₁** offspring receive one of each allele *(Pp)*, but all are purple, the expression of the dominant allele. Next, when these offspring are crossed to produce the **F₂** generation, these gene alleles segregate into gametes (eggs and sperm) but genetically the gametes are of only two kinds—those with the dominant allele *P* and those with the recessive allele *p*. Thus, genetic combinations at fertilization are limited to three genetic possibilities: *PP, Pp, pp*. But in appearance, only two flower colors are produced: purple flowers *(PP, Pp)* and white flowers *(pp)*. This is a statistical outcome based on large numbers of offspring, not a guarantee in a separate breeding cross. On average, then, out of every 4 offspring, 3 will be purple and 1 will be white—a 3:1 ratio. This ratio was Mendel's prediction. His experiments confirmed this and, hence, supported his underlying hypothesis of particulate inheritance.

Let's summarize Mendel's results again in modern terms. Organisms are said to be **homozygous** when they carry two identical alleles for a character—in our example, two alleles for purple flowers *(PP)*, or two gene alleles for white flowers *(pp)*. Crossing parent plants homozygous for purple flowers with homozygous plants for white flowers gives only **heterozygous** hybrids *(Pp)*—purple flowers produce gametes of only one kind, namely alleles *P* for purple flowers, and white-flower parents produce only gametes with alleles *p* for white flowers. However, crossing the heterozygous offspring gives more combinations at fertilization. Some offspring receive a dominant allele *P* from each parent and so are homozygous dominant *(PP)*, some receive a recessive allele *p* from each and so are homozygous recessive *(pp)*, and still others receive a dominant allele *P* from one parent and a recessive allele *p* from the other and so are heterozygous *(Pp)*.

The genetic makeup, or **genotype**, of offspring would be predictable. On average, out of every 4 off-

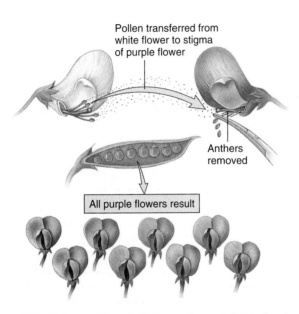

Pollen transferred from white flower to stigma of purple flower

Anthers removed

All purple flowers result

FIGURE 3.4   Mendel's Experimental Method In this experiment of a cross between true-breeding white- and purple-flowered plants, Mendel pried open the surrounding petals of the purple-flowered plant and removed the male part, thus preventing self-fertilization. Then he dusted the anther with pollen he had selected from the white-flowered plant. The resulting seeds were planted, and all produced purple-flowered plants.

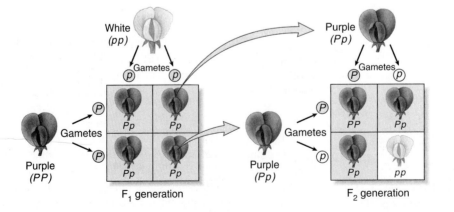

White *(pp)*

Gametes *p*   *p*

Purple *(Pp)*

Gametes *P*   *p*

Purple *(PP)*

Gametes *P*   *P*

Pp   *Pp*   *Pp*   *Pp*

F₁ generation

Purple *(Pp)*

Gametes *P*   *p*

PP   *Pp*   *Pp*   *pp*

F₂ generation

FIGURE 3.5   Independent Segregation—Single Trait, Flower Color Mendel's cross of pea plants for flower color started with true-breeding white-flowered (recessive) and purple-flowered (dominant) plants. All **F₁** offspring of this cross were purple-flowered and genetically heterozygous *(Pp)*. When these were crossed, the resulting **F₂** offspring averaged 3 purple-flowered for every 1 white-flowered plant, a 3:1 phenotypic ratio. However, the ratio of genotypes is 1:2:1 (1 *PP*: 2 *Pp*: 1 *pp*).

spring, there would be 1 homozygous dominant *(PP)* offspring, 1 homozygous recessive *(pp)*, and 2 heterozygous *(Pp)*. The outward appearance of a trait, or its **phenotype**, would be predictable as well, but has a different ratio. Because the purple-flower allele is dominant, heterozygous plants would look phenotypically like the homozygous dominant plants. Thus, on average, every 4 offspring would genetically have a 1:2:1 ratio of *PP:Pp:pp*. But phenotypically, for every 4 offspring, 3 would produce purple flowers *(PP, Pp)* and 1 would produce white flowers *(pp)*, for a ratio of 3:1. The actual numbers Mendel obtained in the **F₂** generation were 705 purple and 224 white, for a ratio of 3.15:1. This was close enough to give him confidence that his hypothesis of the particulate nature of inheritance was correct.

Testcross

To examine predictions based on his hypothesis about particulate inheritance, Mendel devised a test (figure 3.6). His hypothesis predicted that when heterozygous plants are crossed with homozygous plants, there would be a 1:1 ratio of dominant to recessive traits in the **F₁** generation resulting from such a pairing. Today, this same heterozygous/homozygous cross is called a *testcross,* but because Mendel's basic hypotheses are so well substantiated, there is little value in repeating these particular crosses as tests of simple Mendelian principles of inheritance. Instead, the testcross is now used to determine whether an individual expressing the dominant phenotype is genetically homozygous dominant or heterozygous. Certainly, a purple flower cannot be genetically homozygous reces-

sive *(pp)*, or it would be white! But simply by looking at a purple flower it is impossible to tell whether it is homozygous dominant *(PP)* or heterozygous *(Pp)*. Both genotypes produce purple flowers. The genetic composition of an individual with a dominant phenotype is determined by crossing it with an individual known to have a recessive phenotype. Working back from the results of this testcross, the parental individual with dominant phenotype is deduced to be homozygous dominant if all **F₁** progeny also have the dominant phenotype; heterozygous if **F₁** progeny are 50:50, dominant to recessive.

But for Mendel, the heterozygous/homozygous cross was a way to test predictions of his hypothesis that alleles segregate when gametes are formed. Mendel prepared for the testcross by crossing white-flowered with white-flowered plants. He predicted they would breed true: that only white-flowered plants would be produced. This happened. Now he was ready to perform the testcross. He crossed a white-flowered plant (homozygous) with a hybrid purple-flowered plant (heterozygous) from the **F₁** generation. Note how this testcross differs from his original experiments. In his first set of experiments, Mendel first crossed homozygous with homozygous, then heterozygous with heterozygous, giving predicted ratios of 3:1 in the **F₂** generation. But in this testcross, he crossed homozygous and heterozygous plants. If his understanding of inheritance was correct, then this testcross should give different ratios. Specifically, his hypothesis predicted that such a cross should average half purple-flowered and half white-flowered offspring, for a ratio of 1:1 in the **F₁** generation. This too happened (figure 3.6).

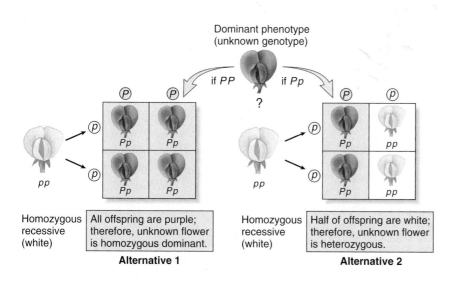

Dominant phenotype
(unknown genotype)

if *PP*   ?   if *Pp*

	P	P
p	Pp	Pp
p	Pp	Pp

pp

	P	p
p	Pp	pp
p	Pp	pp

pp

Homozygous recessive (white) — All offspring are purple; therefore, unknown flower is homozygous dominant.

Alternative 1

Homozygous recessive (white) — Half of offspring are white; therefore, unknown flower is heterozygous.

Alternative 2

FIGURE 3.6 Testcross By just looking at a dominant phenotype, here a plant with purple flowers, you cannot know if it is homozygous or heterozygous for the dominant allele. To determine its genotype, Mendel performed a testcross. In this illustration, the dominant phenotype (unknown genotype) was crossed with a plant known to be homozygous recessive, here white-flowered. If all offspring are purple (Alternative 1), then the unknown flower is homozygous dominant; if offspring are half and half, purple and white (Alternative 2), then the unknown flower is heterozygous.

Mendelian Principles of Inheritance

Mendel's work is summarized in two formal hypotheses, so well established now that they are spoken of as laws or principles. Again, I present these using modern terminology.

The Principle of Segregation

Mendel's first principle concerns single traits. Its central assumption is that pairs of a gene's alleles segregate from each other into separate gametes. At fertilization, when two gametes fuse, each contributes one allele to the offspring. Expression of the gene in the offspring is not a blend of its two alleles, but results from the action of the dominant allele. Although masked and unexpressed, the recessive allele is still present. Only when paired with another recessive allele for that trait is the recessive trait expressed in the phenotype.

The Principle of Independent Assortment

Mendel's second principle deals with the inheritance of multiple traits. Essentially, Mendel showed that alleles for one trait pass simultaneously to offspring independently, without affecting the segregation of alleles for another trait. To illustrate, Mendel looked in one experiment to just two traits—seed shape and seed color—again, capital letters for dominant, lowercase letters for recessive alleles (figure 3.7). The first trait, seed shape, may be round *(R)* or wrinkled *(r)*. The second trait, seed color, may be yellow *(Y)* or green *(y)*.

Mendel hypothesized that the first-generation cross of true-breeding plants involved *RRYY* × *rryy* parents. The F_1 generation offspring would receive an allele for each trait from each parent, and thus all would be heterozygous for each trait *(RrYy)*. The F_2 generation would produce varied offspring of a predictable phenotypic ratio. On average, for every 16 offspring, 9 would show the two dominant traits (round and yellow); 3 would show combinations of dominant and recessive (round and green); 3 would show the reverse combination, recessive and dominant (wrinkled and yellow); and 1 would show the two recessive traits (wrinkled and green). The predicted ratio was 9:3:3:1. Mendel's actual numbers (315:108:101:32) confirmed his prediction, having an actual ratio of 9.8:3.3:3.1:1. Statistically close again.

The results of Mendel's multiple-trait crosses did not contradict his expectations regarding single traits. Taken separately, each trait still segregated close to the 3:1 ratio expected for single traits—seed shape

round (423):wrinkled (133); seed color yellow (416):green (140). (Adding data across all his experiments, over the eight years, the ratio taken separately was very close to 3:1—see figure 3.2.) Together, the alleles for the two traits did not hinder or affect each other as they shuttled into gametes. Thus, the crosses involving multiple traits simultaneously demonstrated quite another feature of inheritance: Alleles for different traits are not influenced by each other; instead, they segregate independently of one another. This became Mendel's second principle of inheritance: Each gene distributes independently into gametes without interaction or interference between genes. Here he was lucky, or perceptive. As will be discussed shortly, some alleles are linked and travel together into gametes. But the alleles Mendel picked independently sorted into gametes.

Mendel's Achievement

I asked earlier if Mendel was merely lucky, and answered no. He seemed to understand the basis of simple inheritance of characteristics, and he used carefully crafted experiments to demonstrate the hereditary principles involved. Had Mendel changed his experimental approach slightly, he might have run into subtleties of heredity that would have made his study less clear, his case more elusive. What were the obstacles Mendel dodged to set up his pea experiments in such a straightforward manner? One was mathematical. He realized that to test and illustrate his ideas on inheritance, he should expect *statistical* results, and not *exact* results. The effects he expected were not fixed ratios, but statistical probabilities. A 3.18:1 outcome was sufficient, close enough to the predicted 3:1 ratio.

The other obstacle Mendel faced was an absence of knowledge about the anatomical arrangement of genes within the cells. How genes were coupled and joined would affect the sorting of genes and, in turn, affect inheritance as well. Mendel followed 7 traits. The pea plant has 7 pairs of chromosomes, but of course, then no one knew this, certainly not Mendel. If he had chosen to follow only one more trait, then inevitably two traits would have ended up on the same chromosome. Like passengers on the same boat, both would have traveled together into the same gametes—*not* sorted independently—and remained linked in their inheritance generation to generation. Mendel picked seven characters, all on separate chromosomes, and no more.

The pattern of inheritance summarized in these two principles is sometimes called "simple" Mendelian

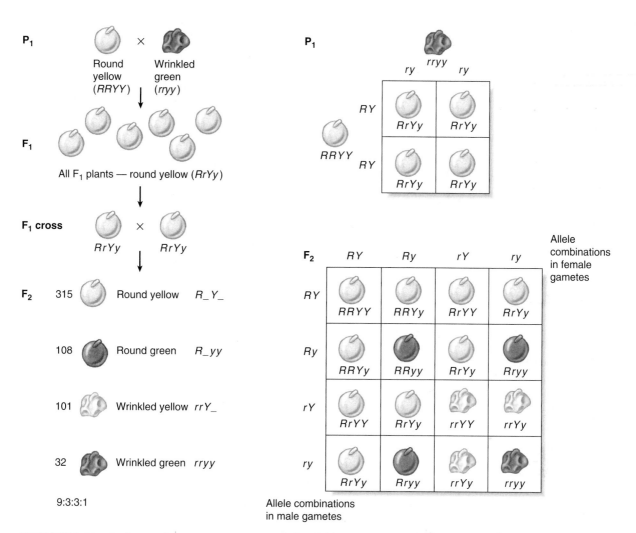

FIGURE 3.7   Independent Assortment—Multiple Traits, Seed Shape and Seed Color   Mendel followed two traits together to see if they influenced each other. Vertically at left, the phenotypic outcomes into the **F₂** generation are followed for seed shape—round (dominant) and wrinkled (recessive); and seed color—yellow (dominant) and green (recessive). A blank space in a genotype indicates that either a dominant or a recessive allele is possible. If the alleles assorted or moved into gametes without affecting each other, then the predicted ratio is 9:3:3:1, which is about what Mendel observed. Vertically at the right, the allele combinations resulting from each successive cross are mapped, showing the genotypes.

inheritance because the detailed pattern of inheritance has proven to be much more intricate. Some traits result from incomplete dominance. In these cases, heterozygotes are neither completely dominant nor completely recessive phenotypically. Some alleles come in many forms, not just one for dominant and one for recessive. Mendel envisioned one gene for one phenotypic character, but more usually, multiple genes contribute to the expression of one phenotypic characteristic. And inheritance of characteristics is even affected by how genes are strung together into chromosomes.

For our purposes, we do not need to explore all these amendments to simple Mendelian inheritance. Modern genetics confirms that Mendel was, and remains, correct about inheritance being particulate, and not blending. Therefore, Darwin was, and is, on solid ground. Favored characteristics preserved by natural selection are not blended into oblivion. They persist in successful offspring from generation to generation. Examining a few of these discoveries of twentieth-century genetics will help us better understand the importance of genetics in evolutionary studies.

Let's begin with chromosomes, their discovery, and why the arrangement of genes on chromosomes is such an important feature of heredity. This will also help us appreciate why Mendel's intuition was such an important part of his scientific discovery.

CHROMOSOMES

In the early twentieth century, Mendel's work was re-discovered. The particulate theory of inheritance gained in respectability. But even by then it was still un-referenced to cell structure. No one had seen Mendel's proposed particles. They were ideas. They accounted for statistical results of experimental crosses. But where were they physically located within cells?

The answer came from two persons independently and in separate papers in the same year, 1903. One was Walter S. Sutton, the other T. Boveri. The Sutton-Boveri hypothesis proposed that Mendel's particles were carried on chromosomes within the cell nucleus (figure 3.8). W. Johansen, another biologist, was first to call the particles, carried on chromosomes, "genes." Collectively, the genetic material of an organism is its genotype. Not only were Mendel's particles located within cells, but these concepts united two fields—genetics, the study of inheritance, and cytology, the study of cells. If genes, or really their alleles, segregated independently into gametes, and if genes resided on chromosomes, then it was important to understand the pattern of chromosome movement during cell division. Attention centered on the chromosomes housed in the cell nucleus, and on how they got into gametes during cell division (see Appendix I).

Cell Division

Two types of cell division occur in eukaryotic organisms, mitosis and meiosis (figure 3.9). Except for a few specialized cells, each body, or **somatic cell**, contains two nearly identical copies of each chromosome. The cells are termed **diploid**, represented "2n." The two companion copies of each chromosome are called *homologous chromosomes*, or just *homologues* (*homologia*, Gr., agreement)—one of maternal origin, the other of paternal origin (figure 3.9). Each gamete or **germ cell** contains only one copy of each chromosome, and these cells are termed **haploid** (1n). During cell division, chromosomes are apportioned into the newly forming "daughter" cells—a full set (mitosis) or half a set (meiosis). As daughter cells separate, each receives its share of constituents of the original dividing cell. Mitosis is involved

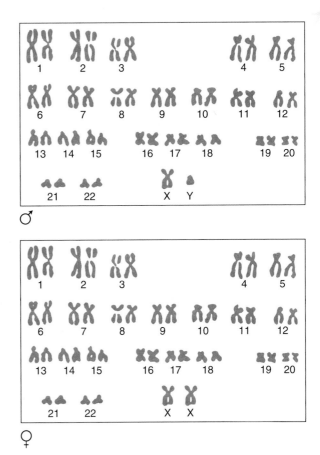

FIGURE 3.8   Homologous Pairs of Chromosomes from Humans   The 23 pairs are shown for a male, top; and a female, bottom. Note that the male sex chromosomes are *XY*; the female *XX*.

in growth in size of the organism and in repair of tissues; meiosis produces gametes for reproduction.

Mitosis

Before beginning to divide, each chromosome replicates, producing identical copies termed *chromatids*, joined at a centromere. As **mitosis** begins, chromatids part—one enters each forming daughter cell, thereby distributing chromosomes equally to each. Separation of the daughter cells completes the process, endowing each of the two cells with a full set of chromosomes, and thus diploid (2n) (figure 3.9).

Meiosis

Similarly, **meiosis** begins when each chromosome replicates, producing identical copies, the chromatids, joined by a centromere. Unlike mitosis, two divisions

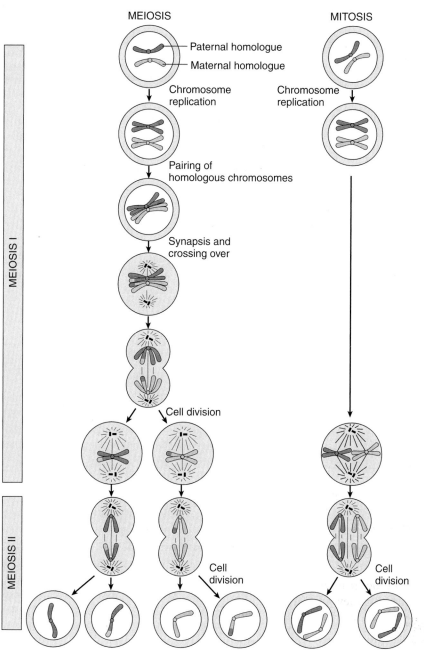

MEIOSIS

Paternal homologue

Maternal homologue

Chromosome replication

Pairing of homologous chromosomes

Synapsis and crossing over

Cell division

Cell division

MEIOSIS I

MEIOSIS II

MITOSIS

Chromosome replication

Cell division

FIGURE 3.9   Mitosis and Meiosis In meiosis, chromosomes replicate, homologous pairs align, and each duplicated homologue separates during cell division; then a second cell division separates replicated chromosomes; four haploid daughter cells (gametes) are produced. In mitosis, chromosomes replicate, but one cell division separates replicated chromosomes into two diploid daughter cells.

of the genetic material occur. In the first division, homologues separate. In preparation for the first division, homologues pair up in the center of the cell. Because each has replicated, this pairing brings together four chromatids: two from each homologue, to form a *tetrad*. Momentarily zipped together, segments of one chromosome interchange with corresponding segments of its homologue, an exchange called *cross-ing over*. Homologous chromosomes then part, one homologue entering each of the forming daughter cells, which separate. In the second division, chromatids separate. In these daughter cells, the chromatids part, one entering each of the two cells formed from each of the daughter cells. From a single cell, four gametes result, each endowed with half a set of chromosomes, and thus haploid (1*n*) (figure 3.9).

MENDEL AMENDED

Gene Linkage (*Many genes, few chromosomes*)

Each diploid cell contains many genes (perhaps several thousand) but only a few chromosomes (perhaps a few dozen)—many genes, but few chromosomes. Thus, understandably, some genes must reside on the same chromosome. During meiosis, these genes ride together into the same gametes. They thus travel together, linked, into the same individual and into the next generation. Their assortment is *inter*dependent, as compared to Mendelian independent assortment.

To illustrate gene linkage, let's look at an example (figure 3.10). In reality, each plant or animal cell holds many chromosomes and each chromosome is a paired chain of thousands of genes. But, to keep our example simple, let's look at a cell with only one chromosome pair and follow only two genes (for wing length and eye color) on the chromosome. Because the two genes reside on the same chromosome, they ride together into the same gamete, linked in their segregation. Even though the two genes control different traits, genetically they are transferred together from generation to generation. If the traits Mendel followed had been controlled by genes linked on the same chromosome, then expression of the characters would have occurred together in the same individual, violating his predicted ratios based on the assumption of independent assortment.

The structural arrangement of genes into chromosomes can change the pattern of heredity. As we saw in our discussion of meiosis, the exchange of chromosomal sections between adjacent pairs—crossing over—can further complicate the pattern of "simple" Mendelian inheritance. Even more subtleties of genetics could have defeated Mendel.

Multiple Alleles (*One gene, many alleles*)

Like Mendel, we have so far envisioned each gene as having two possible allelic expressions. Within one individual, that is true because each individual has two homologous chromosomes, one allele on each. The site of residence of a particular gene on its chromosome is the gene's **locus**. Each individual has only two alleles for a trait. But within a large population of organisms, there may be many alleles for one gene. In one diploid

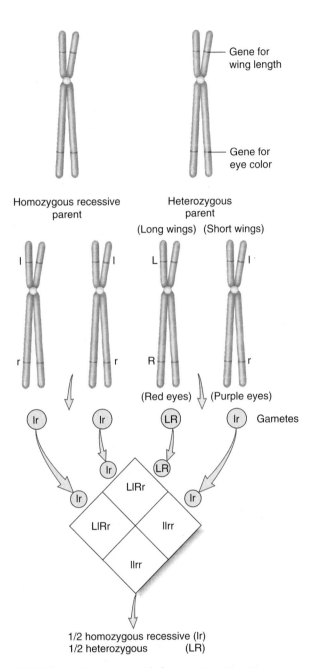

FIGURE 3.10   Gene Linkage   Two fruit-fly genes —for wing length and eye color—reside on the same chromosome. Because they are located on the same chromosome, they ride together into the gametes, thereby reducing the number of genotypes and phenotypes possible.

organism, the two sites may be occupied by only two alleles at a time. But with multiple alleles available for a particular gene, the two particular combinations of alleles carried by an individual can be many. In humans, for example, there are three alleles controlling blood type—A, B, and O alleles (figure 3.11). Possible diploid genotypes include A/A, A/O, B/B, B/O, A/B, and O/O. The phenotypic expression of each genotype determines the individual's blood group. Blood groups are medically important in blood transfusions as they determine blood compatibility between different individuals. Four blood groups are recognized: A, B, O, and AB. O is recessive; A and B are codominant (neither masters the other, both are expressed). As a consequence, persons with *A/A* or *A/O* genotypes belong phenotypically to A blood group; *B/B* or *B/O* to B blood group; *A/B* to AB blood group; and *O/O* to O blood group, as summarized in figure 3.11. Besides their medical importance, blood groups illustrate the possible diploid genotypes that can arise with three alleles. One gene, multiple alleles.

Multiple Genes—Polygenes (*Many genes, one trait*)

One gene, one trait. Essentially, that's the direct relationship Mendel established. More usually, many genes affect single traits. The term *polygenic inheritance* applies where two or more genes govern the same trait. Consider the example of seed color in wheat. In wheat, three genes at three different loci affect color; thus, up to six alleles may be involved. This is represented in figure 3.12. An allele contributing to color is represented by capital letters; one not contributing is in lowercase. Each contributing allele has an additive effect on seed color, so seed color ranges from white (no allele contributing color) to dark red (all six contributing), with various phenotypes in between. Graphing the frequency of each phenotype produces a bell-shaped curve (figure 3.12). When many genes affect one trait, its phenotypic expression occurs in graded differences between extremes—*continuous variation.*

Several traits in humans are polygenic: height, for one (figure 3.13); perhaps intelligence for another. Inheritance is still particulate—alleles may be dominant or recessive. But because several genes are involved, phenotypic variation in offspring is continuous and graded between extremes.

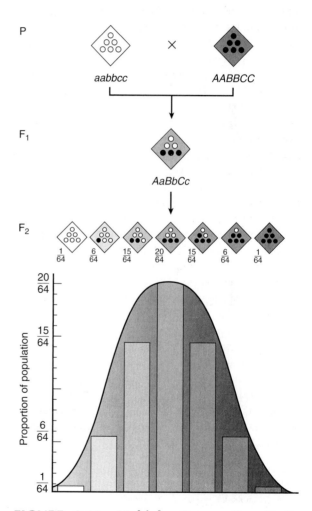

FIGURE 3.12   Multiple Genes, One Trait Polygenic inheritance is illustrated with three genes (A, B, and C), and hence six alleles, for wheat seed color. Alleles contributing to the color are indicated by a capital letter; alleles not contributing are in lower case. In each generation, the six alleles are shown by circles—solid if contributing to seed color and open if not. In the **F$_2$** generation, the phenotypes expressed are additive, producing a continuous range of seed color. If graphed by frequency, they form a bell-shaped curve.

Blood Groups	
Phenotype (group)	Genotype
O	O/O
A	A/A, O/A
B	B/B, O/B
AB	A/B

FIGURE 3.11   Multiple Alleles, One Trait   In humans, the gene that controls blood type has three different alleles: *A*, *B*, and *O*, resulting in four different blood phenotypes: A, B, AB, O, specifically type A (*A/A* or *A/O*); type B (*B/B* or *B/O*); type AB (*A/B*); and type O (*O/O*).

FIGURE 3.13   Polygenic Trait in Humans—Height   Aligned by height, the students show a range of continuous phenotypic variation, with most in the middle.

If we had only the appearance of the phenotypes to guide us, it would look as if characters blended producing this continuous variation. Careful genetic analysis confirms, though, that we are still dealing with an example of particulate inheritance of the sort discovered by Mendel, but one that is much more complicated. Perhaps it is now easy to see why so many early scientists working on the problem of inheritance were confused by the phenotypic patterns they observed. And perhaps we can appreciate why Mendel's scientific breakthrough was so remarkable.

OVERVIEW

As Darwin proposed it, natural selection on average preserves individuals with favorable traits, and these become the basis for subsequent descent with modification—evolution. But if traits are blended away in the next generation, then evolution goes nowhere. Species remain fixed. Clearly, variation existed and persisted—that was obvious in artificially bred animals and plants. Variation did not wash away in blended offspring. But

the mechanism of inheritance awaited discovery. Mendel accomplished this against considerable odds.

Mendel's achievement was substantial, next perhaps only to the discovery of natural selection as one of the major scientific discoveries of nineteenth-century biology. Part luck? Perhaps. But most likely, Mendel had worked it out in his head first, and then coupled clever hunches and insights to design well-crafted crossing experiments that nicely confirmed what he had already guessed. In this, we can only admire the simple genius of this obscure monk toiling patiently in an abbey garden. Mendel's concept of paired particles, independently distributed to offspring during reproduction, is as elegant as it is a simple solution to the mystery of heredity. Specifically, Mendel proposed the following: (1) Inheritance is particulate, not blending. Genes pass undiminished from generation to generation. (2) Each individual carries a pair of alleles for each trait. (3) Alleles exhibit dominant or recessive expression. (4) During formation of gametes, each allele of the pair travels into its own gamete (principle of segregation). (5) Alleles controlling a trait reach gametes independently of alleles controlling other traits (principle of independent assortment).

Subsequent discoveries in the twentieth century amended and extended these principles and clarified inheritance. The discovery of chromosomes, cell division, complex inheritance, and DNA added depth to the genetics we meet today. In modern terms, we can summarize the genetics of the last century and a half: The *genotype* of an individual is the individual's full assortment of genetic material. Today, this totality of characteristic genetic material within each cell is popularly termed the organism's **genome**. The Human Genome Project, for example, successfully labored to produce a description of the molecular anatomy of human DNA. A *chromosome* in a cell is the DNA bundled and packaged in other molecules, mostly proteins.

Gene—trait A *gene* is a region of the genome dedicated to building a particular phenotypic trait; *gene expression* describes the time when a gene is activated—turned on to carry out its role. Genes have roles as *structural genes* that act more or less directly to produce the products composing the trait; some genes have roles as *regulatory genes* that act more indirectly and distantly to synchronize the action and timing of the expression of other genes. It is easier to think in terms of one gene, one trait. But more commonly, multiple genes (*polygenes*) control the expression of a trait.

Allele—particular expression of a gene An allele is an alternative version of the way a gene is expressed. For example, for the trait of plant height, different alleles might express a tall or a short height. In a diploid organism, only two alleles are present at a time—Mendel's two "elemente." But within a population, many different alleles affecting the same trait may be carried in different individuals. Human blood types are one example. There are three blood-type alleles in humans generally, but in one individual only two pair together to produce that individual's particular blood type.

Locus—reserved seating Along a chromosome, the site occupied by a gene is its *locus*, its site of residence. At a given locus, the particular occupant can be any one of the alleles of that gene. Conse-

quently, on homologous chromosomes, the alleles for the same trait reside at the same respective sites. Gene mapping, as was done in the Human Genome Project, describes the molecular anatomy of genes and their particular loci, characteristic sites of residence along the DNA.

For scientists of his day, immersed in vague ideas of blending inheritance and unpracticed in statistics, Mendel's ideas were elusive, even tedious, concepts. Think of your own reaction to the simple genetics of this chapter—single traits through F_1 and F_2 generations, testcrosses, inheritance of two traits, statistical ratios. It's tempting to skip ahead and look for the larger significance. When Mendel first presented results of his pea crosses before the Brunn Natural Science Society, many probably nodded off. What a shame. In these ordinary plants, Mendel had discovered an extraordinary concept—particulate inheritance. Specifically, he discovered that the mechanism of inheritance is particulate, not blending. Later scientists realized that here was the basis for preservation of traits favored by natural selection; here was the source of new traits, via mutations in the genes; here was the raw material of evolution, variation; and here was the basis for inheritance of favored traits, carried by genes.

Selected References

Henig, R. M. 2000. *The monk in the garden: The lost and found genius of Gregor Mendel, the father of genetics.* Boston: Houghton-Mifflin.

Peters, J. A. 1959. *Classic papers in genetics.* Englewood Cliffs, N.J.: Prentice Hall.

Learning Online

Testing and Study Resources

Visit this textbook's website at **www.mhhe.com/ evolution** (click on the book's title) to take advantage of practice quizzing, study/writing tips, timely news articles, and additional URLs for research on the topics in this chapter.

4

"To say that a man is made up of certain chemical elements is a satisfactory
description only for those who intend to use him as a fertilizer."

Hermann Joseph Muller, geneticist, Nobel prize (1946)

Emergence of Life

INTRODUCTION

Life was tediously slow in gaining momentum on planet Earth. At a few places around the world today, rocks of ancient age survive, having escaped the churning and melting and reworking of the crustal surface. But these rocks, several billion years old, contain no skeletons, no skulls, no shells, no fossil leaves, no petrified trunks of great trees. A moment's thought will answer why: Several billion years ago, none of these large, multicellular animals and plants had yet evolved. The Earth was young and life was still microscopic. But these rocks from several billion years ago speak of life, microscopic life, because within these rocks are traces of bacteria and other microorganisms.

With favorable conditions of preservation and some luck, the death of a large dinosaur may leave body impressions or bits and pieces of their more durable hard parts such as bones and teeth. When microorganisms die, they too may leave impressions in the rocks and, sometimes, the telltale chemical debris leaves a signature of information about their composition. The organic chemicals pristine and phytane occur at significant levels in some early rocks. These are chemical fossils, the breakdown products of chlorophyll, which itself is important in directly trapping the energy of sunlight and passing it along to the machinery of the cell that builds these microorganisms. Rocks approximately 1 billion years old from Australia (central Australia) yield a rich set of single-celled microfossils. Still older rocks from about 3.5 billion years ago (western Australia, southern Africa) include impressions of filamentous (figure 4.1) and single-celled microorganisms, and even chemical fossils. Although scant and microscopic, the fossil record of early Earth history speaks of life.

MAJOR TRANSITIONS OF LIFE

Life on Earth is built on carbon. Writers of science fiction delight in speculating on life's character if, instead of carbon, life were built on silicon, a chemically close counterpart to carbon. Who knows? Carbon is stable but reacts favorably with useful elements such as hydrogen and nitrogen, and it easily forms stable complexes, making it

Medium Diameter (2-5μm) Filaments, Cylindrical Cells

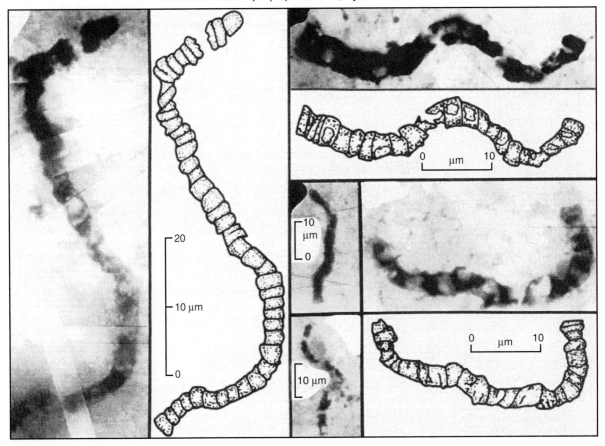

Primaevifilum amoenum

FIGURE 4.1   Microfossils   Microscopic views of these filamentous microorganisms (*Primaevifilum amoenum*) recovered from rocks in Western Australia dated to about 3.5 billion years of age.   (Photo courtesy of J. W. Schopf.)

the molecular backbone of proteins, carbohydrates, and fats. Perhaps silicon-based life forms exist elsewhere in the universe; for the present, we shall leave it to writers of sci-fi to speculate on the consequences. But on Earth, we are carbon creatures. Because carbon is the basis of life here, *organic (life) chemistry* is based on carbon, compared to *inorganic chemistry*, which is based on the rest of the elements.

The history of life on Earth includes major transitions. The first was the development of organic compounds (figure 4.2).

Inorganic to Organic Evolution (4 billion years ago)

Between 4 and 4.6 billion years ago, the Earth had cooled sufficiently that larger molecules were stable

and not heated to levels that quickly dismembered them back into smaller elements. The early Earth's atmosphere essentially lacked free oxygen (O_2). Instead, the atmosphere included some hydrogen, methane gas (CH_4), carbon dioxide (CO_2), and nitrogen (N_2). Under the right conditions, could these form the basic organic building blocks of life?

To answer this, Stanley Miller and Harold Urey in 1953 simulated Earth's early atmosphere within an enclosed laboratory system along with water (H_2O), and heated the mixture with electric sparks to simulate lightning (figure 4.3). In short order, organic molecules, some rather complex, spontaneously appeared within this simulation of Earth's primitive atmosphere. Follow-up experiments by others produced many more organic molecules, including compounds found in the basic structure of genes, deoxyribonucleic acid (**DNA**). Other

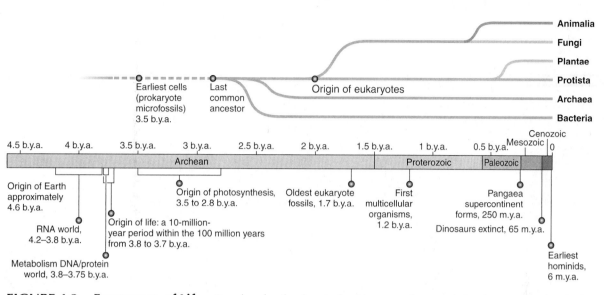

FIGURE 4.2   Emergence of Life   Based on fossils, chemical evidence, and extrapolation from molecular clocks, the first appearance and relationships of the major domains of life are indicated against a 4.6-billion-year time scale.

biochemists reversed this experiment, freezing the mixture of primitive atmospheric compounds, using the crystalline structure of frozen water to force chemicals together. Again organic compounds formed spontaneously. In fact, it has proved to be remarkably easy to get organic compounds to form within these simulated, primitive atmospheric conditions, hot or cold. The young Earth was ripe for the production of the basic building blocks of life.

Of course, these experiments do not prove that this is what happened, but they do establish that it is quite plausible. In particular, such experiments establish that a transition from inorganic to organic evolution likely marked an early moment in life's debut on Earth, a natural genesis of life within Earth's primitive chemical atmosphere. In the young Earth, the exposure to heat, radiation, pressure, and even cold caused a slow accumulation of organic chemicals. Some argue that organic compounds arose elsewhere in the universe, and seeded the Earth like "cosmic spores." Perhaps, but this just begs the question as to how these molecules formed in the first place.

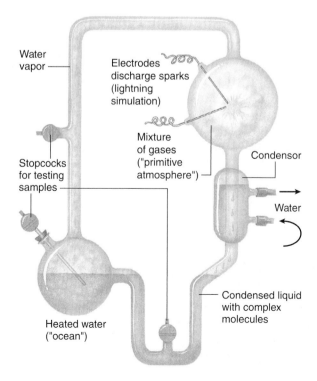

FIGURE 4.3   Life in Test Tubes, The Miller-Urey Experiment   Heated water produced water vapor circulating through the closed system of glass chambers. Into the upper chamber, Miller and Urey placed gases thought present in Earth's early atmosphere, and applied a spark. Condensers cooled any gases, causing molecular products to collect in the water. From this water, samples were taken over the next week and analyzed. Among the organic molecules formed were amino acids, basic building blocks of protein. Subsequent follow-up trials, by many other biologists, using various combinations of "primitive atmospheres," produced even more complex organic compounds.

Whatever their source, this transition on Earth to organic evolution occurred most likely about 4 billion years ago, a best guess, extrapolating backward from cellular forms that appeared later. Much earlier than this, and the Earth was still being pummeled by large chunks of rocks left over from the formation of the solar system. These huge impacts generated heat that vaporized surface water, preventing seas and lakes from persisting. Surface movements and reworking of the Earth's crust have since obliterated these inflicted craters. But the large surviving craters on our geologically inactive companion moon bear witness to this early episode of cosmic bombardment.

The earliest "microorganisms" were more like macromolecules, large chains of replicating, carbon-based molecules. Strictly speaking, these organisms were **heterotrophs** (nourished by others), meaning that they did not directly manufacture ingredients needed for their own maintenance and duplication (reproduction), but instead subsisted on organic chemicals available in their environments. This is one of the least understood transitions, but certainly the most important. This is when life was initiated, setting in motion the remarkable cascade of events that followed.

Cell—Prokaryotic, Heterotroph (3.5 billion years ago)

The oldest fossils of microorganisms are found in rocks 3.5 billion years old in western Australia. These microfossils resemble bacteria that exist today. The important organic molecules and DNA are safely isolated from direct contact with the environment by an encasing plasma membrane that defines the boundaries of the cell. The *plasma membrane*, which defines the boundaries of the cell, protects the activities within, and acts as a selective gatekeeper with the outside environment. Also present is a semirigid *cell wall*, usually outside the plasma membrane, which stabilizes the shape of the bacterium. Within the cell, proteins are synthesized on stabilizing *ribosomes*, and the surrounding matrix is the *cytoplasm*. These early cells were *prokaryotic cells*, meaning that their cellular DNA was free within the cytoplasm—no nuclear membrane wrapped it (figure 4.4). Also absent were *organelles*, specialized bodies within the cell. These cells were also heterotrophs, soaking up available organic molecules in the environment to feed their cellular appetites.

Sexual reproduction, based on the fusion of male and female sex cells (gametes), is absent in prokaryotes. Instead, they reproduce **asexually** by means of *binary fission* whereby a single prokaryotic cell divides in two; these cells then divide again repeatedly, forming an expanding colony.

Cell—Prokaryotic, Autotroph (2.7 billion years ago)

Photosynthesis converts the sun's energy into chemical energy used to meet the metabolic needs of the cell. Several versions of photosynthesis probably evolved early. One of the most successful was based on splitting water to liberate oxygen, taking up the energy released to fuel the cell. Modern plants employ this form of photosynthesis. Rocks about 2.7 billion

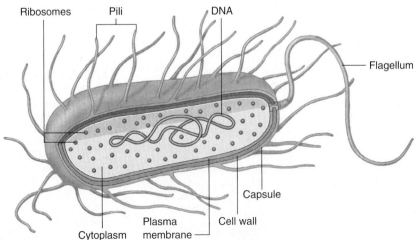

FIGURE 4.4   Bacterial Cell, Generalized   Note the circular DNA, ribosomes, flagellum (for movement), pili (grappling lines to attach to structures), capsule (jelly-like protective coat present in many bacteria), and the plasma membrane contained within a cell wall.

years old indicate that free oxygen (O_2) had already begun to accumulate in the atmosphere. Indirectly, this suggests that photosynthesis was well established and had debuted earlier. These first photosynthesizing, prokaryotic cells were **autotrophs** (self-nourishing); their energy came not from consuming other organisms or organic molecules (heterotrophs), but by the direct conversion of the sun's energy into the manufacture of the cells' own ingredients. Today, the only photosynthetic prokaryotic cells that generate O_2 as a by-product are *cyanobacteria*, leading us to believe that cyanobacteria evolved no later than this and perhaps as early as 3.5 billion years ago.

The Earth was not born with atmospheric oxygen. Instead, free oxygen was of biological origin from the water-splitting step of photosynthesis. The free oxygen produced by the first cyanobacteria dissolved in the surrounding water, where it reacted with iron to form iron oxide (rust). When all the dissolved iron had precipitated out, oxygen reached saturation levels in the bodies of water and bubbled out into the atmosphere, where it then combined with terrestrial iron to again form rust. These rust deposits are found in marine rocks and later in terrestrial rocks. All of this would have taken time, which suggests that cyanobacteria evolved earlier, for which we may yet find more direct fossil evidence.

Cell—Prokaryote to Eukaryote (2 billion years ago)

At 1.7 billion years ago, microfossils first appear that are noticeably different in appearance. They are larger than the earlier prokaryotes, exhibiting evidence of internal membranes and thick walls. These are presumably the first eukaryotic cells. However, chemical traces unique to eukaryotic cells are found in rocks 2.1 billion years of age, suggesting at least this earlier time for the appearance of the first eukaryotes (see figure 4.2).

Eukaryotic cells have more complex interiors than prokaryotic cells. The cellular DNA of a *eukaryotic cell* is enclosed in a specialized membrane—the **nuclear envelope**—and the cell contains numerous organelles, membrane-bound compartments dedicated to particular cellular functions (figure 4.5). One such organelle is the **mitochondrion**, a small power factory within the cytoplasm where oxygen is consumed in organic fuels to obtain energy used by the cell. Plant cells, in addition, have a central vacuole and chloroplasts (figure 4.6). The *central vacuole* is a storage depot, collecting proteins, pigments, waste materials, and defensive toxins; the **chloroplasts** are organelles dedicated to photosynthesis. Further, plant cells have cell

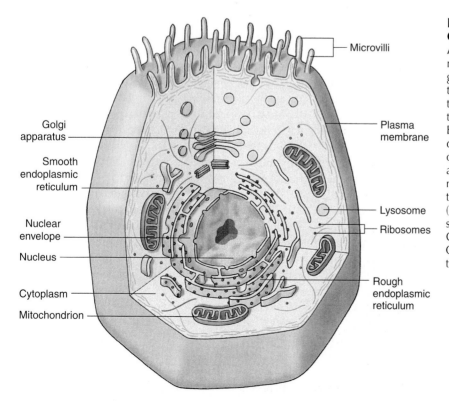

Golgi apparatus

Smooth endoplasmic reticulum

Nuclear envelope

Nucleus

Cytoplasm

Mitochondrion

Microvilli

Plasma membrane

Lysosome

Ribosomes

Rough endoplasmic reticulum

FIGURE 4.5 Eukaryotic Cell, Generalized from an Animal Note the plasma membrane containing various organelles, membrane-bound structures, within the cytoplasm, and the nuclear envelope bounding the nucleus, which holds DNA. Besides the mitochondria (production of energy molecules), the organelles include compartments and channels involved in the manufacture, processing, and transport of synthesized products (rough endoplasmic reticulum, smooth endoplasmic reticulum, Golgi apparatus, and lysosomes). Often microvilli, small projections, occur on the cell surface.

FIGURE 4.6 Eukaryotic Cell, Generalized from a Plant Note the same basic components and organelles as the animal cell (Figure 4.5) plus the addition of a firm cell wall of cellulose, a central vacuole (which sequesters various chemicals), and chloroplasts that carry out photosynthesis.

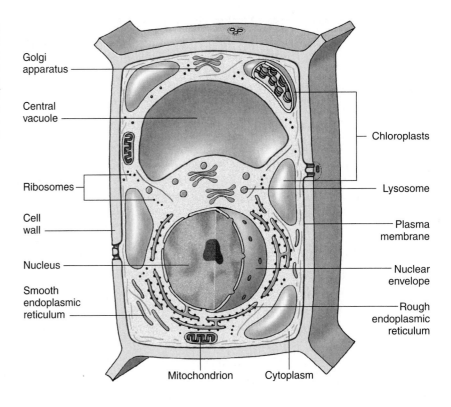

Golgi apparatus

Central vacuole

Ribosomes

Cell wall

Nucleus

Smooth endoplasmic reticulum

Chloroplasts

Lysosome

Plasma membrane

Nuclear envelope

Rough endoplasmic reticulum

Mitochondrion   Cytoplasm

walls that lie outside the plasma membrane to encase the cell. Unlike the cell walls of prokaryotes, the firm plant cell walls are made of *cellulose*, a carbohydrate, which encloses individual cells and collectively provides the structural support of the plant body.

Like prokaryotic cells, eukaryotic cells reproduce asexually by cell division through *mitosis*, a derived form of binary fission (see chapter 3 and Appendix 1). But eukaryotic cells can also effectively reproduce sexually through fusion of two specialized sex cells (gametes). Each sex cell, formed during *meiosis* (see chapter 3 and Appendix 1), contains one copy of the double-stranded DNA. At fertilization, the fusion of the male and female gametes produces offspring that are a genetic mixture of both, rather than the simple duplication or copy produced by binary fission. Sexual reproduction leads to frequent genetic recombination, in turn generating variation, the raw material for evolution. Therefore, by about 2.1 billion years ago, sex had evolved and things were looking up for life on Earth.

Multicellularity

Individual prokaryotic cells may find themselves in the company of other prokaryotic cells within a

colony, the result of binary fission. Although the colony may afford a favorable environment in which each cell operates, the colony is not itself an "organism." After binary fission, some duplicate cells may not fall away but instead, remain in contact with other cells, producing a complex of joined cells. Strictly speaking, such prokaryotic groups are "multicellular" assortments. However, true multicellularity, wherein individual cells are specialized for a particular function in the organism, is found in many eukaryotic organisms, where multicellularity arose independently several times.

Each single-celled microorganism must be a "jack of all trades." Each must be able to meet its own metabolic needs for maintenance and growth, respond to the environment, maintain structural integrity, and prepare for and complete successful reproduction. Multicellular organisms take advantage of cell specializations wherein some cells are involved in growth, others are involved in energy processing and delivery, others are sensory cells responding to environmental challenges, some provide support, and others are dedicated to reproduction. Large organisms living today—plants, fungi, animals—are built on the theme of multicellularity.

Consider This—

Brushing Up on Dental Caries

Tooth decay, or dental caries, is a bacterial disease. Several bacterial species are involved, but most are various species of *Streptococcus* and *Actinomyces*. These bacteria divide and proliferate, forming tight mats or colonies on the surface of the teeth. These patches of bacterial growth are called "dental plaques." From sugars in our diets, these bacteria derive energy for proliferation and release lactic acid as a by-product onto the tooth surface. The compacted mat of bacteria concentrates this acid, leading to local erosion of tooth calcium, the caries. If left unattended (tooth brushing, flossing), the growing plaque impinges on the living gum tissue at the base of the tooth, producing irritation and causing this supportive tissue to retreat from the tooth, further aggravating the dental disease.

MAJOR TRANSITIONS OF LIFE AND CONSEQUENCES

Ozone

Just as Earth had an impact on life, life had an impact on Earth. Today we breathe air and our cells necessarily consume the free oxygen available. Without it, we suffocate. Ozone (O_3), a derivative of O_2, circles the Earth high above in the atmosphere like an encasing shield, filtering out much of the incoming radiation that would otherwise bombard life on the surface of the Earth with harmful ultraviolet rays that penetrate the skin and, over time, shred key biological molecules such as proteins and DNA. Yet the atmosphere of the young Earth included no significant O_2 and hence no O_3.

Pollutant

The first prokaryotic cells to evolve did so in the absence of oxygen. Later, photosynthesis made its debut, and O_2 began to accumulate. Ironically, as it accumulated, oxygen became a pollutant to these first prokaryotes. Theirs was an **anaerobic** environment (*an-* without; *aer*, oxygen; *bios*, life)—without oxygen. In such an environment they evolved, and to such an environment they were adapted. Chemically, oxygen is highly reactive. For prokaryotes adapted to an anaerobic environment, oxygen was destructive. Today, in our oxygen-rich environment, these prokaryotes have retreated to ancient anaerobic environments—swamps, deep ocean trenches, hot springs. Cells and multicellular organisms that evolved later did so in an atmosphere relatively rich in oxygen and came to depend on oxygen's special energetic contribution to new metabolic pathways that used the energy. Oxygen became indispensable.

Eukaryotic Origins

Organisms cooperate in mutually beneficial associations. Why not cells? In fact, that happened. Prokaryotic cells were simple—plasma membrane, DNA, uncomplicated chemical metabolism. The plasma membrane apparently folded inward into the cytoplasm, moving it closer to the active sites involved in cell metabolism. Some of these membranes came to surround and protect the important genetic material of the cell, the DNA. The mitochondria and chloroplasts evolved from prokaryotes that initially lived independently but were incorporated into eukaryotic cells, providing them, respectively, with energy factories and photosynthesis (figure 4.7). In fact, both mitochondria and chloroplasts retain modest amounts of their own DNA, remarkably similar in size and character to the DNA of bacteria, but independent from the eukaryotic cell's own DNA enclosed in the nucleus.

CHEMICAL CODING—FROM GENOTYPE TO PHENOTYPE

DNA

One of the most important groups of molecules in the cell is the *nucleotides*. These consist of a *sugar* (deoxyribose, ribose) joined on one end with the chemical *phosphate* and on the other with an organic *base*, a nitrogen-containing chemical ring (figure 4.8). The genetic material of the cell is made of such chains of nucleotides—two strands wound in parallel. Fundamentally, this genetic material is *DNA* (deoxyribonucleic acid). It contains the information to build the basic organism. Like blueprints for a building, the DNA incorporates materials and information outside

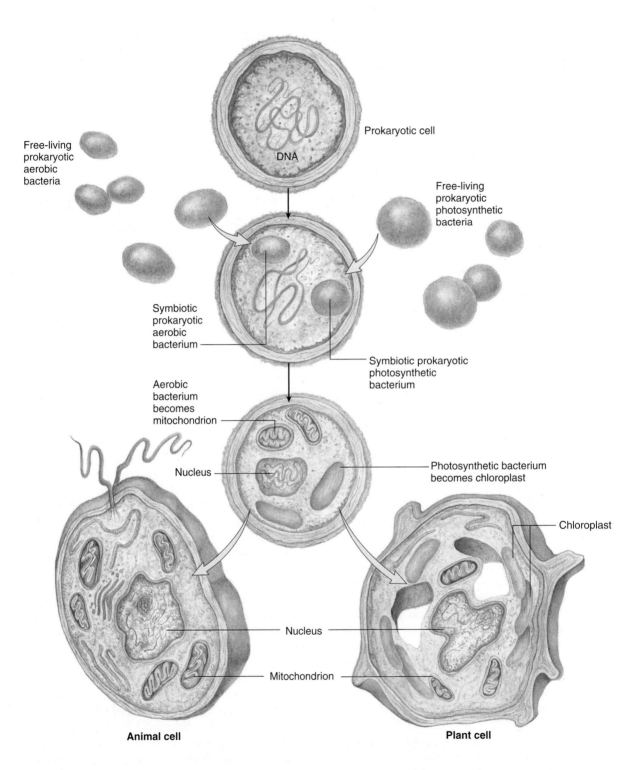

FIGURE 4.7   Origin of Eukaryotes   Free-living bacteria developed mutually beneficial relationships with a host prokaryotic cell. Some aerobic bacteria developed into mitochondria and others developed into chloroplasts, eventually producing the eukaryotic cells of animals and plants.

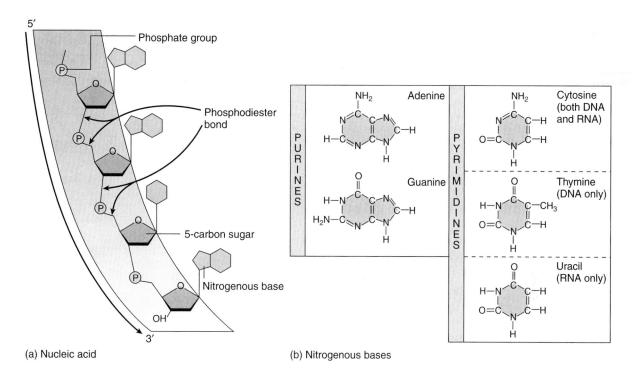

(a) Nucleic acid

(b) Nitrogenous bases

FIGURE 4.8   Structure of Nucleic Acids   DNA and RNA are built of nucleotides chained together (a). Each nucleotide is made up of one of the nitrogenous bases shown in (b) plus a phosphate group (P) plus a 5-carbon sugar (deoxyribose in DNA; ribose in RNA). These nucleotide units are linked together through special chemical bonds formed between phosphate groups (phosphodiester bonds). The sugar-phosphate complex forms the backbone of the nucleic acid; the nitrogenous bases project to the interior of the double helix, where they pair with nitrogenous bases of the other nucleic acid strand (see figure 4.9).

itself. Nutrition, environmental influences, and eventually learning contribute to the finished phenotype. But the DNA blueprint presides over this process of phenotype assembly. The two strands of DNA turn together in a helical pattern—corkscrew fashion—forming a double-stranded molecule wrapped in proteins to form a **chromosome**.

Each nucleotide in the DNA strand has one of four possible bases: thymine (T), cytosine (C), adenine (A), and guanine (G). Between the strands, *basepairs* naturally form loose chemical bonds, A-T and C-G, thereby holding the strands together (figure 4.9).

RNA

The DNA is a template. To carry out its coded instructions, it produces—**transcribes**—a complementary messenger molecule of **RNA** (ribonucleic acid), written *mRNA* (messenger RNA) to distinguish it from *rRNA* (ribosomal RNA), which is involved in a helper role in pro-

tein synthesis. However, it is mRNA that directly serves as a template to guide the assembly of chemicals into useful products. The structure of this mRNA is matched to DNA and so carries forward the DNA's coding instructions. RNA, too, is composed of nucleotides—cytosine, adenine, and guanine, but with uracil (U) in place of thymine. The coding of DNA is preserved in RNA because each nucleotide of DNA pairs with one particular nucleotide of RNA: adenine to uracil; cytosine to guanine; thymine to adenine; guanine to cytosine.

The mRNA now **translates** its coded information into proteins. Each set of three mRNA nucleotides forms a **codon**, a triplet sequence of mRNA nucleotides that specifies a particular amino acid, one of the building blocks of protein. Translation begins at one end of the mRNA and proceeds sequentially, codon by codon, to the other. The various codons of mRNA order their respective amino acids and assemble them into a connected sequence to form a protein, a chain of amino acids (figure 4.10).

FIGURE 4.9   Base-pairing   DNA is composed of two specialized strands of nucleic acid wound in a spiral helix. Between the double strands of DNA, the nitrogenous bases pair specifically and preferentially: adenine (A) with thymine (T), and guanine (G) with cytosine (C). In large part, this is the basis for coding of information within DNA, and later transcription to RNA. Each end of a DNA strand is polarized—one end is designated the 3' end, the other the 5' end. During DNA duplication, the new strand of DNA lengthens in the 5' → 3' direction.

At the molecular level, a *gene* is simply a section of the DNA that codes for a particular protein through the mRNA intermediary it transcribes. When switched on, the gene becomes active; its coding sequence is decoded and its message is translated into a particular product. This is referred to as *gene expression*. Gene expression is controlled in complex ways. Other sections of DNA may initiate gene expression in a timely fashion, or inhibit expression of particular genes until their protein products are required.

In prokaryotes, the DNA is a circular double-stranded helix, the mRNA transcribed is ready to function, and the protein produced is manufactured within the cytoplasm (figure 4.11a). In eukaryotes, the DNA is a linear double-stranded helix that resides within the nucleus. The transcribed mRNA is first modified or processed before moving out of the nucleus into the cytoplasm, where it is then ready to assemble proteins (figure 4.11b). This

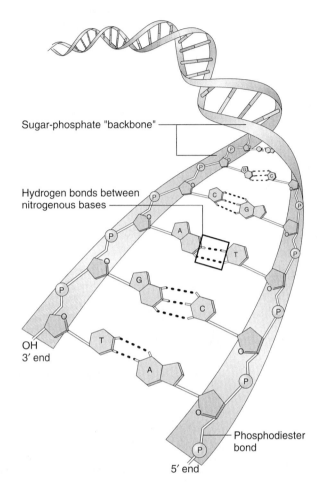

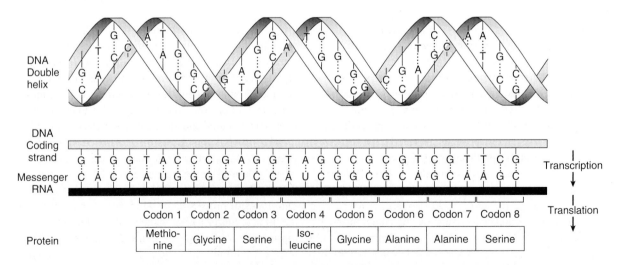

FIGURE 4.10   Information Transfer—DNA to RNA to Protein   The organized sequence of bases in DNA is used in *transcription* to produce a complementary, chemically matched, mRNA molecule. In turn, three adjacent bases form a codon, which specifies a particular amino acid. Sequentially, codon by codon, the mRNA is a cipher program used in *translation* to produce a connected chain of particular amino acids, the protein molecule.

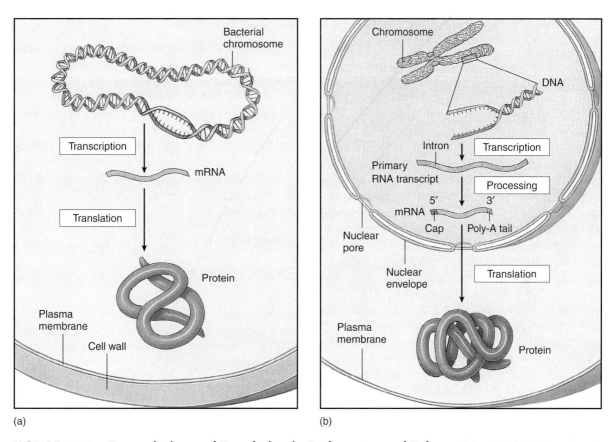

FIGURE 4.11   Transcription and Translation in Prokaryotes and Eukaryotes   (a) Bacterial genes are transcribed into mRNA, which is translated immediately into a protein. (b) Eukaryotic genes typically contain long stretches of nucleotides called introns that do not code for proteins. Introns, whose function is poorly known, are removed from mRNA before mRNA directs the synthesis of the protein.

nuclear DNA is not the only genetic material in eukaryotic cells. Recall that some of the organelles, such as the chloroplasts and mitochondria, carry their own short sections of DNA, a likely holdover from when these organelles were independent microorganisms. This *chloroplast DNA* and *mitochondrial DNA* contain several genes that produce supplemental proteins essential to the respective roles of these organelles in the cell.

In addition to proteins, carbohydrates and fats (fatty acids) are produced indirectly in chemical pathways within the cell through promoter enzymes (proteins) produced by the DNA. Together, proteins, carbohydrates (sugars), and fats are the basic building blocks of cells and, in turn, of the organism.

CELL METABOLISM

Metabolism is the term for the general chemical processes of breaking down molecules to capture the energy released, or the building up of complex molecules, which requires energy input. Whether breaking or making molecules, the common feature of each process is energy—extracting it or adding it. Energy is present in the chemical bonds that bind atoms together. As large molecules are broken down, these bonds split and the released energy is captured and stored in specialized molecules. One of these major "storage batteries" that captures this energy is the chemical ATP (adenosine triphosphate), whose chemical bonds receive and securely hold the captured energy; another is NADH (nicotinamide adenosine dinucleotide). Energy-consuming processes recover this stored energy from ATP, changing it to ADP (adenosine diphosphate), and from NADH, reducing it to NAD+.

Metabolic Pathways

In animals, there are several major **metabolic pathways**, a series of connected chemical steps that gradually capture, in increments, energy from molecules. These pathways are *glycolysis*, the *Krebs cycle*, and the *electron transport system*. Digestion breaks down the food consumed into its basic end products that will fuel metabolism—carbohydrates, fats, and proteins. As these are subsequently broken down within cells, the released energy is captured and stored in ATP (figure 4.12), used later to power the cell's activity. Animals eat plants or each other. Digested, these foods eventually enter these metabolic pathways to yield up their energy. This harvesting of energy from food is termed *cellular respiration*. Foods also yield organic chemicals, the basic end products of digestion, which are used in turn as chemical building blocks to construct the basic tissues of the animal.

In plants, sunlight provides the energy that supports cellular metabolism, but indirectly through a series of complex processes (figure 4.13). First, solar energy is converted into chemical energy. This energy capture occurs in chloroplasts, plant organelles that hold highly organized stacks of membranes that focus incoming light and convert the solar energy into the chemical bonds of chlorophyll molecules. Through a series of metabolic pathways, this energy is passed from chlorophyll to ATP and its chemical cousin NADPH (nicotinamide adenine dinucleotide phosphate) for storage. In the process, water (H_2O) is broken down for its hydrogen ions (H^+), and to oxygen (O_2), which is released as a waste product.

Carbon Fixation

Unlike animals, wherein the building blocks of their tissues come from food consumed, plants are able to directly manufacture (synthesize) the various building blocks of their tissues. Plants do this in the *Calvin cycle*, a specialized metabolic pathway. In the Calvin cycle, carbon dioxide (CO_2) is snatched from the air and used to synthesize an intermediate sugar. ATP and NADPH provide the energy. This intermediate sugar is then released into the cytoplasm, where it is incorporated into more complex sugars that make up the plant tissues. This direct incorporation of carbon, from CO_2, into stable organic compounds is termed **carbon fixation**.

Photosynthesis

Formally, plant **photosynthesis** includes the overall process whereby solar energy is captured (chloroplasts) and then transformed into organic molecules manufactured from carbon dioxide (Calvin cycle). The primary organic molecules produced in photosynthesis are sugars, destined for use in the cell wall (cellulose), for storage (starch), as cellular compounds, and in support of cell metabolism.

FIGURE 4.12 Cell Metabolism Food is broken down during digestion, yielding the end products of carbohydrates, fats, and proteins. These enter various metabolic pathways where energy is harvested and stored in ATP or temporarily in NADH. These metabolic pathways include glycolysis, the Krebs cycle, and the electron transport system, represented here diagrammatically. Various chemical intermediates in these pathways are indicated (pyruvate, lactate, acetyl-CoA). Oxygen (O_2) is used up; carbon dioxide (CO_2) is given off.

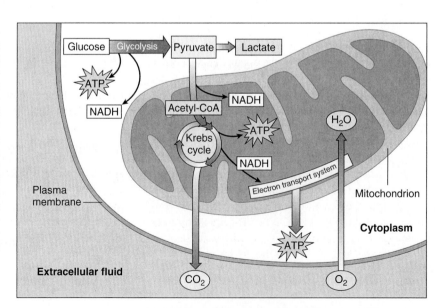

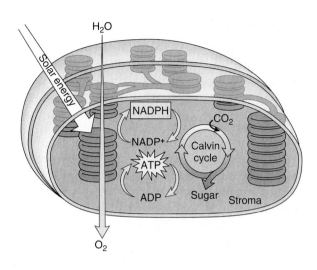

FIGURE 4.13   Photosynthesis Plant cells include numerous chloroplasts, such as the one shown here in this cutaway view. In chloroplasts, energy from light is used eventually to produce energy storage molecules, ATP and NADP + . These fuel the Calvin cycle, which in turn produces PGAL, an intermediate on its way to becoming a sugar used in cellular respiration and the manufacture of plant tissues. Note that water (H_2O) is taken up and oxygen (O_2) is given off as a by-product.

OVERVIEW

The world shaped life, and life shaped the world. It has been so from the very beginning, more than 4 billion years ago. The first major transition of life was from inorganic to organic evolution. Organic molecules, some of surprising complexity, were born and self-assembled under natural conditions from simple gases in the young Earth's atmosphere. Protective cell walls sheltered the organic machinery within the earliest of these cells. Photosynthesis debuted in microorganisms, endowing them with the ability to manufacture by themselves needed sustaining chemicals, taking solar energy directly and fueling their own cellular factories. Their waste product was oxygen, released into the nearby water and air, eventually transforming Earth's atmosphere and changing the environment into which all subsequent species evolved, lived, and strived. Sex in eukaryotic cells expanded genetic diversity, the raw material of evolution. All of this had been completed by about a billion years ago. The major transitions had occurred, save for one more—multicellularity. Although multicellularity was not new, plants and fungi and animals became multicellular specialists. This method of body organization in these larger organisms made it possible for them to develop specialized structures—new implements to address old survival challenges. In chapter 5, we will look at some of the great diversity that developed.

Selected References

Davies, P. 1999. *The 5th miracle: The search for the origin and meaning of life*. New York: Touchstone.

Dyson, F. 1999. *Origin of life*. 2nd ed. Cambridge: Cambridge University Press.

Enger, E., and F. Ross. 2003. *Concepts in biology*. 10th ed. New York: McGraw-Hill.

Johnson, G. B. 2003. *The living world*. 3rd ed. New York: McGraw-Hill.

Lewis, R., M. Hoefnagels, D. Gaffin, and B. Parker. 2002. *Life*. 4th ed. New York: McGraw-Hill.

Lurquin, Paul F. 2003. *The Origins of Life and the Universe*. New York: Columbia University Press.

Mader, S. S. 2003. *Inquiry into life*. 10th ed. New York: McGraw-Hill.

Raven, P. H., and G. B. Johnson. 2003. *Biology*. 6th ed. New York: McGraw-Hill.

Learning Online

Testing and Study Resources

Visit this textbook's website at **www.mhhe.com/ evolution** (click on the book's title) to take advantage of practice quizzing, study/writing tips, timely news articles, and additional URLs for research on the topics in this chapter.

5

"Just think how much deeper the ocean would be if sponges didn't live there."

"After eating, do amphibians need to wait an hour before getting OUT of the water?"

"Do fish get cramps after eating?"

"If a tree falls in the forest and no one is around to see it,
do the other trees make fun of it?"

Steve Wright, comedian

Diversity of Life

INTRODUCTION

Life is built upon a deep, underlying molecular plan—DNA. A nucleotide backbone makes up this fundamental, common molecular motif throughout all life. The same four nitrogen-containing bases are paired into a long and spiraled double helix, a universal standard in all living organisms. Yet life is also complex and varied (figure 5.1). Life is diversity on a theme. The bases of this diversity include fundamental changes in molecular metabolism, in the cellular unit itself, in the assembly of single-celled into multicellular organisms, and in differences in basic lifestyle. Some of these changes are summarized in figure 5.2, which represents major transitions in the history of life. Notice that multicellularity in plants, fungi, and animals accompanies three different, respective lifestyles. Plants incorporate *photosynthesis* into their biology, fungi *absorb* environmental nutrients, and animals *ingest* energy—three different survival strategies.

PROKARYOTES

Prokaryotes are *bacteria*, which in human terms is a nasty word. Bacteria are associated with disease, and in many cases they certainly deserve our contempt. But not all bacteria are harmful to humans. Some actually provide benefits. Let's look at the groups.

At one time, most microorganisms were lumped together into just "bacteria," which included prokaryotic microbes, mostly single-celled. However, newer techniques of analysis reveal that there are really two major groups of microbes. One is the "true" bacteria, sometimes called eubacteria or now more commonly just *Bacteria*; the other is the *Archaea* (*Archaebacteria*), or "ancient" bacteria (figure 5.1). Members of both groups are prokaryotes. The groups differ primarily in their chemical composition and the structure of their cell walls.

Bacteria (Eubacteria)

The bacteria include the *cyanobacteria*, sometimes called "blue-green algae." Cyanobacteria are photosynthetic; their chlorophyll is

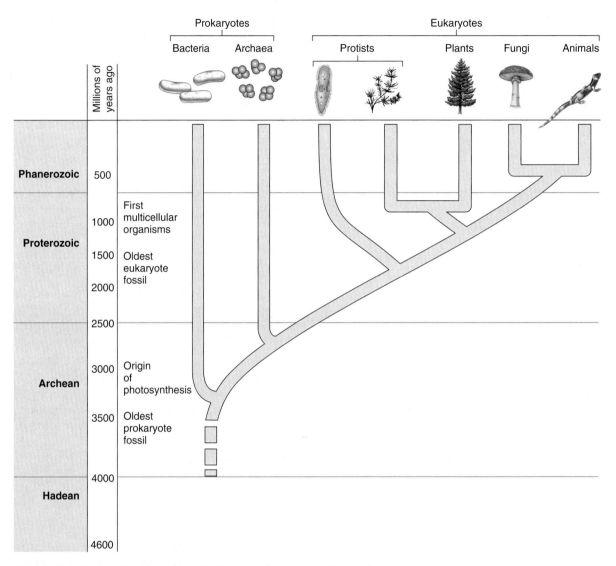

FIGURE 5.1   Major Groups of Organisms   The phylogenetic tree represents the approximate time of appearance and relationships of these groups.

similar to that in plants and algae today, and they similarly produce free oxygen as a by-product. Cyanobacteria are *photoautotrophs*—they carry out photosynthesis using solar energy (*photo-*) to build directly for their own use (*autotrophs*) organic molecules from carbon dioxide. Their appearance about 3.5 billion years ago coincided with the onset of free oxygen buildup in the Earth's atmosphere, which changed the environment in which subsequent life evolved.

Some nasty bacteria are found amongst the eubacteria. Pathogenic forms include bacteria responsible for the plague, cholera, lyme disease, dental caries (tooth decay),

anthrax, and the sexually transmitted disease chlamydia. But also included are the usually harmless *Escherichia coli*, or *E. coli* for short. Some eubacteria are *chemoautotrophs*—they obtain their energy not from sunlight but instead from the chemical bonds (hence, "chemo-") of inorganic molecules. Some of the most important chemoautotrophs convert nitrogen (N_2) to ammonia (NH_3), a process termed **nitrogen fixation**. Through several sequential steps, nitrate (NO_3^-) is produced, which is then taken up by plants. This is critical because nitrogen is an important component of proteins and nucleic acids, yet only these eubacteria—and no other organisms—have

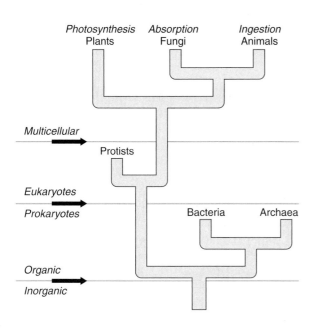

FIGURE 5.2   **Major Evolutionary Transitions and Lifestyles**   The basic domains of life represent major changes in structure, function, and basic strategies of existence. The first major transition was from inorganic to organic existence, followed by the prokaryotic cell as heterotroph and autotroph, and prokaryote to eukaryote. Although clumps of cells occurred earlier, plants, fungi, and animals represent specialists that build on multicellular organization, respectively, in photosynthesis, absorption, ingestion.

the ability to take nitrogen and fix it in usable compounds. In so doing, these single-celled prokaryotes make nitrogen available to other organisms, mostly plants.

Archaea (Archaebacteria)

Archaea are the extremists of the bacterial world, perhaps evolutionary products of the harsh environmental conditions under which they emerged on the young, inhospitable Earth more than 3 billion years ago. They grow only in oxygen-free (anaerobic) environments like the atmosphere of the young Earth. Today, they have withdrawn into the remnants of such sheltered anaerobic environments. Some archaea are *thermophiles* (heat-loving); these thrive in hot springs in Yellowstone National Park (United States) at 70–75°C (160°F) or next to deep-ocean hydrothermal rift vents

at temperatures above 100°C (212°F). Others such as *Halophilic* (salt-loving) archaea prosper in water with extremely high salt concentrations such as occur in the Great Salt Lake or the Dead Sea. The *methanogens* produce methane gas. Some methanogens are denizens of swamps and marshes, where the methane they produce bubbles out as "marsh gas"; others live in the guts of animals such as ruminants (e.g., cattle, bison, wildebeests), where they break down cellulose but produce methane as a by-product.

EUKARYOTES

Eukaryotes have nucleated cells filled with organelles. Many (e.g., plants, fungi, animals) are multicellular, but not all. Perhaps the most diverse of the eukaryotes are the protists.

Protists

The protists are a motley lot, an artificial assemblage, but one of useful convenience. Some are unicellular, others colonial, and some even multicellular. In early classifications, photosynthetic protists were placed with plants; those that ingested food, with animals;

Consider This—

E. coli—Friend and Foe

The human gut holds about 2 pounds of bacteria of various kinds. A small proportion of these are *E. coli*. Like different breeds of dogs, different strains of *E. coli* may be docile or fierce. Most *E. coli* synthesize vitamin K and B-complex vitamins, which are then beneficially absorbed by the body. However, some strains of *E. coli* can become pathogenic, producing urinary tract infections, diarrhea, and even life-threatening blood infections. Why different strains should switch from friend to foe is not known. But the outbreak of such rogue bacteria seems to result from opportunity (contaminated food, poorly cooked food) and from some genetic mutations that promote the release of virulent toxins.

others stood alone. Most likely, protists will eventually be split into smaller, natural taxonomic units. For now, let's just look at the diversity. You may know of some of them already.

Algae.   Algae are one of the photosynthetic groups of protists, including red, brown, and green algae. Some red algae live in fresh water, but most are marine protists. The long, leafy "kelp" along shallow ocean shores is a brown alga that can form massive underwater groves. In green algae, the close chemical similarities of their chlorophylls betray their close relationship to plants. Along with cyanobacteria, the green algae make up the community of aquatic organisms known as **phytoplankton**. The phytoplankton are at the base of most marine and freshwater food webs, as well as being responsible for much of the world's O_2 production.

Sarcodina.   Amongst the Sarcodina are the amoebas, which move by means of fluid cell extensions, the pseudopods. Most are harmless, but one species causes amoebic dysentery. Foraminifera, or just "forams," are part of this protist group. Each unicellular foram secretes a shell, usually of calcium, in which it lives.

Flagellates.   The flagellates include protists with a whiplike *flagellum* that, like a cellular tail, beats to move the microorganism along. Sudden, explosive population buildup can produce "red tides," a bloom of flagellates toxic to many marine organisms. The flagellate *Trypanosoma*, the vector carried by the tsetse fly, is responsible for sleeping sickness common in tropical regions. Another flagellate, *Giardia*, present in clear waters of mountains streams, causes the hiker's diarrhea.

Plants

Plants are denizens of the land. They arose from aquatic seaweeds—specifically from green algae, perhaps leafy or filamentous in shape—and colonized terrestrial environments. Some, such as water lilies and cattails, have returned to a watery environment. But plants are designed for life on land and carry adaptations to serve them there.

Terrestrial Adaptations

Life on land poses special problems and opportunities, compared to life in water. On land, sunlight is unfiltered by sea or fresh water, and its full energy can be tapped through photosynthesis. On the other hand, the buoyancy of surrounding water that floats aquatic algae is gone in thin air. Remaining upright, alone against gravity, is a major problem. So is the threat of desiccation—dehydration

from exposure to the hot sun and drying wind. For most plants, these problems are addressed with *root* and *shoot systems* (figure 5.3). The root system taps into one environment, the soil, where it anchors the plant and finds water and minerals to support the plant. The shoot system takes advantage of another environment, the air, where it gathers carbon dioxide and spreads its leaves out into solar collectors, taking full advantage of photosynthesis.

The leaves and other parts of the shoot are coated with a waxy layer, the *cuticle*. You may have noticed this shiny cuticle on the surface of polished apples. This watertight sealant on the aerial parts of the plant reduces water loss from interior tissue. Gas exchange between the atmosphere and interior tissues of the leaf, a necessary part of photosynthesis, occurs through numerous tiny pores, the *stomata* (sing., *stoma*), passing through the cuticle. Stomata open and close to permit entry of

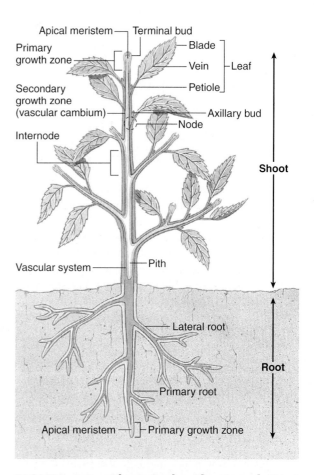

FIGURE 5.3   Plant Body, Shoot and Root
Primary growth in length is within apical meristems at the tips of stems and tips of roots.

carbon dioxide and exit of oxygen. Walls of cells supporting the plant are often reinforced with *lignin*, a chemical that hardens to strengthen the shoot projecting above the ground and help hold it erect. Most plants also have a plumbing system made up of **vascular tissue**, which is composed of tubular cells running throughout the plant. Some of the vascular tissue conducts water and minerals from roots to shoots; other tubes conduct sugars from shoots to roots. Bundles of vascular tissue also form pillars that help support the plant. The vascular system is easily seen as *veins* that spread along the undersides of leaves.

Growth

A plant's growth in height occurs at the shoot tip, not by elongation within its stem or trunk. However, increase in girth—widening of woody plants—occurs throughout the shoot and root systems by adding material laterally, along the plant's length. Some plants are **annual plants** that live only one year and then die. Others are **perennial plants**, those that live for more than a year. In trees, which are perennial plants, much of the growth in girth occurs in an active ring of cells just beneath the bark, the *vascular cambium*. In environments imposing stressful dry (tropical) or cold (temperate) seasons, the vascular cambium ceases growth during these stressful seasons and enters *dormancy*. When rains return or spring brings warm temperatures, growth resumes, producing a new layer of large, water-swollen cells. As the tree approaches the next dormant phase, growth slows, cells in the vascular cambium are small, as the tree enters dormancy once again. Each annual pulse of growth plus dormancy therefore produces one **growth ring**—an annual episode of growth marked by a ring of large and compact cells. Preserved in the trunk of the tree from year to year, the age of the tree is represented in these yearly growth rings. Foresters and others use core-sampling tools on living trees to take a small plug of the tree and count the rings on the extracted sample to determine the age of the tree.

Whether slow or swift, plants grow continuously by simple cell proliferation, termed **vegetative growth**. If a plant part is nibbled or broken off, new shoots grow out; parts or "runners" also may extend out from the main plant, take root, and give rise to whole new plants. Some animals are capable of similar growth via part replacement and can produce a new organism from body fragments or by budding. But in plants, vegetative growth is highly accommodating, so much so that gardeners can clip off "cuttings," bits of plant shoots, to start new growth or even new plants. Some plants prop-

agate by such vegetative means. In fact, this natural method of proliferation, released from natural controls, can become an ecological problem. The watermilfoil is a freshwater plant introduced into North America from Eurasia. Largely through vegetative growth, it has spread to lakes, rivers, and ponds in most states and provinces, where it forms dense floating mats that choke off light to native plants below.

Reproduction

Plants reproduce by both mitosis and meiosis. But unlike animals, plants have a life cycle that includes an **alternation of generations**, wherein diploid (2n) individuals and haploid (n) individuals successively alternate with each other. Recall that during meiosis, the pairs of chromosomes in a cell divide—each half of the pair moving into one of the gametes (eggs or sperm) produced (see chapter 3 and Appendix 1). Just before meiosis, the cell holds a full complement of paired chromosomes and is *diploid* (2n); after meiosis, the chromosome number is halved in the gametes, which are consequently *haploid* (n). Within plants, the life cycle alternates between **sporophyte individuals** (2n; produce spores) and **gametophyte individuals** (n; produce gametes) (figure 5.4).

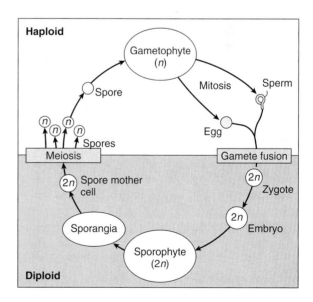

FIGURE 5.4 Alternation of Generations, Plants The life cycles of plants are different from ours and those of all animals. Animals are diploid (2n) and the only haploid (n) stage is found in their eggs and sperm. In plants, diploid individuals (sporophytes) alternate with haploid individuals (gametophytes), although the prominence of each might be quite different in different groups of plants.

Plant Diversity

The transition from green algae to plants—from water to land—may have been an easy one. Even today, many green algae live in shallow water around the edges of lakes and ponds, areas subjected to occasional drying. If plants adapted to such drying, then the colonization of land required only an extension of the time exposed to air until the transition became permanent.

Estimates of the time of this transition recently were pushed back. Fossils speak to the presence of plants at about 450 million years ago. However, extrapolating back from molecular changes in current plants, the much deeper date of 700 million years ago is now tentatively proposed for the transition to land plants (see figure 5.1).

Bryophytes (mosses).   Some of the earliest land plants were the *bryophytes*, mosses being the most common example (figure 5.5). Mosses mostly lack both vascular tissue and true roots for water and nutrient transport. They thrive in damp, shady places, although they can withstand prolonged periods of drought. They also lack lignin to strengthen their cell walls, keeping them short and low to the ground. Mosses are common in near-polar regions, forming great ground-covering mats.

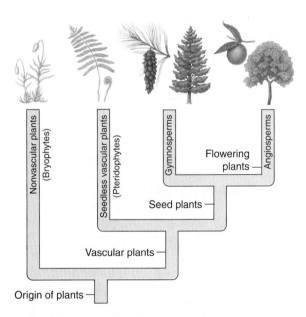

FIGURE 5.5   Plant Evolution   The major groups of plants are shown. Note that adaptive transitions evolved at different points: water to land from ancestral green algae (not shown); nonvascular to vascular; seedless to seed; flowers.

Pteridophytes (ferns).   Ferns are the most abundant member of the *pteridophytes* (ter.rid.o-fites), which also include the horsetails and the much lesser known club "mosses," not to be confused with true mosses, which are properly members of the bryophytes (figure 5.5). Fossils mark the presence of ferns at 350 million years ago (Devonian period). Ferns are characterized by an underground specialized stem, the *rhizome*, which is equipped with short roots. The leaves are aerial *fronds* that unfurl from coiled-up, young leaves ("fiddleheads") at the tip of the rhizome. No seeds are produced, but ferns show alternation of generations. Spores are produced in clusters on the underside of the mature frond (sporophyte individual). After meiosis, the shed spores (now haploid) produce a small, heart-shaped form (gametophyte individual) that produces eggs and sperm, which, following fertilization, grow directly into the rhizome and fronds of the young fern.

Pteridophytes are vascular plants. During the mid to late Paleozoic, they became predominant, forming great forests in tropical regions. With lignin-reinforced cell walls and added support from the vascular tissue, some reached great heights. Distant relatives of the club mosses produced giant woody trees more than 120 feet (40 m) tall. One period of the Paleozoic era was especially favorable to coal formation and was named, in recognition of this, the Carboniferous. In this period, the dead plants did not completely decay in waters of tropical swamps. Instead, their organic debris accumulated, forming peat, as occurs today in peat bogs. Later, when the peat was inundated and covered by overriding marine seas and sediments, the heat and pressure converted the peat into coal. These great coal deposits literally fueled the Industrial Revolution of the nineteenth and twentieth centuries.

The *seed plants* are vascular plants that include the gymnosperms (e.g., conifers) and angiosperms (flowering plants). They introduce new innovations especially suited to address the problems of terrestrial life. One innovation is the seed, a protective case for the young plant embryo; the other innovation is pollen, a hardy transport package for sperm. Specifically, a **seed** is a small capsule composed of a protective *seed coat*, the *plant embryo*, and a supply of *nutrients* to support the young plant up until it germinates. Seedless plants, such as mosses and ferns, produce resistant spores that can develop into new organisms. Spores survive extremes of temperature and moisture to which mosses and fern plants themselves might succumb. The seed represents a different solution to environmental challenges. The seed is a protective package in which the tiny plant embryo develops and is sheltered.

The other innovation of seed plants is **pollen**. A pollen grain is a much-reduced male gametophyte, composed of just a few cells. Transported to the female, some of these cells develop into sperm, which eventually reach eggs of the female plant, resulting in fertilization. Pollen grains are resistant to extremes of temperature and moisture, a further adaptation to reproduction in a terrestrial environment.

Gymnosperms (Conifers, allies). The earliest of the seed plants to evolve are the *gymnosperms*. They are present in the Carboniferous along with the then more abundant seedless plants. By the end of the Carboniferous, the climate began to dry, most swamps disappeared, and the gymnosperms began a rise to prominence. The most common among them today are the *conifers*—cone-bearing trees—including pine, fir, redwood, spruce, cedar, and juniper. The conifer tree is a sporophyte with the tiny, much-reduced gametophyte living in the cones. The cones are made up of specialized, compacted leaves called *scales*. In female cones, the scales protect the **ovule**, in which the tiny female gametophyte develops forming eggs. In male cones, the male gametophyte produces pollen grains that hold sperm. Pollen grains are usually carried on the wind to the ovule. The successful trip from male to female is termed **pollination**. When the pollen enters the ovule, the pollen discharges sperm that fertilize the eggs.

Following fertilization, the ovule develops into the seed, including the protective coats, food supply, and young plant embryo within. Released from the cone, the seed, often attached to a propeller-like extension, may be carried by the wind a long distance and survive for months or years until favorable conditions promote germination. The food supply enclosed in the seed sustains the young sprout until it can set roots and begin photosynthesis.

Angiosperms (Flowering plants). The **angiosperms** are also vascular seed plants. They debuted after gymnosperms during the mid Mesozoic, but came into prominence during the late part of the Mesozoic. Today, the gymnosperms supply us with lumber and paper; the angiosperms supply us with most of our plant foods—cereal grains (wheat, barley, oats, corn) and fruits (apples, berries, oranges, and the like). Flowers distinguish angiosperms from all other plants and advertise their sex organs for all to see. The flower is derived, in part, from modified leaves (figure 5.6). During the reproductive season, flowers open to surround and protect

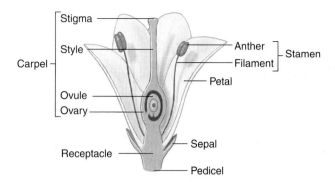

FIGURE 5.6 Flower Structure The female structures (carpel) and male structures (stamen) are shown. Note the ovule within the ovary of the female.

the reproductive organs of the plant—stamen and carpel (pistil)—and often to catch the attention of animals, the **pollinators** (sometimes called animal vectors) that are attracted to these plant parts. Pollen is produced in the anther, the male part of the flower that is perched atop the thin filament (figures 5.6 and 5.7). Pollen grains (male gametophyte) are carried or transported by pollinators, by wind, or sometimes by water and gravity, to other flowers where they settle on the sticky tip of the female stigma; sperm travel down the style into the ovule, which holds the female gametophyte, and fertilize the eggs within (figure 5.7). Oddly, a second fertilization occurs, a feature unique to angiosperms. A second sperm fertilizes another female cell, which develops not into an embryo, like the first, but into the *endosperm*—the nutrient tissue that provides food for the young embryo. Some plants may *self-fertilize* (fertilization of one plant by its own pollen); others *cross-fertilize* (fertilization of one plant by pollen from a different plant).

The swollen base of the carpel is the **ovary**, a protective chamber that may hold one to hundreds of ovules. Following fertilization, the ovary ripens into a *fruit*, enclosing the seeds derived from the ovule or ovules. The fruit surface often develops structures that help the seeds disperse. The fluffy parachute of dandelion seeds helps in their wind-borne journeys; the burrs or prickly processes of some seeds help them catch on the fur or feathers of passing animals. The fruit may be fleshy, as in apples, peaches, and cherries, or dry, as is the pod of a soybean. A single flower with multiple carpels produces an aggregate fruit, such as a blackberry. A pineapple, like some other angiosperms, forms from an *inflorescence*, a tightly packed group of multiple flowers that fuse together to form the single fruit.

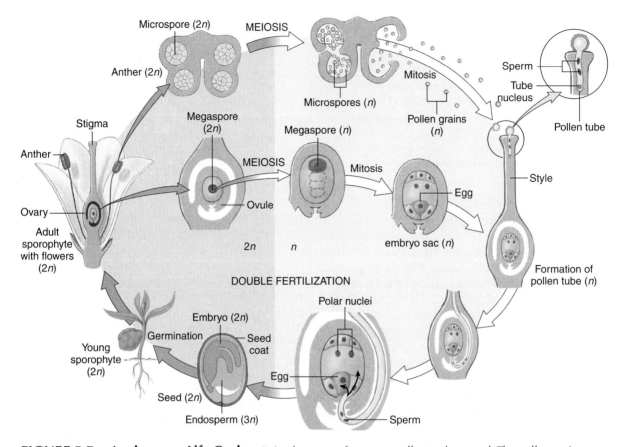

FIGURE 5.7   Angiosperm Life Cycle   Animal vectors often carry pollen to the carpel. The pollen grains contain generative cells that produce the sperm and, after alighting on the stigma, travel down and within the growing pollen tube to reach the ovule and its eggs. There, sperm fertilize the egg, producing an embryo, and fertilize other cells, producing endosperm. Specifically, in the male anther, diploid spores develop (microspores, $2n$), which after meiosis develop into pollen grains (n) that reach the female part of a flower and grow down to and into the ovule in a pollen tube. In the female, diploid spores develop (megaspore, $2n$), which after meiosis develop into an egg within the embryo sac. An arriving sperm fertilizes the egg, which develops into the plant embryo ($2n$); other sperm join with the polar nuclei to produce the nutritive endosperm ($3n$).

Fungi

The fungi are not plants. In place of cellulose, the cell walls of fungi are built of chitin, a tough, nitrogen-containing polysaccharide (chain of sugars). Fungi have no chlorophyll and do not photosynthesize. Instead, fungi absorb nutrients from the substrate to which they are attached—old logs, moist ground, damp bark. They secrete digestive enzymes into the substrate, then soak up the organic molecules released. This makes fungi, along with various bacteria, the principal decomposers in forests. As decomposers, fungi feed on the dead bodies of other organisms, breaking them down and eventually returning nitrogen, carbon, and other elements to the soil where they become available for recycling to plants and animals.

Truffles, morels, and yeasts are fungi. Mushrooms, with their familiar umbrella-like tops, are the temporary reproductive structures. Fungi are made up of thin, long filaments called **hyphae** (figure 5.8). The tubular walls of the hyphae are composed of chitin and other polysaccharides, which enclose the fungal cells. Within these walls, the cytoplasm flows freely, carrying along nutrients to the far-flung parts of the fungus. Bundles of hyphae form the body of the fungus and expand into the specialized reproductive structure, the *fruiting body*. The mass of hyphae permeating the soil, wood, or other substrate is the *mycelium* (figure 5.8), which is the feeding network of a fungus.

Consider This—

Fungus among Us

The state of Oregon currently holds the record for the world's largest single organism. It is a fungus. The interwoven mycelium of this fungus spreads over 2,200 acres (about 1,660 football fields), is 3.4 miles in diameter, and weighs collectively several tons. It is pathogenic, infecting primarily coniferous forests, where it has been lodged and growing for an estimated 2,400 years. Similar, but smaller, great disks of this fungus are known from other forested regions as well. The girth and heft of this fungus qualify it as the Earth's largest known organism.

Fungi can be parasitic and destructive. The American elm tree has been almost eliminated by fungus infections across North America. Agricultural products are susceptible to rusts, fungi that attack seeds of rye, wheat, and oats. The skin disease ringworm is caused not by an animal worm, but by a fungus, as is itching athlete's foot. Fungi (and some bacteria) produce what we call "rot." A wood-digesting fungus does not distinguish between blown-down oak tree trunks and oak tree planks in a wooden ship. Fungal rot was often more of a problem to wooden warships than enemy cannon fire. Molds are odd types of fungi that appear on unattended bread, fruits, and other foods. In moist regions, canvas tents, clothing, leather boots, and even shower curtains support growths of molds.

Some fungi form cooperative associations with green algae, producing **lichens**. The fungal hyphae embrace and supply the algae with a favorable surrounding of moisture and nutrients; the photosynthetic algae return organic compounds to the fungi. Fungi also form beneficial associations with the roots of plants. Such root-fungus combinations are *mycorrhizae*. The root supplies sugars to the fungi; the fungi facilitate transfer of essential minerals from the soil to the roots.

Animals

Animals meet their metabolic needs by eating other organisms. If they eat plants, they are **herbivores** or **phytophages** (*phyto-*, plant; *-phage*, eat); if they eat each other, they are **carnivores**. This survival strategy—ingestion—sets animals apart from plants, which photosynthesize, and from fungi, which absorb. The cells of animals lack the supportive cellulose cell walls of plants, so animals depend for support upon the buoyancy of water or upon specialized skeletons. Animals also move under their own power, not just when buffeted about by a breeze.

Animals arose from protists, although the specifics still elude us. The likely ancestor of animals was a single-celled *choanoflagellate* that lived in clumps or colonies. These microscopic cells came equipped with flagella, long whiplike processes that can whirl about, imparting motion to the cell, or draw in currents of food-bearing water. These animal ancestors were heterotrophs, a fundamental characteristic upon which animal descendants are based. As in most other organisms, both sexual and asexual reproduction are found in animals. *Sexual reproduction* is based on the fusion of gametes, wherein haploid egg and sperm unite; *asexual reproduction* involves budding where diploid cells proliferate (see Appendix 1). Animals that reproduce sexually pass through a **life cycle**, typically beginning with fertilized *eggs*, which grow into *larvae* or *juveniles* (sexually immature stage), which become *adults* (sexually mature stage).

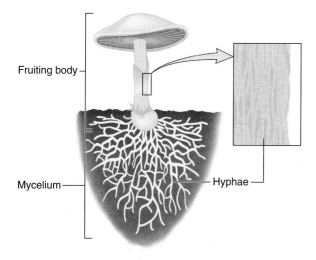

FIGURE 5.8 Fungus Filamentous hyphae produce a feeding mat, the mycelium, that permeates the food source. The hyphae typically continue aboveground as a reproductive structure, the fruiting body.

Fruiting body

Mycelium

Hyphae

Sponges (parazoa)

The simplest of the animals are the sponges. Connected needlelike spicules provide support to the body, which is **sessile**— attached to a secure aquatic substrate. Like most marine animals, sponges pass through a larval stage that is **planktonic**, free and unattached, carried about by tides and currents. A few sponges live in fresh water; most are marine. The bodies surround an internal cavity lined by flagellated cells, reminiscent of their protist ancestry. Some have tubular bodies, but most sponges are *asymmetrical*, lacking defined shape and instead conforming to rock shape or local space. This lack of symmetry sets them apart from all other animals, providing them with their alternative name, **parazoa** (beside animals). All other animals make up the **eumetazoa** ("true" animals), exhibiting well-defined body symmetry (figure 5.9).

Eumetazoa

Radiata. Jellyfishes, corals, and sea anemones are the most common examples of cnidarians, an almost exclusively marine group. The name *cnidaria* means "nettles," a reference to the *stinging cells* (cnidocytes) that cover their bodies and are especially abundant on feeding tentacles. When making contact with food or foe, these cells discharge, lashing through the victim's skin with threads laced with toxins. An unwary swimmer brushing against a jellyfish receives the same treatment, coming away with a "burning" sensation— a reaction to these toxins. Corals are important reef animals, as they secrete a calcium base on which they reside. Found in tropical waters, these are the protective coral reefs that encircle small islands, raise physical barriers that buffer strong wave action, and create a complex environment upon which other reef animals and plants depend.

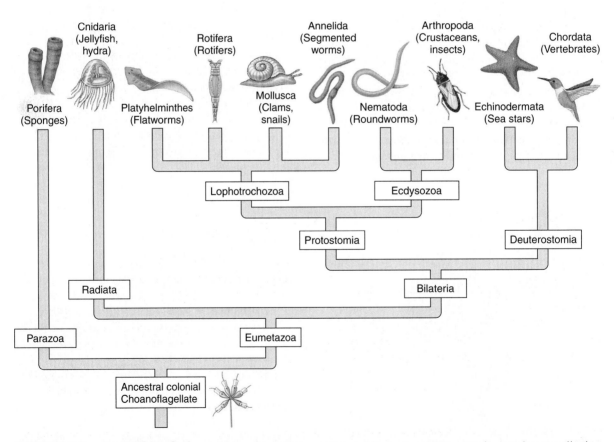

FIGURE 5.9   Animal Evolution   The major groups are shown. After sponges (Parazoa) diverge, leaving all other animals (Eumetazoa), differences in symmetry reveal two groups (Radiata, Bilateria). Embryonic differences within the Bilateria are diagnostic for the Protostomia and Deuterostomia. The subgroups of protostomes are the Ecdysozoa and Lophotrochozoa. Within the deuterostomes occur the chordates, where we as vertebrates are placed. These divisions are based on molecular similarities with anatomical correlations.

Bilateria.   Cnidarian body symmetry is *radial*, meaning that a central axis can be found around which all the rest of the body is equally arranged. All other eumetazoans are characterized by a body symmetry that is *bilateral*, meaning that they can be divided into left and right halves—mirror images. Most bilateral animals also have a definite front end (anterior), back end (posterior), top (dorsal), and bottom (ventral) (figure 5.10a,b). The bilateral animals divide into two major groups, originally defined by two distinctive types of embryonic development (figure 5.10 c,d). One group is the **protostomia** and the other, which evolved late in the Precambrian, is the **deuterostomia**.

The protostomia are made up of two sublineages: the lophotrochozoa and the ecdysozoa (see figure 5.9).

Protostomia—Lophotrochozoa.   The **lophotrchozoa** take their name from a lophophore, a specialized crown of feeding tentacles, characteristic of many. Some of the most important amongst them are the following:

Flatworms (platyhelminthes).   The flatworm body is compressed and ribbonlike. Some are **free-living**, moving about in the environment; others are **parasitic**, living within an animal host. Planarians are free-living forms that glide about on cilia that cover their ventral surfaces. Parasitic forms include liver flukes and tapeworms. Tapeworms are *segmented*, wherein body sections are repeated to produce the long adult.

Roundworms (nematodes).   Roundworms are unsegmented and their bodies are round in cross section.

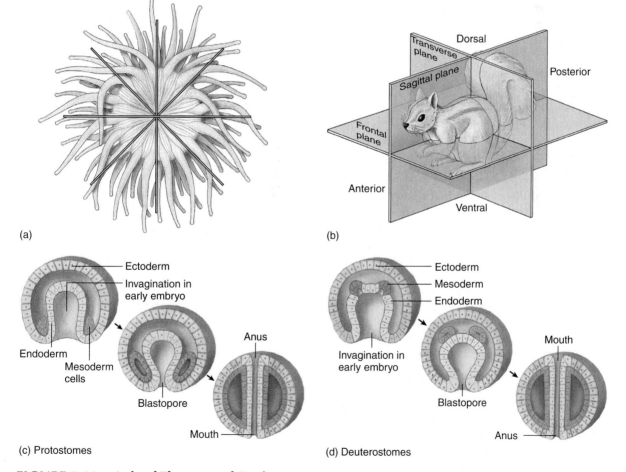

(a)

(b)

(c) Protostomes

(d) Deuterostomes

FIGURE 5.10   Animal Themes and Designs   *Symmetry*. (a) Radial symmetry, illustrated by a sea anemone. (b) Bilateral symmetry, illustrated by a squirrel. Other planes of symmetry are also recognized. *Embryonic development*. (c) Embryonic development of protostomes, wherein the embryonic blastopore becomes the mouth of the adult. (d) In deuterostomes, the opposite occurs, and the blastopore becomes the anus. In both patterns, the basic embryonic body layers—ectoderm, mesoderm, and endoderm—are laid down as well. These generally give rise primarily to the adult skin, muscles and blood vessels, and gut, respectively.

They are one of the most abundant animal groups, occupying land and water in tropical to subarctic habitats. Some are parasitic, such as pinworms, hookworms, and *Ascaris*, which is especially common in farm animals.

Annelids.   The common earthworm is an annelid, exhibiting the basic segmental body plan of the group. As they burrow, earthworms feed on and pass soil through their bodies, digesting the organic material it contains and leaving the indigestible material in castings (feces). Leeches are also annelids, feeding on aquatic insects or more often attaching to a passing animal and dining on its blood.

Mollusks.   Many soft-bodied mollusks possess a shell or hard case into which they can withdraw. Some burrowing forms, such as clams, have a pair of closing shells. Others, such as snails, carry about a single shell for easy retreat. Others are active predators, with a reduced shell, tentacles with suckers that reach out to grasp prey, and a pair of large, keen eyes; these include the squid and octopus.

Protostomia—Ecdysozoa.   The **ecdysozoa** take their name from the characteristic **molting** (*ecdysis*) wherein the growing animal sheds its external skeleton and grows a new one. The most important amongst them are the following:

Arthropods.   By far, arthropods (*arthro-*, jointed; *pods*, legged) are the largest group of animals, an estimated 10 million species and still counting. They are segmental animals, with an **exoskeleton**—an external skeleton made of a tough, nitrogen-containing sugar, *chitin*. The exoskeleton is jointed, making movement possible. The exoskeleton also serves as protective armor against assaults by predators and abrasion from the environment. One of the oldest arthropods, the *trilobites*, arose in the late Precambrian. They occupied shallow marine seas, lasted throughout the subsequent almost 300 million years, and then perished, along with many other groups, during the mysterious mass extinctions at the end of the Paleozoic. The *arachnids* include the daddy longlegs, mites, ticks, scorpions, and spiders. The *crustaceans* are the crabs, lobsters, and shrimps.

By far, the largest group of arthropods is the *insects* (uniramians), most of which are winged arthropods. Half of all named animal species are insects. Their varied and diverse lifestyles place them on almost every continent, moving through complex life cycles. Generally, the larva is a feeding stage, the adult a dispersal stage. Aphids to zygoptera (damselflies), butterflies to beetles are included.

Deuterostomia.   The deuterostomia include the echinoderms and chordates (figure 5.9). The echinoderms include starfishes and allies; the chordates include several small groups and one very central group, the vertebrates—which includes us, human beings.

Echinoderms.   Besides starfish (sea stars), echinoderms include the sea urchins, sand dollars, brittle stars, sea cucumbers, and ancient crinoids. As larvae, echinoderms are bilateral in symmetry. As adults, they are five-armed or some derivative thereof and move along with sweeps of their arms or on the synchronized motion of tube feet (tiny suction cups on extending processes). At face value, echinoderms are unlikely taxonomic companions to the chordates. But their similar embryology and chemical structure make the connection undeniable.

Chordates.   Included in the chordates are several unfamiliar groups such as sea squirts and amphioxus, but also a group central to our story, the **vertebrates**. The vertebrates include various groups of fishes, plus amphibians, reptiles, birds, and mammals. Bone tends to replace soft supportive tissues, and it leaves traces in the fossil record. A stiffening rod, the *notochord*, runs down the long axis of the body of chordates. Against its action, segmental blocks of muscle act to impart side-to-side swimming motions. Within many vertebrates, the notochord is replaced by a chain of bony or cartilage blocks, the *vertebrae*, collectively the vertebral column or "backbone." This vertebral column, from which the group takes its name, functionally forms part of an **endoskeleton**, an internal skeleton.

Upon the vertebrate body plan, a great many and diverse groups evolved (figure 5.11). The earliest vertebrates were the *jawless fishes*, including the now extinct *ostracoderms*, which were encased in protective bony armor, and their modern descendants such as *lampreys*, which lack bone and today are restricted to specialized, scavenging or parasitic lifestyles (figure 5.12). Jaws are the bony or cartilaginous supports that rim the mouth. Two early groups first to possess jaws were the *placoderms* and the *acanthodians*. Both groups are extinct, but in their time they commanded a formidable oceanic, predaceous lifestyle, guided by simple forms of paired fins. Major fish groups followed, such as the *cartilaginous fishes* (chondrichthyes) and *bony fishes* (osteichthyes). The skeleton of cartilaginous fishes is just that, cartilage, the same material that gives your ear and nose tip flexibility. Sharks and their allies have skeletons made of such material. The bony fish have extensive skeletons of bone. Today, this is the largest single group of verte-

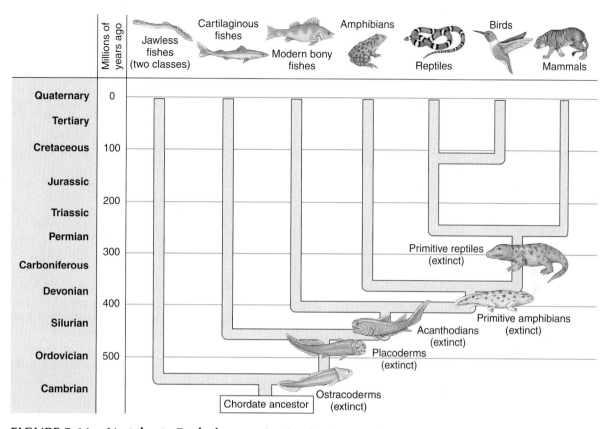

FIGURE 5.11   Vertebrate Evolution   Within the chordates (see figure 5.9), the vertebrates arise from a primitive chordate ancestor. Notice the sequence of appearance; first, various fishes; then amphibians, reptiles, and mammals. Birds evolve within the reptile radiation.

brates and includes most fishes, from those seen in pet shops to tuna, trout, and many others. The *amphibians* were the first vertebrates to employ limbs, and therefore the first **tetrapods**. Arising later within the tetrapods were the reptiles, birds, and mammals. *Tetrapod* means literally "four" *(tetra-)* "footed" *(pod)*, but is understood to be a taxonomic term, not an anatomical descriptor—legless forms, such as snakes (reptile), and finned forms, such as whales (mammal), are included.

Today the amphibians include salamanders and frogs as well as a legless, burrowing group, the caecilians. Adult amphibians often venture onto land, although a moist location, a stream or pond, is usually close by. To breed, most amphibians return to water, where eggs are laid, fertilization occurs, and embryos then typically develop into small aquatic larvae or tadpoles. These larvae feed and mature, eventually undergoing a rapid and radical anatomical change—**metamorphosis**—turning them from larvae into ju-

veniles. In most adult amphibians the gills of their fish ancestors are lost, but they still gather oxygen from the air in lungs or oxygen diffuses directly across the moist skin and is picked up by surface blood vessels. Evolving from amphibians were the **amniotes**, including reptiles, birds, and mammals (figure 5.13). Vertebrates that came before—amphibians and all fishes—are **anamniotes** (*an-*, without; -*amnion*, an amnion) because their embryos are not wrapped in the specialized embryonic amnion.

The collective name *amniotes* is taken from an embryonic innovation. The eggs of anamniotes are typically laid in water and hatched there. The invention of the shelled (cleidoic) egg emancipated the embryos from bodies of water. The amniote embryo is wrapped immediately in a thin membrane—the **amnion**—which floats the young animal in a protective jacket of water. In a sense, amniote embryos carry the water environment with them onto land. The embryo and its enveloping amnion,

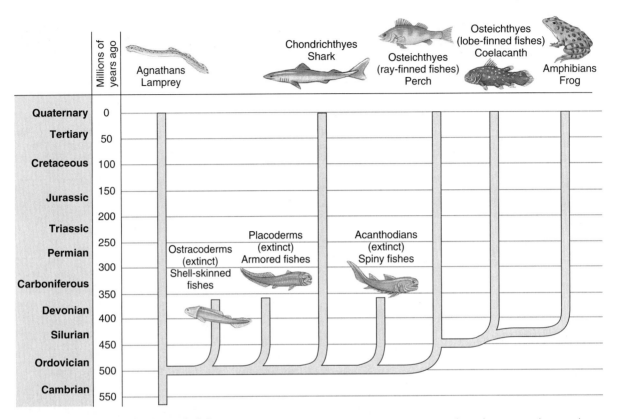

FIGURE 5.12   Evolution of Fishes   Many early fish groups are now extinct, such as the ostracoderms, placoderms, and acanthodians. Amphibians, represented by the specialized frog, arose during the Devonian from a special group of bony fishes, the lobe-finned fishes, illustrated by the coelacanth.

plus any food reserves, are packaged in a leathery shell (reptiles, monotremes) or in a brittle shell (birds), altogether called the **cleidoic egg** (figure 5.14).

Radiation of the amniotes produced some of the best known and most dramatic of all the vertebrates (see figure 5.13). One lineage, *sauropsids*, produced the familiar *reptiles*—snakes, lizards, turtles, crocodiles—but also the extinct plesiosaurs, ichthyosaurs, flying pterosaurs, and, of course, the "dinosaurs." All have scaly skin and a suite of common anatomical features. Most lay eggs, but some are live-bearing. Notice that appearing within this sauropsid lineage are the *birds* (Aves), part of this radiation (see figure 5.13). Strictly speaking, birds are a specialized type of reptile, "reptiles with feathers." All birds lay eggs and are covered with feathers, used also for flight. The other great lineage of amniotes is the *synapsids*. The *mammals* evolved within this lineage. Mammals share unique anatomical similarities of internal structure, but they are characterized by a coat of hair, not scales or feathers, and females nurse their young from milk glands—mammary glands. Today, there are three living groups of mammals (figure 5.15). The smallest group is the *monotremes*, the duckbilled platypus being an example. Representatives are found today in Australia, Tasmania, and limited parts of New Guinea. Monotremes are mammals (hair, lactation), but they lay eggs incubated and hatched outside the body of the female. The *marsupials* are the pouched mammals—kangaroos, koalas, opossums, and their allies. Young are born at a very early age (no cleidoic egg) and then migrate into the female's pouch, where they find and nurse from mammary glands until they grow to a large size. Most marsupials live today in Australia, but some biologists place their center of origin in North or South America from whence they spread to current locations. The *placental mammals* are the largest group of mammals, worldwide in distribution. Neither eggs nor early birth characterized their reproduction. Instead, tissues of the fetus (young embryo) cooperate with specialized tissues along the inside wall of the female uterus ("womb"). These specialized fetal-maternal tissues

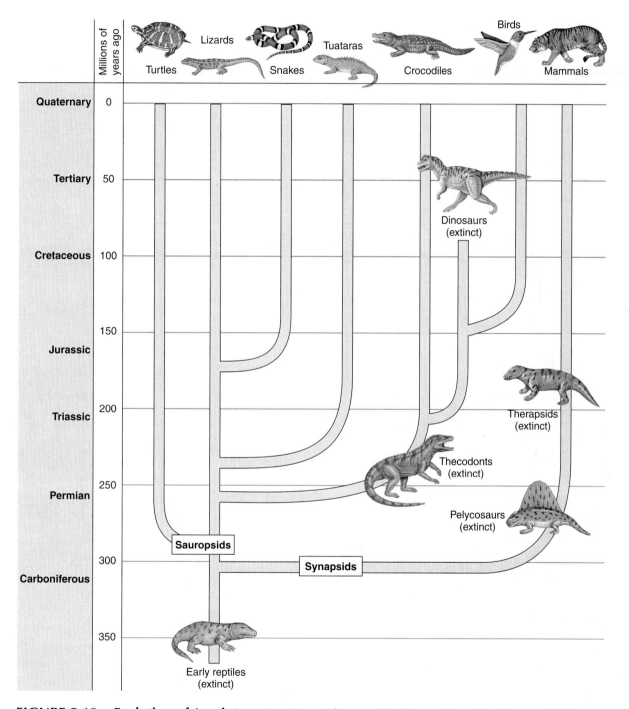

FIGURE 5.13   Evolution of Amniotes   Primitive reptiles were the first amniotes arising from amphibian ancestors (not shown). From these early reptiles arose all later groups, the sauropsids and synapsids. Within the synapsid lineage, mammals arose. Within the sauropsid lineage, great diversity occurred, including modern groups of reptiles and birds, as well as the extinct dinosaurs.

form a **placenta**, connected by major blood vessels traveling to and from the fetus as the *umbilical cord*. The placenta passes oxygen and nutrients from the mother's blood to the fetus and fetal CO_2 to the mother for elimination. At birth, the placenta detaches and, along with its umbilical cord, is passed out along with

FIGURE 5.14   Cleidoic Egg   Sometimes called an "amniotic egg," the cleidoic egg includes the embryo floated in a water jacket formed from a thin membrane, the amnion, and several other embryonic membranes. One is the chorion, just under the outer shell, which serves respiration. The other, the yolk sac, contains energy-rich yolk upon which the embryo draws to meet its nutritional and growth needs. All is wrapped in a leathery (reptiles, monotremes) or hard (birds) shell.

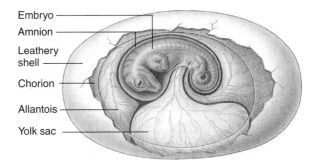

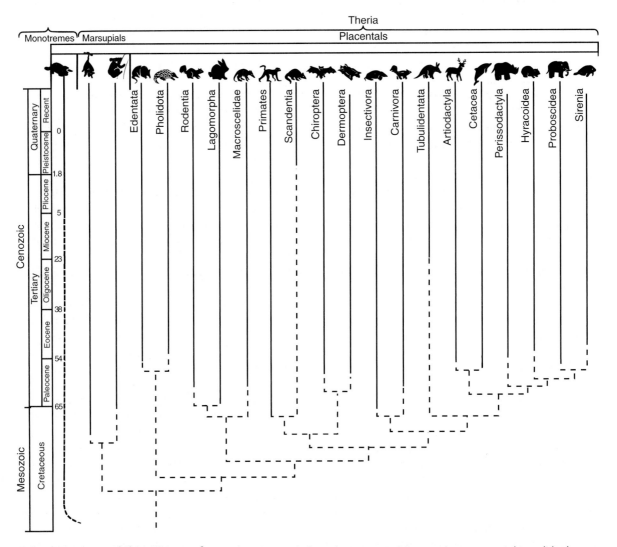

FIGURE 5.15   Living Mammals   Today, mammals form three groups: monotremes, marsupials, and the largest, the placentals. The marsupials and placental mammals are sometimes placed together in the Theria, and living monotremes plus allied fossil forms in the Prototheria (not indicated).

the newborn infant mammal. All placental mammals support their growing fetus by means of a placenta, but this reproductive device is not unique to them. Placen-

tas have been independently invented in other groups such as some lizards, some snakes, and even in a few amphibians and fishes.

ENVIRONMENT

All organisms live in an **environment**—the external world where they reside, find food, contend with foes, attract mates, elude predators, and deal with harsh climates. Here they will survive or not; here they will breed or not. It is not their characters standing in splendid isolation that determine their success, but their characters measured against the challenges of the environment. Interaction is the word—interaction between an organism and its immediate environment. The future success of an organism cannot be determined in isolation. Adaptations are not independent features living alone. All must serve, and serve well, within the environment where the organism lives. Yet environments are different, so organisms are different.

Globally, organisms live in warm *tropical* regions; adjacent *subtropical* regions; *temperate* regions; and, at the poles, extreme *polar* regions. These regions lie, generally, from the equator (tropical) to the poles (polar). The position within latitude, equator to poles, affects the character of the environment, determining the seasons and thus day length, heat/cold, and weather. In temperate regions, the year passes between warm summers and cold winters. Tropical regions lack such extreme swings in temperature but often experience extremes of moisture alternating between wet and dry seasons. Local conditions of geography, such as mountain ranges, lakes, and nearby bodies of water, affect these latitudinal caprices of climate, producing *biomes*—distinctive *ecosystems* with distinct local communities of plants and animals adapted to the particular climatic conditions. A desert ecosystem would be one example; African savannas would be another. Where oceans wash up against the land, they support *intertidal communities*. The pull of the moon's gravity sloshes the marine waters about within their large oceanic basins. Their characteristic maximum flooding up the beach during the day marks the *high tide* level; their maximum retreat the *low tide* level. Tidal patterns depend upon the size of the ocean basin and shape of the continental coastline, so tidal ebb and flow varies around the world. The consequence for marine animals living within an intertidal zone is that they are successively exposed to drying air and direct exposure to the sun as water retreats, and re-immersed when the tide returns. Dense water filters the penetrating light. The level penetrated, and thus made available for photosynthesis, is the **photic zone**; below that, where significant light does not reach, is the **aphotic zone**.

All life on Earth depends eventually on the sun. The *producers* (cyanobacteria, algae, and plants) harvest light energy in photosynthesis. Producers use this energy, along with carbon dioxide and water, to grow and proliferate. Producers are consumed by *herbivores*, which are in turn eaten by *carnivores*. The *decomposers* (bacteria, fungi) break down dead organisms, thereby returning nutrients to the soil or water, where they are again gathered up by producers and reincorporated into the tissues of living organisms.

As once succinctly put by the ecologist Eugene Odum, an organism's **habitat** is its ecological address; its **niche** is its ecological profession. The habitat is where it lives; the niche is how it lives. The habitat is an organism's specific site of residence within the environment. The niche includes the organism's sum total of immediate available resources within the environment—food, predators, mates, temperature, humidity, living space, sunlight, wind chill, soil properties, and other factors. How it utilizes these resources defines its profession, its niche. The full range of available resources that could be used describes the **fundamental niche**; what the organism actually uses is its **realized niche**. Both concepts—habitat and niche—are useful when we think about the diversity, abundance, and distribution of species. Habitat loss or habitat recovery helps us understand species loss or return; if resources are available, but no species is present, we can speak of an "empty niche" and look for the reasons why it is empty.

When two organisms attempt to use the same limited resource, they enter into **competition**—specifically, *inter*specific competition if the organisms belong to different species, *intra*specific competition if to the same species. During the fall rutting season, persons venturing into the northern woods may hear male moose trumpeting with other male moose; racks of antlers may lock in combat. But more often, a trip into the wild is not attended by the sounds of conflict and competition. You don't usually hear the sounds of beasts going at it red in tooth and claw. You don't hear the rending of tissue, gnashing of teeth, or the sound of limbs being torn from sockets. Competition can occur quietly.

For example, two species of intertidal barnacles, if occurring separately, reach their fundamental niches (figure 5.16). The high tidal areas are the first areas exposed as the tide retreats and the last areas covered when it returns. Consequently, the high tidal area is especially dry (exposed to air) and hot (exposed directly to sunlight). The barnacle *Semibalanus* is susceptible to heat and desiccation, so it reaches no higher than mid-tide; *Chthamalus*, tolerant of drying,

FIGURE 5.16 Competition Among Barnacles In the absence of competition, *Chthamalus* lives in low- to high-tide regions. *Semibalanus* lives in low- to mid-tide regions. These are their fundamental niches. But, together and in competition, *Semibalanus* overrides and excludes *Chthamalus* from their areas of overlap, reducing it to a smaller realized niche.

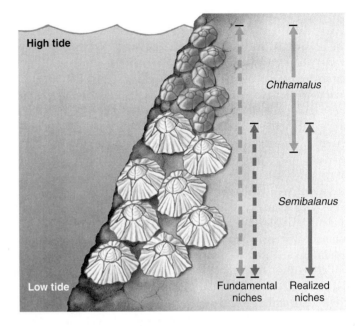

reaches all the way from low to high tide. However, when both occur together, *Semibalanus* more successfully competes for space, excluding *Chthamalus*, whose realized niche becomes restricted.

Warblers are a group of brightly decorated birds, sporting various combinations amongst the species of yellow, red, and white with black accents.

In northeastern parts of North America, five species were closely observed as they foraged in and around trees in search of insects. So closely related, competition could have been intense. But each of the species frequents different parts of the tree where it searches for food (figure 5.17). Ecologically, the species are separated, thus reducing any challenges from direct competition.

Forest plants compete for nutrients in the soil, a space in the stand of trees, and for access to sunlight. In temperate and tropical forests, major trees grow to great heights, spread their branches into a broad canopy, and intercept the sunlight for photosynthesis. Young, small tree seedlings and ground-hugging plants are in a shaded world, choked off from sunlight. Shade-tolerant plants grow slowly; others perish, outcompeted for the resource of sunlight.

OVERVIEW

The diversity of life on Earth has passed through major transitions, producing changes in complexity and

major new organization of parts that function to ensure survival and reproduction of the whole organism. There is nothing inevitable about such changes, nor are they necessarily progressive. Bacteria today are probably no more complex than were their ancestors of 2 billion years ago. And bacteria today have just as tight a hold on their place in nature as more complex organisms that came later. These major transitions produced changes in complexity that afforded different ways of finding successful strategies of survival. The first of these transitions was the origin of life itself. Prokaryotes represent a stage wherein the DNA and the machinery of life it directs were housed within the walls of a protective cell. The ancestors of mitochondria and chloroplasts were once free-living prokaryotes, but they became incorporated into a host cell, thus producing eukaryotes. With eukaryotes also comes the origin of sex. Experiments with a multicellular organization occur in protists, but multicellularity is the foundation for plants, fungi, and animals, where, respectively, photosynthesis, absorption, and ingestion are characteristic lifestyle strategies.

Each organism within an ecosystem confronts survival challenges from the physical environment and from competitors of its own species and from other species.

Competition may be seldom observed directly because it can be nonviolent, as with plants vying for sunlight; or competition may be seasonal, or occur only in infrequent years of harsh weather.

| Cape May warbler | Black-throated green warbler | Yellow rumped (Myrtle) warbler | Bay-breasted warbler | Blackburnian warbler |

FIGURE 5.17   Wood Warblers   Five species of wood warblers occur in spruce forests of the northeastern United States. Their foraging efforts are localized in different parts of the tree, represented by the shading.

And what of humans? What is our contribution to this evolution and diversification of life? Stasis and stagnation are not part of this history. The history of life on Earth is summarized in one word—*change*. We emerged from this history, as did the other organisms with which we share this planet. Stamped on our character are the consequences of that history. If we are to understand our contribution and ourselves, then it is that history we need to understand. We'll consider the evidence for evolution in chapter 6.

Selected References

Enger, E., and F. Ross. 2003. *Concepts in biology.* 10th ed. New York: McGraw-Hill.

Johnson, G. B. 2003. *The living world.* 3rd ed. New York: McGraw-Hill.

Lewis, R., M. Hoefnagels, D. Gaffin, and B. Parker. 2002. *Life.* 4th ed. New York: McGraw-Hill.

Mader, S. S. 2003. *Inquiry into life.* 10th edition. New York: McGraw-Hill.

Maynard Smith, J., and E. Szathmáry. 1995. *The major transitions in evolution.* Oxford: W. H. Freeman.

Raven, P. H., and G. B. Johnson. 2003. *Biology.* 6th ed. New York: McGraw-Hill.

Learning Online

Testing and Study Resources

Visit this textbook's website at **www.mhhe.com/ evolution** (click on the book's title) to take advantage of practice quizzing, study/writing tips, timely news articles, and additional URLs for research on the topics in this chapter.

6

"There are some people that if they don't know, you can't tell 'em."

Louis Armstrong, jazz musician

"When a thing is new, people say: `It is not true.' Later, when its truth becomes obvious, they say: `It is not important.' Finally, when its importance cannot be denied, they say: `Anyway, it is not new.' "

William James, psychologist and philosopher

Evidence of Evolution

INTRODUCTION

If we were to mark a moment when the world changed, it would be 1859, the year Charles Darwin reluctantly published *The Origin of Species, by Means of Natural Selection.* Those who read and understood what Darwin had discovered could never again see the world in the same way. It was a fundamental shift in our view of ourselves as human beings and of our place in the universe. Just as Newton (1642–1727) earlier provided a natural set of laws governing the motion of planets, Darwin (1809–1882) supplied a natural explanation for the evolution of life—the world and its creatures all moved by natural causes. And if that were true, then humans too came out of a long history shaped by natural selection, and bear the character of that evolution rather than the stamped image of a divine Creator. The implications were at once stark and troubling, and also liberating and fresh. As we have seen, the view that organisms change through time is a view that preceded Darwin by 2,500 years, reaching back to Greek philosophers. Since the ancient Greeks, others had stepped forward to proclaim that evolution, in fact, occurred. But no one marshaled the case as comprehensively as Darwin and without resorting to discredited ideas of spontaneous generation or dubious views of inheritance. Darwin was the first to notice that natural selection was a plausible mechanism, an engine behind these evolutionary changes.

Unlike other scientists whose insights changed our view of the world, Darwin was dealing with deep historical events that unfolded over an immense amount of time. No television reporter from CNN was present to watch these events firsthand, scribble notes as species evolved over the millions of years of Earth history, then report back to us. By human speeds, the processes were too slow to comfortably observe in a single lifetime. Darwin conducted no single, definitive experiment, rubbed no test tubes together, dropped no objects from towers to test the natural laws. His was an insight derived from careful observation and original deduction from events already at hand. As we have also seen, he was not baffled by the distracting obstacles and misunderstandings of the science of his day—the underestimates of geologic time and the confusions about heredity. What Darwin saw, and what we must see, is the compelling evidence *overall*, rather than the clues arising from one particular line of argument. In fact, Darwin recognized this. He noted that what he proposed rested on not a single, isolated critical piece of evidence, but on a body of proof, "one long argument" as he called it. But for us, as a matter of convenience, we will look at particular arguments, evaluate each critically, and then end by summarizing what these arguments say together, as a cohesive and comprehensive view of evolution.

THE FACT OF EVOLUTION

The Fossil Record

Old rocks hold the remains of old organisms. Dating techniques allow us to place these rocks and the fossils they bear in a chronological sequence, often with actual estimates of their age in years. Such estimates trace a history of life back over several billion years, through the fossils to simpler forms the farther back we go. A dated sequence of morphological change emerges. Single-celled organisms come before dinosaurs; fish precede amphibians, which precede reptiles, mammals, and birds. The bits of organisms or the impressions they leave in rocks represent direct contact with organisms of the past, and bear witness to the fact of evolution.

Even more compelling is the evidence of sequential order to the fossil record. If species were stamped out one at a time—independently, without reference to each other—then there would be no order to the fossil record (figure 6.1a). Elephants might appear first, then bacteria, then giraffes, then some fish, then flowering plants, and

FIGURE 6.1 Paleontology Appearance of animals and plants, early (bottom) to recent (top). Random (left) appearance of organisms—humans, some plants, more mammals, fishes, microorganisms, birds. Ordered (right) appearance from single-celled, to multicelled, to fishes, amphibians, reptiles, birds, and mammals, including humans.

Time

(a)   (b)

so on, until the world was filled up with species. This did not happen. Simple bacteria were the first to evolve, then eukaryotes, then multicellular organisms. Within vertebrates, fish arose first, then amphibians, then reptiles, within which arose birds and, separately, mammals. Evolution is change through time, building on those that came before—an ordered sequence. The fossil record holds evidence of this sequence. But if separately and specially created, this would not be the pattern. Kangaroos might appear first, then fish, then bacteria, then birds, then amphibians, and so forth. No order, just a special creative process randomly speckling the fossil record with freshly minted species. Species, if specially created, would appear without connection or relationship to those appearing before.

But the fossil record does not testify to indiscriminate appearance. It documents order (figure 6.1b). An unfolding of life characterizes the fossil record, not random appearance of species. Each species or group is preceded by a logical and related ancestor: descent with modification.

Deduction

Formally reasoned, if species were created separately, one at a time, then they would have no evolutionary continuity or connection to one another. Consequently, we would predict that the fossil record would preserve this random appearance of new species. This does not happen.

The same applies for groups. If all mammals were specially created at once, all birds at another time, all sharks some other time, plants at another time still, and so on, then groups should be randomly distributed through the fossil record. This, too, is not shown in the fossil record. Instead, there is a sequence, an order, to the presence of such groups, not a scattered distribution within the rocks of the past.

On the other hand, if species evolved—descendant species emerging out of prior ancestors—then all species are sequentially connected. Evolution, in fact, has an order: simple before more complex, amphibians before birds, reptiles before mammals, bacteria before all the rest, and so on. This is the evidence of the fossil record. For instance, we do not have some mammals appearing 4.6 billion years ago and others appearing yesterday. No bird species appeared before bacteria, some later, the rest yesterday. Birds evolved during the Mesozoic, then radiated during the Cenozoic. Prokaryotic cells gave way to more complex eukaryotic cells, which gave way to multicellular organisms. The simple preceded the complex.

We conclude that the fossil record overall is ordered, not random, and therefore does not support a view that species (or groups) were specially created one at a time. Instead, the fossil record supports the prediction that all species are part of an unfolding evolutionary past, producing a lineage of ancestors and descendants, connected through time in an ordered sequence.

Comparative Anatomy

Comparison of organisms reveals their shared similarities. The lion and domestic cat, although different species, exhibit basic underlying similarities. Even quite distinct species, such as humans and birds and lions and seals, show correspondence between similar parts (figure 6.2). Organisms as different as bats

FIGURE 6.2 Comparative Anatomy Correspondence between parts can be seen in a comparison of forelimbs among four vertebrates. Although all use the forelimb differently—human (grasping), bird (flight), lion (running), seal (swimming)—all of these forelimbs have the same basic underlying structure of bones and soft tissues.

(flying), moles (burrowing), and dugong (swimming) retain the same, underlying basic pattern of their architectural design (figure 6.3). Although suited to different functions, all possess a humerus (upper arm bone), forearm, and wrist with digits. Suppose we were intelligent human engineers and were asked to build flying, burrowing, and swimming machines. We would do so from scratch. Each machine would have different functions and so there would be no reason to build them alike and no reason to borrow the basic design from one upon which to build the other. Our digging machine would not be a transformed boat, or our boat a transformed airplane, or our airplane a modified digging machine. Each would be created separately, independently, by intelligent designers and builders, without a common underlying

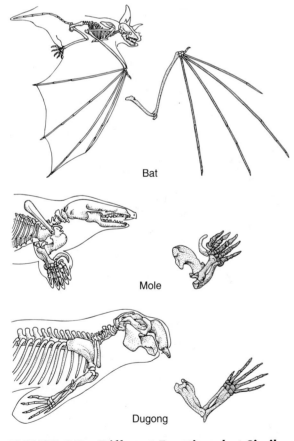

FIGURE 6.3   Different Functions but Similar Underlying Forelimb Anatomy   Forelimbs of bat, mole, and dugong. Each limb performs a different function—flight, digging, and swimming, respectively—and all are superficially different, but all three share a common, underlying anatomical plan.

plan uniting them (figure 6.4). The wings of the jet plane are not modified oars from the racing shell; the hull of the shell is not modified from the body or lugged treads of the mining machine; the mining machine is not a modified airplane.

In the biological world, unlike the intelligently engineered world, organisms are modified from ancestors. They retain the basic, underlying features of their ancestors even as those features are modified for different functions—flying, burrowing, swimming.

Further, a *morphological series* can be constructed within a group that identifies corresponding parts in different species. Parts correspond among related species, but the morphological series tracks their subsequent modification through time, from one species to the next. Within horses, the digits become reduced to a single element (figure 6.5). The gill arches of fishes correspond to the jaws of other vertebrates (figure 6.6a). From fish to first tetrapod, the shared similarities of each are apparent (figure 6.6b). Fossil chronology enables us to track these changes through time, confirming their ordered sequence based upon comparative anatomy.

Adding time and the chronology of the fossil record provides a cross-check on the morphological series alone. But remember, here we are taking each piece of evidence independently one at a time—examining it in isolation on its own merits. So, examined independently, comparative anatomy itself allows us to construct a transformation series, noting what is basic and what is a derived plan.

Correspondence between parts extends down to the molecular level. All life, from bacteria to multicellular organisms, is based on a common genetic code. Base pairs in the DNA and amino acids in proteins are universal to all life from bottom to top. The harmless bacterium in your intestines, *Escherichia coli* (*E. coli* to its friends), and all organisms up to you yourself, have genes made up of exactly the same four molecular building blocks—nucleic acids and their bases. These common triplet codes preside over the manufacture of the phenotype from microorganism to multicellular organism.

Why should such a code be universal? Perhaps there is some chemical determinant, a molecular imperative imposing a universal convention on all life. Alternatively, and more plausibly, DNA based on triplet codes evolved early, within the first simple microorganisms. This elegant solution to protein coding proved to be beneficial, and its success was incorporated into later organisms, providing the genetic foundation for complex lifestyles and body organizations. The universality of the genetic code testifies that all

FIGURE 6.4   By Human Design   Functions vary and designs vary. Unlike biological organisms, these humanly designed and built machines show no correspondence of parts from planes (flight), to mining machines (burrowing), to boats (swimming).

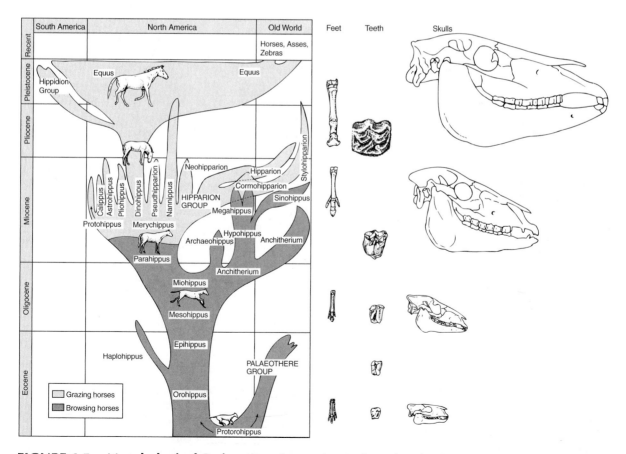

FIGURE 6.5   Morphological Series   From four-toed to single-toed modern horses, this morphological series illustrates the correspondence between parts (feet, teeth, skull) and their modifications. Here, the stratigraphic position of these species is added.

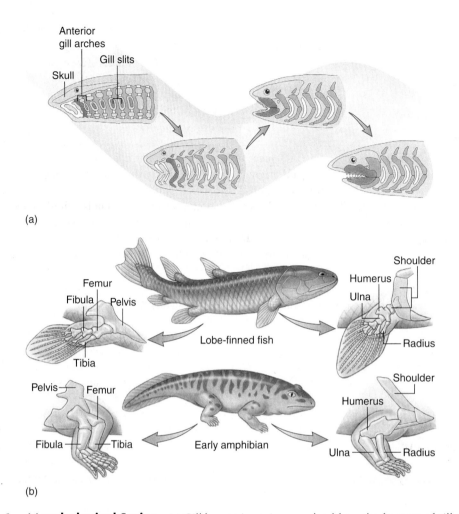

FIGURE 6.6   Morphological Series   (a) Gill bars to jaws. Jaws evolved from the front set of gill arches of jaw-less ancestors. (b) Fish fin to tetrapod limb. Note that the bones of the fish hip and shoulder correspond to bones in this early tetrapod (amphibian). Here, the morphological series carries from one type of vertebrate (fish) to another (amphibian), and from one environment (water) to another (land).

life shares a single origin. Like morphological correspondence, molecular correspondence reaches all the way back to the origin of life on Earth.

Deduction

If species were constructed or created separately, each stamped out individually and without relationship to any other, then there would be no reason to expect underlying similarities or the presence of morphological series. But we do find similarities, and because series can be constructed, we deduce that evolution of one

form to another would account for this historical source passed from ancestor to descendant derivatives. Common origin with common ancestors explains the presence of shared similarities in descendants.

Related Concepts

Similarity of corresponding parts in different organisms is determined by three criteria: ancestry, function, or appearance. The term **homology** applies to two or more features that share a common ancestry; the term **analogy** applies to features with a similar function; and the term **homoplasy** applies to features that simply look alike (figure 6.7a). These terms date back to the early nineteenth century but gained their current meanings after Darwin established the theory of common descent.

Homologous structures, more formally, are features in two or more species that can be traced back to the same feature in a common ancestor. The bird's wing and the mole's arm are homologous forelimbs, tracing their common ancestry to reptiles. Homology recognizes similarity based upon common, ancestral origin.

Analogous structures perform similar functions, but they may or may not have similar ancestry. Wings of bats and wings of bees function in flight, and are therefore analogous as flight devices, but neither structure can be traced to a similar part in a common ancestor. On the other hand, turtle and dolphin forelimbs function as paddles (analogy) and additionally can be traced back historically to a common source (homology). Analogy recognizes similarity based upon similar function.

Homoplastic structures look alike and may or may not be homologous or analogous. In addition to sharing common origin (homology) and function (analogy), turtle and dolphin flippers also look superficially similar; they are homoplastic as well as homologous and analogous. The most obvious examples of homoplasy come from mimicry or camouflage, wherein an organism is, in part, designed to conceal its presence by resembling something unat-

tractive. Some insects have wings shaped and sculptured like leaves. Such wings function in flight but not in photosynthesis (they are not analogous to leaves), and certainly such parts share no common ancestor (they are not homologous to leaves), but outwardly they have a similar appearance to leaves—they are homoplastic.

Analogy, homology, and homoplasy are each separate contributors to biological design. Dolphins and bats live quite different lives, yet within their designs we can find fundamental likenesses—hair (at least some), mammary glands, similarities of teeth and skeleton. These features are shared by both because both are mammals with a distinct but common ancestry. Dolphins and ichthyosaurs come out of quite different vertebrate ancestries, yet they share certain likenesses—flippers in place of arms and legs, and streamlined bodies. These features appear in both because both are designed to meet the common hydrodynamic demands of life in open marine waters. In this example, convergence of design to meet common environmental demands helps account for likenesses of some locomotor features (figure 6.7b). On the other hand, the webbing between the fingers of a bat wing and the webbing between the fingers on penguin arms have little to do with common ancestry (bats and penguins are not closely related) or with common environmental demands (the bat flies in air, the penguin swims in water). Thus, structural similarity can arise

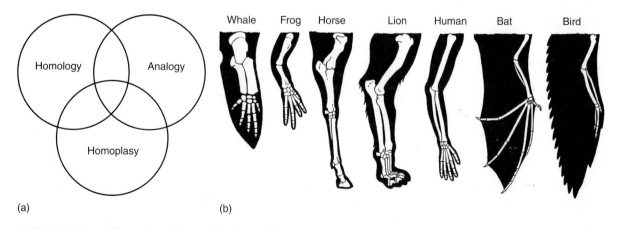

(a)   (b)

FIGURE 6.7   Diversity of Type, Unity of Pattern   (a) Similarities. Parts may be similar in ancestry, function, and/or appearance. Respectively, these are defined as homology, analogy, or homoplasy. (b) Although the vertebrate species differ, the underlying pattern of the forelimb is fundamentally the same.

in several ways (figure 6.8). Similar function in similar habitats can produce convergence of form (analogy); common historical ancestry can carry forward shared and similar structure to descendants (homology); occasionally, accidents or serendipity or other events can by chance lead to unrelated parts that simply look alike (homoplasy).

Comparative Embryology

Embryos from different species retain similarities within their development (figure 6.9). Within very early fish embryos, perforations form within the sides of the throat—future gill slits of the adult-to-be. As the embryo matures, these gill slits become invested with

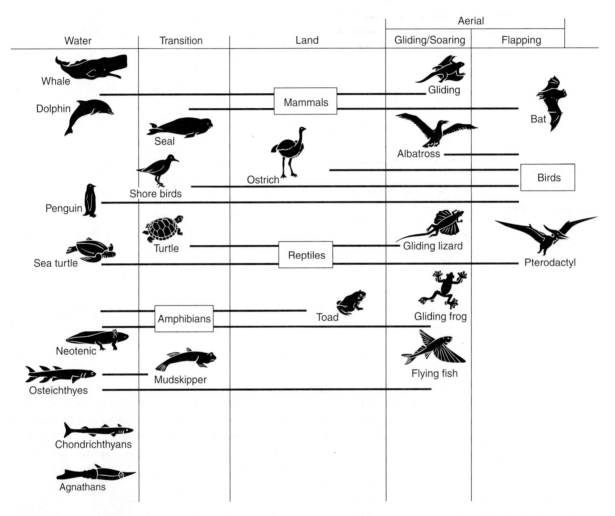

FIGURE 6.8   Convergence of Design   Groups of animals often adapt to habitats that differ from those of most other members of their group. Most birds fly, but some, such as ostriches, cannot, and live exclusively on land; others, such as penguins, live much of their lives in water. Most mammals are terrestrial, but some fly (bats) and others live exclusively in water (whales, dolphins). "Flying" fishes take to the air. As species from different groups enter similar habitats, they experience similar biological demands. Convergence to similar habitats, in part, accounts for the sleek bodies and fins or flippers of tuna and dolphins, because similar functions (analogy) are served by similar parts under similar conditions. Yet tuna and dolphins come from different ancestries and are still fishes and mammals, respectively. Common function alone is insufficient to explain all aspects of design. Despite current similar habitats, each design carries evolutionary features of the past into the present.

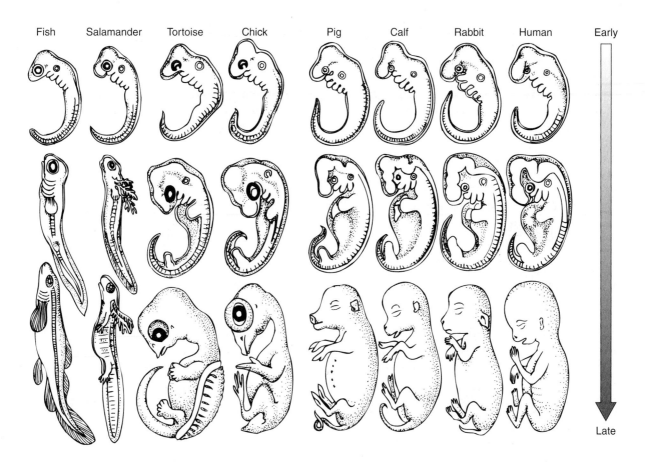

FIGURE 6.9   Comparative Embryology   Embryonic retention of ancestral characteristics. Note the persistence of gill slits in early stages of embryonic development in the mammals (right) that of course, do not possess functional gills in the adult. Also note that a tail is present early in humans, but lost before birth.

capillary beds and are supported internally by gill arches. Next, in a young fish fry, these gill slits are ready to serve in water breathing, taking up oxygen from the water flowing across the gills. The oxygen enters the circulating blood as carbon dioxide is released back to the passing stream of water. What is surprising is that gill slits also form in embryos of animals that, as adults, live in air. For example, in mammals, gill slits appear in the sides of the throat of the young embryo. They never form functional gills for water breathing, but seal over and disappear before birth, because they have no function to perform.

Deduction

We have no reason to expect gill slits in mammals because, as adults, lungs, not gills, are the basis for breath-

ing. But remarkably, gill slits are present. If there is historical continuity between vertebrates, then we might expect to find remnants of structures carried forward by evolution as fish evolved into terrestrial organisms. That is what we find. Within the embryology, the remnants of ancestors persist in the embryos of descendants.

Related Concepts

Early events of embryonic development retain current clues to distant evolutionary events. Ernst Haeckel, a nineteenth-century German biologist, stated this boldly in what became known as the **biogenetic law**. Gill slits, numerous gill arches, and other fish characteristics even appear in the early embryos of reptiles, birds, and mammals, but they are lost as these tetrapod embryos proceed to term (figure 6.9). Although lost as tetrapod development unfolds, these and many similar structures are remnants of fish features carried forward to air-breathing animals out of a water-breathing evolutionary past. Haeckel argued that from egg to complete adult body, the individual passes through a series of developmental stages that constitute a brief, condensed repetition of stages through which its successive ancestors evolved. The biogenetic law states that the sequence of embryonic stages in current species is a rerun of former evolutionary steps. In short, ontogeny (embryology) in abbreviated form recapitulates (repeats) phylogeny (evolution).

Haeckel certainly recognized that recapitulation was approximate. Comparing it to an alphabet, he suggested that the ancestors of a current species might have evolved through a series of stages: A, B, C, D, E, F, G, H, etc., but not all these historical steps would necessarily be repeated in current individuals. On its way to the adult stage, the embryology of a current species might pass through, or recapitulate, an apparently defective historical series: A, B, . . . , D, E ,. . . , G, H, etc., before reaching the adult stage. In this example, several evolutionary stages have fallen out of the developmental series. Although the ancestry of an organism might include an entire series of steps, Haeckel did not believe that all these would necessarily appear in the ontogeny of a later individual. Evolutionary stages could disappear from the developmental series. Nevertheless, he felt that the basic series of major ancestral stages remained the same and thus, the biogenetic law applied.

Development certainly exhibits a conservatism wherein ancient features persist like heirlooms into modern groups. Ontogeny (embryology), however, is not so literally a repeat of phylogeny (evolution) as Haeckel supposed. A contemporary of his, Karl Ernst von Baer, cited examples from embryos of descendant animals that did not conform to the biogenetic law—chick embryos lack the scales, swim bladders, fin rays, and so forth of adult fishes that evolutionarily preceded them. Furthermore, the order of appearance of ancestral structures is sometimes altered in descendant embryos. Haeckel allowed for exceptions; von Baer did not. Von Baer said that these exceptions and "thousands" more were too much. He proposed alternative laws of development. Foremost was von Baer's proposal that development proceeds from the general to the specific: Development begins with undifferentiated cells of the tiny embryo that become simple tissue layers, then organs, then young

organisms, then juveniles, and finally adults. Young embryos are formative or undifferentiated (general), but as development proceeds, distinguishing features (specific) of the species appear. Each embryo, instead of passing through stages of distant ancestors, departs more and more from them. Thus, the *embryo* of a descendant is never like the *adult* of an ancestor and only generally like the ancestral embryo. Other scientists since von Baer have also dissented from strict application of the biogenetic law. What can be made of all this?

We should recognize that embryos are adapted to their environments. Anamniote (fish and amphibian) embryos usually mature in salt- or freshwater surroundings. Amniote (reptiles, birds, mammals) embryos grow in an aquatic environment as well—the amniotic fluid. Thus, similarities among embryos might be expected. But, as von Baer pointed out, what we observe at best is correspondence between the *embryos* of descendants and the *embryos* of their ancestors, not between the descendant *embryos* and the ancestral *adults*. Haeckel was certainly mistaken to see a correspondence between descendant embryos and ancestor adults. Organs are adapted to their current functions. As a fish embryo approaches hatching, its "limb" buds become fins, a bird's become wings, a mammal's become paws or hooves or hands, and so forth. There is, however, an element of deep conservatism in ontogeny, even if it is not an exact telescoping of evolutionary events. After all, the young embryos of mammals, birds, and reptiles do develop gill slits that never become functional as breathing devices. Is this recapitulation? No. It is better to think of this as **preservationism**, for reasons not too difficult to imagine.

Each adult part is the developmental product of prior embryonic preparation. The fertilized egg divides to form the blastula, a hollow ball of cells; gastrulation rolls up and moves embryonic cells around to position them at strategic locations within the embryo, forming basic tissue layers; these tissue layers interact to form organ rudiments; tissues within organ rudiments differentiate into adult organs. Skip a step, and the whole cascade of ensuing developmental events may fail to unfold properly.

In mammals, the supporting notochord of the embryo is replaced almost entirely in the adult by the bony backbone, the vertebral column. For the young embryo, the notochord provides an initial axis, a scaffolding

Consider This—

Human Appendix—Out of a Job

The human appendix receives bad press. Like an old sock, it is thought of as being unemployed and therefore expendable. Its full name is *vermiform appendix*. Certainly, a person can get along without an appendix. For reasons not particularly clear, it occasionally becomes infected and inflamed. When this happens, the appendix may rupture, spilling out intestinal contents (digestive juices, bacteria, partially digested food, and pus from the inflammation) into the surrounding viscera and there create a life-threatening condition. So if your appendix becomes infected, a reasonable surgeon will quickly relieve you of it. But what of its function?

No longer does our appendix (cecum) offer a spacious chamber for microbial digestion of plant material. But the walls of the appendix are richly endowed with lymphoid tissue, part of our immune system, and involved in neutralizing foreign bacteria riding in with our food. Like the walls of the gut generally, the lymphoid tissue of the appendix monitors the passing food, detecting and responding to harmful foreign materials and potentially pathogenic bacteria. In short, the human appendix is part (although a small part) of our immune system, clinging to this little function, refusing to become completely redundant. Its glory days as a major digestive compartment are over, in humans, and it is reduced to a relic, a vestigial leftover from a time in its distant history when it served so well our plant-eating primate ancestors.

---

along which the delicate body of the growing embryo is laid out. The notochord also stimulates development of the overlying nerve tube. If the notochord is removed, the nervous system does not develop. In the adult, body support is taken over by the vertebral column, but the notochord performs a vital *embryonic* role before disappearing—namely, it serves the growing embryo as a central element of embryonic organization.

A notochord that persists in the mammalian embryo should not be interpreted as a sentimental memento of a distant phylogenetic history. Instead, it should be seen as a functioning component of early embryonic development. Structures and processes intertwine to produce the conservatism evident in development. They are not easily eliminated without a broad disruption of the cascade of all ensuing events. Anatomical innovations—new structures brought into service in the adult—are usually added at the end of developmental processes, not at the beginning. A new structure inserted early into the developmental process would require many simultaneous replacements of many disrupted developmental processes thereafter. Evolutionary innovations thus usually arise by remodeling rather than by entirely new construction.

Terrestrial ancestors came equipped with robust forelimbs that supported the body and allowed the organism to romp over the surface of the land. For vertebrates that have taken to the skies, these terrestrial forelimbs are renovated into the wings that carry bats and birds aloft. We need look no further than our own human bodies to find similar examples of evolutionary remodeling. Our backbone and legs, which carried our distant mammalian ancestors comfortably on all fours, hold us in an upright, bipedal stance. The arms and hands that can control the delicate strokes of a paintbrush come refashioned from ancient forelegs that carried a hefty trunk and helped our mammalian ancestors dash from predators. The past is hard to erase. When parts are already available, renovation is easier than starting over with new construction. Preservation of old structures in new species is the result.

Vestigial and Atavistic Structures

Vestigial structures are rudimentary structures, the remains from a prominent past. **Atavistic structures** are throwbacks, the return of structures typical of ancestors. Whales evolved from four-footed land ancestors (figure 6.10a). But today, whales live exclusively in water, guided by flippers modified from the forelimbs of these ancestors. Whale hips and hindlimbs are meager vestiges, remnants of their former selves

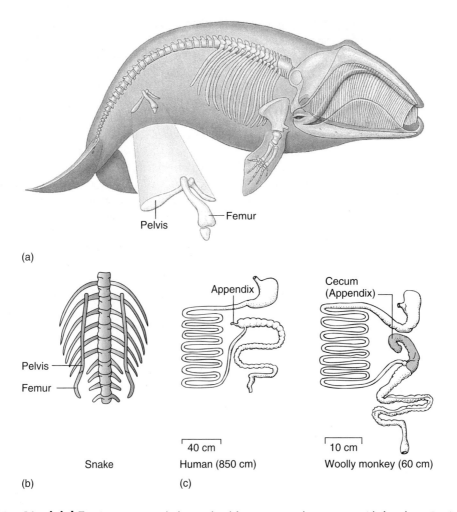

Pelvis

Femur

(a)

Pelvis

Femur

Snake

(b)

Appendix

40 cm

Human (850 cm)

(c)

Cecum
(Appendix)

10 cm

Woolly monkey (60 cm)

FIGURE 6.10 **Vestigial Features** (a) Whales evolved from terrestrial ancestors with four legs. But in whales, the hips and hindlimbs are reduced to small bones with no function. (b) Snakes evolved from lizards with four legs. But in primitive snakes, the remnants of hindlimbs persist (forelimbs are absent). (c) The human appendix is a vestigial structure, reduced from the cecum of primate ancestors, which was involved in digestion of significant plant material.

in tetrapod ancestors, buried under the skin without a major function to perform. Primitive snakes also possess vestigial hindlimbs, leftovers of the supportive hindlimbs from lizard ancestors (figure 6.10b). Humans possess an appendix, a short, blind-ended pouch that apparently lacks any significant digestive function like the intestine to which it is attached (figure 6.10c). In primate species specializing on plant diets, the homologue of this diminutive appendix is a commodious cecum, sometimes of considerable length. Ingested plant material is directed into this cecum for extended action by bacteria that can break down the chemically tough cellulose of plant cell walls. The

vertebrate intestine alone does not release sufficient enzymes of the right kind to digest cellulose effectively. But bacteria can digest cellulose and in so doing make the breakdown products available to themselves and to the herbivore in which they live. The cecum extends the working area of the gut and provides an extra compartment in which much of this digestion occurs. The human appendix is a vestige of that cecum, a reduced relic of that cecal compartment that in herbivorous ancestors digested the fibrous plant material that constituted the diet of ancestors.

Occasionally, atavisms arise, apparently by an embryonic expression of an adult structure long ago suppressed. Atavisms are not just anatomical aberrations, developmental monstrosities from a bungling somewhere within a muddled genetic code. Atavisms are the return, or partial return, of ancestral parts. In the modern species of horse, the lost toes of ancestors occasionally reappear (figure 6.11). Occasionally, single-toed horses of today are born with atavisms, additional toes that might have characterized their ancestors. Today, these extra toes are maladaptive for the fleet and efficient locomotion horses employ across open, grassy plains. But an embryonic mistake has returned these ancient features that millions of years ago equipped an early member of the horse family.

Deduction

If species were created separately, why do we find such vestigial structures? If humans were created new, without an evolutionary history, why is the appendix still around, cluttering up the gut? If we were intelligent designers, we would not leave such scraps and debris lying about in the new species we create. We could start fresh and not work with leftovers. Nor, while building a new organism, would we expect to suddenly find atavisms, the return of ancestral parts long ago discarded.

On the other hand, if organisms are related through evolutionary continuity, then we would deduce that vestigial structures are evolutionary scraps and leftovers—structures reduced from a prominent role in an ancestor and passed in a reduced role to more recent descendants. The human appendix is an abridged cecum evolutionarily passed to humans, but reduced in size as tough plant cellulose decreased in the diet during human evolution. The vestigial hips and hindlimbs of whales no longer serve in swimming, as they might have done in ancestors, but now are reduced to remnants. The vestigial limbs of primitive snakes contribute, in males, to stroking and stimulating the female during courtship, but no longer carry the weight of the body as they did in lizard ancestors.

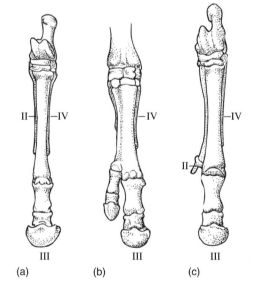

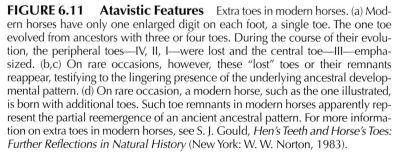

FIGURE 6.11   Atavistic Features   Extra toes in modern horses. (a) Modern horses have only one enlarged digit on each foot, a single toe. The one toe evolved from ancestors with three or four toes. During the course of their evolution, the peripheral toes—IV, II, I—were lost and the central toe—III—emphasized. (b,c) On rare occasions, however, these "lost" toes or their remnants reappear, testifying to the lingering presence of the underlying ancestral developmental pattern. (d) On rare occasion, a modern horse, such as the one illustrated, is born with additional toes. Such toe remnants in modern horses apparently represent the partial reemergence of an ancient ancestral pattern. For more information on extra toes in modern horses, see S. J. Gould, *Hen's Teeth and Horse's Toes: Further Reflections in Natural History* (New York: W. W. Norton, 1983).

Surprise atavisms, the return of ancestral structures, would not be expected if each species were created separately. But if descent with modification is the character of evolution, then occasionally we might expect to see such ancestral parts suddenly reappear. Current mistakes of genetics or embryology reactivate and express old bits and pieces of these ancestral parts because remnants of the original genetic code ride forward within the genetic code into current species.

Related Concepts

Evolution works by remodeling, not from new construction started from scratch. Novel mutations arise to provide fresh variety, but their effect is usually to modify an existing structure rather than to replace it entirely. As a result, the anatomical specializations that characterize each group are modifications upon a common, underlying pattern. It is "unity of plan and diversity of execution," as T. H. Huxley, a friend and contemporary of Darwin, said in 1858. The bird wing is a modified tetrapod forelimb (figure 6.7b), which is a modified fish fin (figure 6.6b). This remodeling feature of evolution accounts for the structural similarities from one group to the next.

Distributional Evidence

Species of plants and animals are not distributed evenly across the globe. Biologists accompanying voyages of discovery and exploration in the eighteenth and nineteenth centuries gradually amassed evidence of an uneven distribution of organisms on the continents. Elephants occurred in Africa, another species in Asia, but elephants were absent from North and South America. Marsupial mammals occurred in abundance in Australia, less so elsewhere, and so on. Based on these distinctive assortments of fauna and flora, **biogeographic realms** were recognized, wherein each continent today sports its defining complement of species (figure 6.12).

But if we step back in time, adding fossil communities, some common extinct species occurred at the same time horizon on continents that today stand thousands of kilometers apart. At the beginning of the twentieth century, some scientists explained this common distribution of fossils on distant continents with the supposed presence of past "isthmuses," bridges that once joined distant continents, across which species

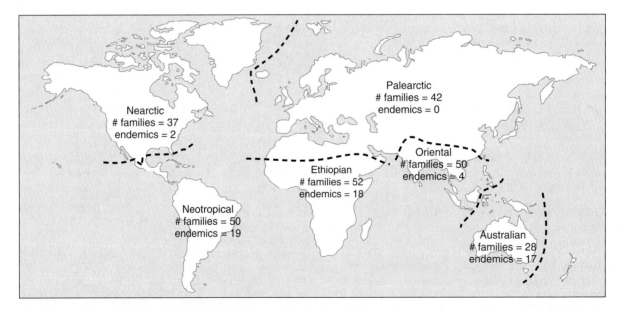

FIGURE 6.12   Biogeographic Realms   Continental land areas support characteristic assortments of plants and animals, which in turn define six biogeographic realms. The number of mammalian families are indicated, along with the number of endemic families.

spread, populating connected continents and leaving their fossil remains on both.

Certainly at major time horizons, the same fossil species are found on continents now distantly placed from each other. But the reason for this is not through connecting isthmuses, now fallen away. Most isthmuses proposed in the early twentieth century proved to be figments of the imagination, never having been present. But how then did species of the past come to populate distant continents of their time?

The answer emerged in the mid twentieth century. Continental drift accounted for these biological distributions. The Earth's solid surface crust wraps a hot mantle and core. The solid crust is a puzzle of plates abutting each other along suture zones (figure 6.13a). The high landmass of the plate, lifting above the surrounding oceans, is the *continent*, part of the plate on which it rides. These plates slide past one another, move apart from each other, or dip one under the other at subduction zones. The energy that drives this shifting of the plates, referred to as **plate tectonics**, is caused largely from activity beginning deep within the Earth's molten mantle. Convection currents of molten material rise to the surface and split through the crust at suture zones, powering the plates apart. As the molten lava wells up and hardens, it adds new rock along the edges of the plates. At their opposite sides where they are pressed into adjacent plates, great mountain ranges might be pushed up, such as the Himalayas in northern India; or one plate may be forced down under the other, as is occurring off the west coast of North America. The overridden plate is remelted as it enters the Earth's hot mantle, and the slow slide of the two, one over the other, produces occasional earthquakes as slipping and settling occur.

Measured by a human tempo, continental drift is an exceedingly slow process, easily outpaced by speedy snails. Occasional slippage of tectonic blocks, one upon the another, may rumble through the land, producing chattering earthquakes. Sudden, local advances in plates may then measure in a few feet. But averaged over the millennia, movement is by inches per year, plenty of time for plant and animal communities to adjust (figure 6.13b). Collectively, it is time again—the vast stretches of time—that accumulates these inches into miles, and moves the continents in stately grace about the planet, through the geologic ages. On board are the communities of organisms, passengers into the future.

Related Concepts

Isthmuses, as first proposed, are not needed to account for the distribution of fossils between continents of the past. Joined continents, now separated by tectonic events, account for the easy spread of ancient fossils of the past. But let's not throw out the proverbial baby with the bathwater. Bridges do exist, or at least connections between continental areas exist today, opening routes to dispersion of species. Some routes of dispersal are open; others restricted. *Corridors* are uninterrupted connections, easy routes of travel across continental areas (figure 6.14a). Bison in North America are examples of a species moving through a large geographic area, along corridors, to occupy vast regions, without interruption by geological barriers such as mountain ranges or unswimmable rivers. However, some species meet restrictive barriers, limiting their continued expansions into new areas. These are *filter bridges*. They are selective with respect to which species pass and which do not (figure 6.14 a,b). Sometimes climate is limiting. Early in the Cenozoic era, cold-blooded snakes and lizards moved easily between Siberia and Alaska across the Bering land bridge, which then stood above the seas that would later cover it. This exposed land bridge offered reptiles a dry crossing with warm, hospitable temperatures, similar to the temperatures experienced today in Los Angeles. But by the middle Cenozoic, climates had cooled and the Bering land bridge, although still present, experienced even cold summer temperatures that limited movement by heat-dependent reptiles. With the onset of the Pleistocene ice ages, only cold-adapted organisms could cross—polar bears, caribou, musk oxen, and, until their extinction, mammoths.

Ecological barriers may restrict dispersal of populations. The African and Eurasian continents, separated from the end of the Mesozoic, again came into contact in the mid Cenozoic, across a narrow land bridge in the Middle East. Tropical species of snakes in both continents could then have moved across this bridge, populating new areas (figure 6.14a). But this did not happen, at least not extensively. Climate did not limit their dispersal, and with the formation of a land bridge, no physical barrier stopped the exchange of species. Instead, between the two continents, in the Sinai Peninsula, was an established community of snakes, well adapted to the intermediate conditions.

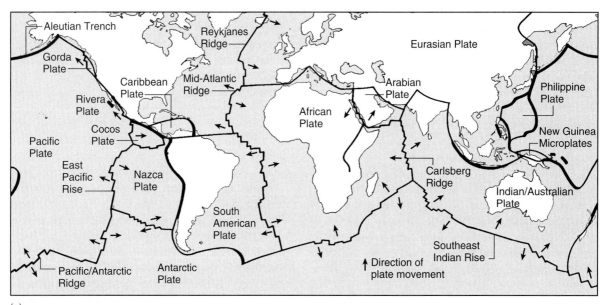

(a)

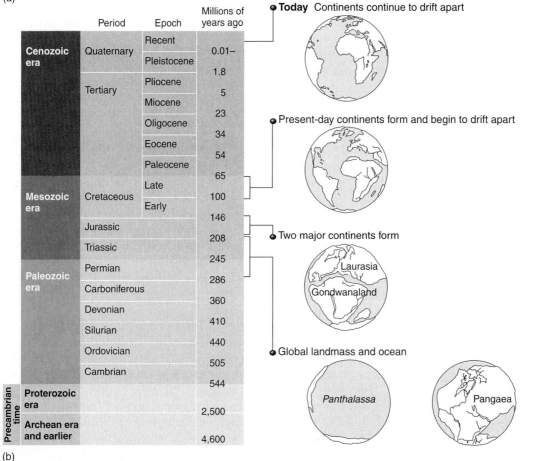

(b)

FIGURE 6.13   Continental Drift   (a) Plate tectonics. Shown are the suture zones of abutting major current crustal plates and their respective movement directions. (b) Shown are the changing continental positions through most of the Phanerozoic.

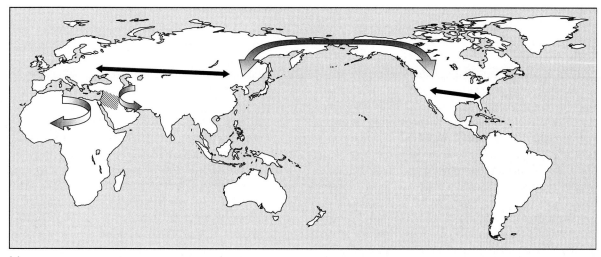

(a)

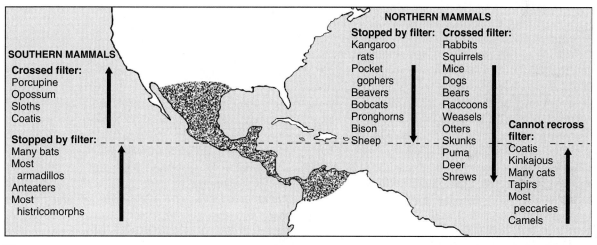

(b)

FIGURE 6.14   Bridges   Connections between geographic areas allow for or restrict dispersal of plants and animals. (a) Corridors and filter bridges. Corridors allow for the relatively uninterrupted spread of organisms. The double-headed arrows indicate such open expanses across Eurasia and North America. Filter bridges permit selective transit of organisms that pass, either because of inhospitable climate or ecological obstruction. One major selective filter has been across the Bering Strait; another is in the Middle East, restricting reptile species. (b) Filter bridges between North and South America occur in the narrow land connection between these continents. Some species have crossed this filter, but others have not.

Apparently, this entrenched community of snakes presented a biological obstacle to any migrating snakes from the adjacent continental areas, restricting successful dispersal and keeping them in place.

Sweepstakes are fortuitous or accidental transportations of individuals to new, suitable habitats. Winds and storms can blow migrating birds off course and into new areas. Hurricanes, tides, and strong currents can dislodge trees and floating debris to which cling terrestrial animals. Hurricanes in the fall of 1998 dislodged from Guadeloupe Island in the Caribbean a raft of floating, waterlogged trees with more than a dozen iguana lizards on board. A month later, these tattered and dehydrated reptilian castaways finally, and fortuitously, reached landfall on the distant island of Anguilla. As the wasted and withered lizard colonists disembarked their disintegrating raft, they were spotted by a local fisherman who recorded their arrival. After inspection by a biologist, these accidental travelers were returned to their new Anguillan home.

The island of Krakatoa in Indonesia, off the western tip of Java, was subjected in 1883 to a devastating volcanic eruption that eliminated all life on the island. The slate was clean, but would not remain so. Within a year, a spider was present, probably blown in on a silken parachute. Later, coconuts washed ashore along the beach, along with strong swimmers such as large lizards and snakes. Flowering plants and ferns returned, likely sprouting from hardy seeds or rhizomes (fern roots) making the oceanic crossing. Within 50 years, more than 1,200 species had made it to the island, although no amphibians had survived the saltwater crossing from nearby islands.

Battered but lucky colonists arrive few in numbers, often not a representative or true statistical sample of the source population from which they departed (see also chapter 11). From these few pioneers grows the new island population, spreading across and eventually inhabiting available landscape. By the luck of the draw, these few founders may carry only a few of the many alleles from the large source population from which they came. Consequently, the island population they establish is built on the limited variety of traits they carry. This phenomenon, in which rare traits are more common in new and isolated locations, is the **founder effect**. It is especially important on oceanic islands, such as the Hawaiian Islands, because the particular character of the flora and fauna on these islands is directly related to the limited sample of species arriving with their restricted genetic variation.

Low numbers may limit reproductive choices. After a perilous open-water crossing, a single individual may make landfall and survive. But who to mate with? If the surviving individual is a pregnant female, then eventually, with some luck, young will follow, grow up, and provide the start to a new population. A single seed, washed ashore, sprouts and may grow vegetatively to propagate. If the plant is self-pollinating, then a new population can become established from a single seed.

Even if organisms do not travel to new locations, genetic variation in their populations may become drastically reduced. Natural events, such as floods, river diversions, and climate changes, may temporarily reduce numbers, leaving a few individuals with limited genetic variation. Build-back of the population is then based on

this restricted sample of the few survivors. As a consequence, even when population numbers recover, genetic variability can be very limited, an outcome termed the **bottleneck effect**. Today, cheetahs in the wild, and those taken from the wild into zoos, show almost no genetic variation. Apparently, sometime in their recent past, cheetahs went through a bottleneck effect, depleting their genetic variability.

Other sweepstake introductions of organisms may be human accidents. Early Europeans entering the northwestern United States to break ground for farms and raise cattle brought crop seeds with them. This seed, especially that from the Russian steppes, inadvertently included the seeds of cheat grass, a naturally occurring grass in parts of Asia. The introduced cheat grass prospered in arid regions of the western United States, apparently well-adapted to disturbed soils worked over by farming and grazing, and in some areas cheat grass displaced the endemic bunchgrasses.

Many species or families of species are restricted to a local area or predominate on a particular continent. The region where the group is localized is referred to as its **center of origin** (table 6.1). For example, blackbirds and close relatives occur naturally in the Nearctic and Neotropic biogeographic realms; pigeons and doves predominate in the Palearctic, as do pheasants; hummingbirds are primarily denizens of the Neotropics; rattlesnakes occupy only the Nearctic and Neotropic realms. The implication is that the group is found in a restricted range because it first evolved there. Consequently, as species evolve from original ancestors, they evolve in the same location, live in the same area, and, as diversification proceeds, come to occupy the same biogeographic realm.

Deduction

The distinctive assemblages of plant and animal species occupying a particular biogeographic realm differentiated on these continents. The drifting continents carried the founding groups "on board" into isolation from other parting continents. The continents were like large islands, on which diversification and evolution of the species proceeded along their own lines.

If species were created one by one, each a separate and independent act of production, then we might expect to find members of common groups evenly distributed around the world. There is no reason to expect the preponderance of large marsupials in Australia if the different marsupials species were created sepa-

Table 6.1   Center of Origin

The numbers of current species occurring in each of three biogeographic realms are plotted. Organisms are not evenly distributed across the globe but usually are predominant in one region—their site of initial origin and subsequent diversification.

	PALEARCTIC (Old World)	NEARCTIC (New World)	NEOTROPIC (New World)
Icterids (blackbirds, etc.)	0	96	43
Columbids (pigeons, doves)	290	40	15
Phasianids (pheasants, etc.)	165	11	0
Trochilids (hummingbirds)	0	18	319
Rattlesnakes	0	31	1

Numbers are approximate. Most subspecies are not included in the counts.

rately. We would expect to find some in the Neotropics, others in Africa, still others in Asia. Certainly there are species of marsupials elsewhere in the world, but most today reside in Australia, not evenly dispersed around the globe.

On the other hand, an evolutionary descent with modification from founding marsupial stocks, in combination with continental drift, accounts for this quite nicely. When the Australian continent separated from the other continents of Gondwanaland (figure 6.13b), primitive marsupial species were the major mammalian group on board the departing continent. Elsewhere in the world, placental mammals would later emerge and diversify. But the marsupials of Australia, carried into isolation from these evolutionary events on other continents, evolved into a variety of independent forms, counterparts to the placental mammals diversifying elsewhere in the world.

On a smaller scale, we would make a similar deduction from centers of origin. If created separately, why are hummingbird species, for instance, not distributed in suitable habitats around the world? Why are they confined to the Neotropical and Nearctic regions? On the other hand, if hummingbirds are products of evolution, then we would expect their distribution to be restricted generally to the region in which the first hummingbird species debuted. From this first species, others would arise. Reasonably, we would deduce that new species of humming-

birds would evolve, not on the other side of the world, but in the vicinity of the ancestors from which they spring. The tendency for restriction of a family to a local area is interpreted as evidence that they evolved there and diversified there and so are most numerous there.

THE COURSE OF EVOLUTION

If, in fact, evolution occurred, as this evidence testifies, then how might we next assemble this evidence to reveal the *course* of that evolution? One way is by direct means of the fossil record. Stratigraphy preserves the chronological order of appearance of species—what came first, what next, what later—the course of evolution. If dating techniques are available, then these dates from different fossils can be used to assemble them into order.

Indirectly, we can use the relationships amongst groups to hypothesize the evolutionary order of their appearance. That is what we did in chapter 5 (also see Appendix 2) when we constructed and examined **phylogenetic trees**, branching systems of hypothesized relationships and evolutionary sequence.

OVERVIEW

Evolution works with parts available—descent with modification. But an intelligent designer (for example, a clever engineer with money to spend) starts with fresh, new parts to create a fresh, new contrivance. For evolution, adaptations are jury-rigged from the limited parts-shelf supplied by ancestors; the intelligent engineer can requisition the unused and original parts to make the perfect device, a wonder of optimum design. But each new biological adaptation is a renovation of the old, testimony to the fact that organisms and their designs are products of history. Parts of ancestors are held over and incorporated, often in new ways in new species.

For example, the giant panda from the forests of China is related to bears. Unlike other bears, which are omnivores and eat almost anything, the giant panda feeds exclusively on bamboo shoots, feasting away for up to 15 hours per day. Pandas strip leaves by passing the stalks between their thumb and adjacent fingers. In addition to diet, pandas are unique among bears in apparently possessing six digits on the forelimb instead of the customary five. The extra digit is the "thumb," which is actually not a thumb at all but an elongated wrist bone controlled by muscles that work it against the other five digits to strip leaves from bamboo stalks (figure 6.15). The panda's true thumb is committed to another function, so it is unavailable to act in opposition to the other fingers. The radial sesamoid bone of the wrist has become remodeled and pressed into service as the effective "thumb."

The availability of parts narrows or enlarges evolutionary opportunities. Had the panda's wrist bone been locked irretrievably into another function, as was its original thumb, the doors for evolution to bamboo feeding might have been closed and this endearing bear might never have evolved (see also Davis, 1964; Gould, 1980).

Whether building up new parts (panda's thumb) or breaking them down (vestigial parts), evolution refashions the parts of ancestors into the refurbished parts of descendants. The resulting continuity from species to species holds the record of an accumulating complexity producing an ordered history (fossil record). Where parts no longer serve in the adult, they may be retained as necessary steps through which to build a delicate embryo (comparative embryology). Underlying structures are similar but modified to particular habitats and lifestyles (comparative anatomy). The geographic location where a new species evolves is likely to be the geographic location where first members of the family appear (distributional evidence). Overall, the evidence speaks to continuity and change through time, descent with modification, evolution.

A century and a half after Darwin, the major scientific obstacles that faced him are now resolved. Time in abundance was available to evolutionary processes, to turn the improbable into the probable, to provide for evolutionary testing and sorting, and to produce the great diversification of organisms we meet today. Heredity is known, down to the DNA molecules that make it happen, including those in humans. Genes carry on, undiminished, the traits preserved by natural selection. Mutations in these genes, together with the results of sexual reproduction, produce the raw variations upon which natural selection acts afresh in each generation. In addition, the science of evolution includes population biology, the result of groups of organisms within a species interacting in their immediate environment. So radically different and advanced are our evidence and understanding of biological change today, that we should call it **neo-Darwinian evolution** in celebration of these important modern breakthroughs.

Natural selection was Darwin's major intellectual contribution to understanding biological evolution, along with the marshaling of the then-available evidence that spoke to change through time. Natural selection remains a centerpiece of modern evolutionary thought, a sophisticated and complex concept to which we turn next in chapter 7.

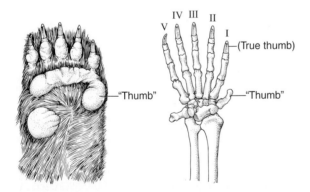

FIGURE 6.15   The Panda's Thumb   The panda has five digits like most mammals; however, opposing these is another digit, a "thumb, " which is actually not a thumb at all but an enlarged wrist bone. The furred and padded panda right paw is shown; next to it, the underlying skeleton of hand and wrist bones are sketched.

Selected References

Davis, D. D. 1964. The giant panda: A study in evolutionary mechanisms. *Fieldiana Zool. Mem.* 3:1–339.

Gould, S. J. 1980. *The panda's thumb: More reflections in natural history.* New York: W. W. Norton.

Gould, S. J. 1983. *Hen's teeth and horse's toes: Further reflections in natural history.* New York: W. W. Norton.

Rudwick, M. J. S. 1985. *The meaning of fossils.* Chicago: University of Chicago Press.

Learning Online

Testing and Study Resources

Visit the textbook's website at **www.mhhe.com/evolution** (click on the book's title) to take advantage of practice quizzing, study/writing tips, timely news articles, and additional URLs for research on the topics in this chapter.

7

"Although they can neither dance nor sing, plants possess a number of potential means of discriminating among pollen donors, and the donors possess several possible mechanisms for rendering themselves discriminable to pollen receivers."

Mary F. Willson and Nancy Burley, *Mate Choice in Plants*

Selection

INTRODUCTION

Herbert Spencer (1820–1903) introduced the phrase *survival of the fittest* to describe, in a few words, the complex struggle between an organism and its environment, and to capture the idea that those organisms with superior traits (the "fittest") were more likely to survive. Darwin's phrase was *struggle for existence*. For both, competition between and predation among individuals was the centerpiece of natural selection. Today we call these **biotic** (*bios,* living) **factors**, the influences of other living organisms on survival. Eventually Darwin recognized that equally important for survival was the outcome of an organism's confrontation with the physical environment. Today these are **abiotic** (*abios*, nonliving) **factors**, physical demands arising from conditions of climate or habitat harshness.

But this is not enough. Since about the 1930s, most evolutionary biologists have recognized further that survival alone is not enough to account for descent with modification. Evolutionary success is ultimately measured not by simple survival but by reproductive achievements. Surviving the assaults of biotic and abiotic factors brings no evolutionary benefits unless it culminates in successful reproduction. Consequently, **fitness** has come to mean the relative reproductive success of individuals, within a population, in leaving offspring in the next generation. Individuals that contribute more offspring to the next generation enjoy greater fitness than those that contribute fewer offspring. Fitness is not an absolute measure but a relative measure of one individual compared to another; it is calculated as a reproductive outcome. At the genetic level, fitness is measured by the relative success of one genotype (or allele) compared to other genotypes (or alleles).

We will explore the significance of fitness, especially fitness of the genotype, as we meet this in later chapters. For now, let's recognize that descent with modification is driven by two processes: one is reproductive success (fitness), the other is its prerequisite, environmental success (survival). Fitness is measured, then, by the relative ability of an individual organism in a population to survive and reproduce in a particular environment. And central to evolutionary outcomes is natural selection, especially as it affects survival success of individual organisms.

THE PHENOTYPE TAKES A BEATING

Natural selection directly assaults the phenotype, not the genotype. It is the phenotype of the organism that immediately receives the batterings from the environment and thus endures, or not, the hazards posed by climate and topography, challenges from competitors, or threats from predators. The intensity of such environmental factors acting upon members of a population is called **selection pressure**. The phenotype experiences selection pressure, and consequently current survival most immediately depends upon the phenotype. The phenotype must provide the insulation to protect the organism in cold climates, the physiological vitality to meet competitors, and the swiftness or cunning to escape predators. Certainly the phenotype itself is largely the product of the genes, so when a poorly designed phenotype is lost, so is the genotype that contributed. Eventually the genotype is affected by the consequences of survival, but only indirectly through the fortunes of the phenotype. It is the phenotype that comes face-to-face with natural selection.

ARTIFICIAL VERSUS NATURAL SELECTION

Selection is about choices. We make choices. We choose our friends, choose our neighborhoods, choose our jobs, choose our cars. We pick and choose, exercise preferences, examine alternatives, and cull from among the choices. Our choices may be limited. Not everyone is rich and wealthy. But in a word, we select. At least, we select from among the available suite of possibilities before us. We select to suit our purposes, human purposes. An equivalent culling process occurs in nature wherein some survive and others perish. It is useful, if carefully done, to compare the effects of human selection on plants and animals with the analogous action of natural selection. By looking at the consequences of human choices on breeding organisms, we can observe how human intervention produces great variety, and thereby we can appreciate the similar consequences of diversity produced over greater periods of time in nature.

Artificial Selection

The weeding out of organisms by humans for human purposes is **artificial selection**. Humans preserve organisms carrying traits that suit human needs, wishes, or just plain fancy. Humans intervene actively and intentionally to select amongst available choices. For example, a dairy farmer seeking to improve milk production might breed just those cows that give high-volume output, and hold back those with low production. Repeating this over successive generations, the farmer can gradually increase the herd's milk production. The dairy farmer does not create whole new cows endowed with increased milk output. Instead, the farmer takes advantage of existing variation within the herd, breeding just high-volume milk producers. Purchase of cows from other herds may bring into the breeding program cows with favorable traits not present in the farmer's own genetically limited herd, and allow the farmer to more quickly breed desired traits into successive generations.

Dogs and Cats

Under artificial selection, the traits amplified through the generations need not be practical traits. They may be capricious and only suit human fad or fancy. Dogs come in great variety (figure 7.1), big and small, long-hair and short-hair, mellow or edgy in personality. Thanks to artificial selection, some are traction animals, some herd, some guard, some hunt, and some just sit patiently in human laps. Yet all dogs derive from a wild, gray wolf ancestor. The first archaeological sites containing dog bones date to 14,000 years ago. Genetic comparisons confirm a divergence from wolves 15,000 years ago, in East Asia. From there, dogs accompanied migrating humans to Europe and across winter ice of the Bering Strait into the Americas. After wolves, today's domestic dogs are genetically closest to coyotes and then jackals.

Why domestication of dogs first occurred is not known, but dogs certainly provide an extra set of acute ears to warn of intruders, sentinels ready to alert. Under domestication, the irritable, high-strung, and aggressive behavior of adult wolves proved incompatible with human social systems. The more submissive traits characteristic of wolf pups eased wolves' assimilation into human company. Humans seem to have preferred juvenile characteristics of wolves, including floppy ears, greetings that include licking, and begging for food. Individuals that retained these puppy-like traits into maturity found favor among humans who nurtured them. In modern times, selective breeding programs have focused on more fanciful features of dogs, making dogs a model for how rapid morphological change can occur.

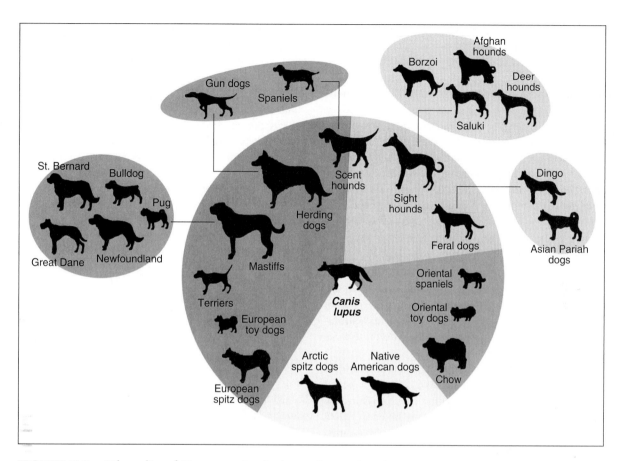

FIGURE 7.1   Diversity of Dogs   Artificial selection has produced numerous breeds of dogs, which descended from wolves, *Canis lupus.*

Domestic cats apparently descend from ancestors of small cats in the Middle East, around the eastern shores of the Mediterranean. The earliest remains of almost all major domesticated animals (goats, pigs, sheep, cattle, birds) are found in this region. But unlike these early domesticated animals, cats give no significant milk or meat or eggs or tilling traction. And unlike dogs, they provide no warning of threat, do not easily bond with humans, and so do not readily join in the defense of home and hearth. But early human settlements were based on agriculture. Stored grain and barns likely provided unintended safe housing for rodents, especially the "house" mouse, *Mus musculus,* which thrived in and around grain stores and debris. Safe from their natural predators, rodents prospered. Perhaps cats followed this food source into farms, and there were tolerated for their favorable effects on keeping the vermin in check. By about 4000 B.C.E., cats had moved up from being tolerated guests to being cult an-

imals in religions in Egypt. During the third millennium B.C.E., cats appear first in the artwork of Crete, mainland Greece, Libya, and then eastward into India and China. Cats spread or were carried in the second millennium B.C.E. to Italy (1900 B.C.E.), Britain (1600 B.C.E.), and Germany (1000 B.C.E.). Not until the seventeenth century C.E. did cats reach North America in numbers. Some were probably escapees from commercial ships, which observed a long tradition of carrying cats onboard as charms for safe voyage and companions to combat rodent vermin in the hold below deck. Others arrived later as favored pets with early human immigrants to New England, New York, and the Canadian Maritime Provinces, respectively from England, Holland, and France.

Cats, like dogs, followed similar routes into domestication. First, wild cats were symbiotic with humans. Both humans and cats benefited from the interaction and neither was harmed. Second, capture of these wild cats would lead to artificial selection for individuals that could adapt to confinement and proximity to humans and, in such an environment, rear young. These were practical traits that made cats more reliable partners in the management of farming affairs (namely, control of vermin). Third, cats were selectively bred for fanciful traits—long or short hair, fur colors, or special features such as blue eyes. Cats arose from a common ancestor in the wild, but under domestication and selective breeding have produced today's extraordinary diversity of breeds.

Agriculture

Artificial selection has had a large and favorable effect in agriculture generally, giving us many enhanced varieties of food crops (figure 7.2). Some artificial selection has been driven by practical and intuitive methods. From wild grasses, unheralded farmers in the past set aside and then later planted seeds from plants with favorable traits. Corn was one agricultural crop to arise by such intuitive methods (figure 7.3).

A deliberate effort to select high- and low-protein content in corn was begun in 1896, on experimental plots of land in Illinois. As with most characteristics, the oil content in the sample kernels of corn varied from low to high, averaging about 4.5%. The scientists measured oil content, and then selected seed kernels at the extremes. Those with low oil content were planted in one plot of land, those of high oil content at another. Over the year, these seeds germinated, grew into corn plants, and produced their own ears of corn, which were harvested. The scientists then repeated the process, selecting seeds (kernels) low and high in oil content to replant the next year. After 90 generations (by 1986), they were able to produce two lines of corn, one becoming successively higher in oil content (increased by 450%) and the other successively lower in kernel oil (decreased to about 0.5%).

Artificial selection has also produced enhanced varieties of plants to excite simple human pleasure. Master gardeners, over the last several centuries, were part of the household staff on wealthy estates, presiding over the grounds around the estates. In greenhouses and garden plots, they crossed plants with characteristics, especially flower color, that would produce the unusual or especially pleasing. Begonias are one example of selective crossing that produced a great variety of colors and shapes satisfying to humans (figure 7.4).

Tulips

Like many plants, such as the potato and tobacco, tulips came to western Europe from other regions of the world. The tulip was first seen by Europeans in Turkey. The Austrian ambassador, on a visit to the Sultan of the Ottoman Empire, spotted the flowers growing in the gardens within Constantinople and Adrianople. The

FIGURE 7.2 Diversity of Tomatoes and Roses Through selective breeding, tomatoes with different shapes have been produced; within the rose family, different colors and flower structures have been artificially selected.

Pear tomato

Cherry tomato

Beefsteak tomato

Hybrid Tea

Floribunda

Grandiflora

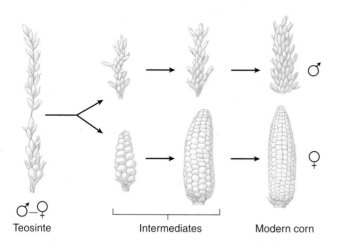

FIGURE 7.3 Corn Artificial selection through the centuries evolved the modern male tassel and female ears of corn from wild grass.

Teosinte   Intermediates   Modern corn

tulips were probably native to areas around the Black Sea and had been cultivated by the Turks since about 1000 C.E. They were introduced to Holland in 1593, where their popularity quickly grew. In fact, by 1636, tulips were a craze out of control. Prices for single bulbs reached exorbitant prices, in today's values, thousands of dollars, as individual speculators raced to purchase a new variety that could be grown, multiplied, and sold for even more. Horticulturists produced new varieties, some striped and elegant, at first affordable to only the wealthiest aristocrats. But some color varieties were quickly produced, such as shades of red and yellow. These became common, the price dropped, and more giddy people bought tulips. The craze grew. Rare varieties promised great prices. The black tulip became the

prize. This was a tulip with not just a deep-purple flower, but a completely black flower. It never happened. Try as they would, the best horticulturists of the day never managed to breed a truly "black tulip." Tulips simply did not carry the natural variation, the bell-shaped curve, to include such a variant plant within their range of possibilities.

The quest for a black tulip, although unsuccessful in nature, did inspire the writer Alexander Dumas to pen a novel, *The Black Tulip*, built around a fictional character who succeeded. In real-life Holland, in 1637, a Dutch farmer paid, for a single tulip bulb of the Viceroy variety, two loads of wheat, four loads of rye, four fat oxen, eight pigs, a dozen sheep, two hogsheads of wine, four tons of butter, a thousand pounds of cheese, a bed,

FIGURE 7.4 Begonias Artificial selection of begonias for flower shape, size, and color has produced distinct varieties.

Camellia   Rosebud   Sunset   Wild begonia

Marmorata   Picotee   Carnation

clothing, and a silver beaker, worth about 2,500 guilders of that day. Unfortunately for that farmer and thousands more, the financial bubble for tulips burst that same year, the government enforced economic controls, and the "tulip mania" ended. Tulips are still a popular flower, and now more reasonably priced. But to this day, the black tulip has appeared only in fiction.

Natural Selection

The weeding out of organisms by biological processes, without deliberate or directed human intervention, is *natural selection*. Organisms less suited to an environment, on average, perish and consequently do not pass on their genes to subsequent generations. Natural selection is an average event, a tendency for the less suited to die and thereby fail to reproduce. It is a process of elimination. The other side of that evolutionary coin is that organisms with traits serving them well fare better, tend to survive, and pass along those favorable traits. The process, played out often over thousands of years, moves slowly by human standards. Fortunately, we have on hand some examples.

Industrial Melanism

The peppered moth (*Biston betularia*) comes in two distinctive color phases: one melanic (dark), the other peppered (light, with black flecks), from which the common name derives. Up to about the mid nineteenth century (1848), the dark, melanic phase was estimated to make up not more than about 1% of the population in the countryside outside of Manchester, England. By the end of the nineteenth century (1898), it made up 99% of the population. These drastic shifts in the frequency of the melanic phase appear also in the insect collections made by naturalists of the time. In the early twentieth century, this strange shift in frequency drew the attention of H. B. D. Kettlewell and several other biologists, including N. Tinbergen. Kettlewell proposed that the change in frequency was due to a change in selection pressure from the environment. In proving this, Kettlewell and Tinbergen performed a series of experiments, showing that the polluting consequences of the growing industrial revolution in the late nineteenth century changed the forests downwind from the manufacturing centers in England. These changes in the forests led to a change in the color phase of the moth that was most advantageous.

Moths are creatures of the night, becoming active when they can go about their business without exposing themselves to the community of daylight hunting predators, mainly birds. During daylight hours, they are usually inactive, finding places to hide where they are inconspicuous and escape the notice of prowling birds. Prior to the industrial revolution, the surfaces of many trees were overgrown with light-colored lichen. Against such a background, the light phase was cryptic (camouflaged) and blended in; the dark phase was conspicuous and stood out (figure 7.5a). Birds tended to take the conspicuous, which were the melanic phase, and overlook the light phase, which, on average, survived more frequently to reproduce. But as manufacturing industries expanded in the late nineteenth century, that changed. The air pollutants killed the lichens encrusting the trees where the moths spent their daylight hours. This left the uncovered bark, which was naturally dark. Against such a dark background, the dark phase blended in; the light phase stood out (figure 7.5b). Now the melanic phase enjoyed the advantages of their dark, cryptic color and survived better than the conspicuous light-phase individuals; more of the melanic moths escaped notice and lived to reproduce, thus accounting for their increasing proportion of the population.

Kettlewell released equal numbers of dark- and light-phase moths in unpolluted and polluted woods. He then recaptured moths (drawn to a night light) to document the surviving percentages in each of the two different environments. During daylight hours, he also observed and photographed from a blind to actually document predation in action. His findings matched his predictions: In unpolluted forests (white, lichen-covered bark), the light-phase moths survived predation significantly more than the conspicuous dark-phase moths; in polluted forests (dark bark), the reverse occurred—the dark-phase moths survived significantly more than the now-conspicuous light-phase moths. Flycatchers, nuthatches, yellowhammers, robins, and thrushes all tended to overlook the cryptic-phase individuals, taking the conspicuous-phase moths. Movies of the birds' foraging behavior show them searching tree trunks and eating the conspicuous moths without noticing adjacent, cryptic individuals.

Critics suggested that for this process to work, moths would have to choose the bark that matched their color. So Kettlewell released the two color phases in a barrel painted alternatively black and white. In fact, melanic moths tended to settle on dark, light moths on white. Critics suggested that the polluted bark itself physiologically changed the light moth into the dark. So Kettlewell tested this too, and found that melanic

(a)   (b)

FIGURE 7.5   Peppered Moth   The moth occurs in two color phases, peppered and melanic. (a) Both phases are displayed against an unpolluted, lichen-covered tree. (b) Both phases are displayed against a dark tree, on which the lichen were killed by pollution.

moths stayed dark even on light backgrounds and light moths stayed light on dark backgrounds. The color phases were genetically determined, not physiologically determined. Some critics today note that a careful search for either phase of the moth on tree bark seldom turns up moths actually on the bark; instead they are hidden in leaves and elsewhere on the tree. This may be an artifact of observation, or an example of behavioral evolution in the moth, accompanying its morphological evolution of an earlier century. We shall just have to see, as with other criticism of Kettlewell's work, where the explanation lies.

But overall, the high correlation between moth color phase and forest type, polluted or unpolluted, along with the careful work of Kettlewell and others, gives us a persuasive example of rapid evolutionary change within a few decades rather than millennia. Humans were present as the changes were underway, and naturalists documented the changes as they occurred. These observations of change were then put to experimental test. These tests clearly document the relationship between a trait (color phase) and the survival service performed (camouflage). We can actually dig into this example and understand why some moths survive (cryptic) and others perish (conspicuous). Kettlewell documented the particular part of the environment that did the culling of moths, namely, the community of insect-eating birds searching for moths against different backgrounds. This leads us to under-

stand why such evolution occurred during the late-nineteenth-century in England, from predominance of light-phase moths in pristine forests to a predominance of the dark phase in polluted forests. Although the change in forests resulted from human activity, the culling process was not mounted by humans for human purposes. Instead, the ongoing process of natural selection did the weeding out of individual moths with unfavorable traits over the decades when the forests changed.

Cliff Swallows

In the spring, cliff swallows return from southern wintering areas to nesting sites in North America. Along banks and cliffs, they build mud nests clustered into colonies, feeding only on aerial insects they catch in flight. In one unusual year, returning swallows were surprised by the assault of a very harsh week of cold, rainy weather. Temperatures were stressful, winds harsh, and insects few. Thousands of swallows died. After the storm passed, biologists collected bodies of dead swallows and captured survivors. They took various measurements of wings, tail, and other parts from both sides of the body, as well as total body weight. They found a significant difference in survivors compared to casualties. Survivors, on average, were larger and their bodies more symmetrical (less difference in wing and tail feathers from left and right sides). Larger birds enjoyed the relative advantage of retaining body heat, so were less temperature stressed. And larger birds can store more body fat, so these could cope better with the reduced availability of insects. The survivors were more symmetrical. Their greater symmetry of wings and tails reflects their greater airborne stability and balance, left and right, thereby conferring more efficient and less-costly foraging flights during the blast of cold weather.

In one week of cold, harsh weather visited on these swallows, selection favored the large and symmetrical, which tended to survive, over the small and asymmetrical, which did not. Such harsh weather greeting the returning swallows is rare. Most springs are less devastating. But for some animals, natural selection acts regularly on an annual, even seasonal, basis.

At a Snail's Pace

Working in the fields and woodlands of England, two scientists, A.J. Cain and P.M. Sheppard, documented the survival of different color phases of the land snail *Cepaea*. Their collective work over several years was extensive, and included some genetics of the traits, but we will just focus on part of it.

Birds, thrushes in particular, are a major predator on *Cepaea*. When attacked, the snail pulls its soft body into its shell, preventing the bird from dislodging it. Thrushes defeat this defense by finding a hard stone, termed an "anvil stone," and flinging the snail against it. Repeated, this action eventually breaks open the shell, breeching the snail's defenses. The thrushes eat the snail, leaving the shell debris behind at the anvil rock.

The snails come in several background colors that may be detailed with up to five encircling bands. The background colors are basically three—brown, pink, and green. (The snail's partially translucent shell looks green when the snail is inside, yellow if the snail has been removed.) To examine the prevalence of the three basic color types, Cain, Sheppard, and associates, down on hands and knees, carefully searched for and collected snails in two different habitats. One was a woodland-beech forest with brown leaf litter coating the forest floor essentially year-round. The other was a meadow environment, coated with grass essentially green the year around.

Their collecting found that *Cepaea* with brown background coloration was common in woodland-beech; green snails were rare. In meadows, the reverse was true—brown was rare, green prevalent (figure 7.6). Pink tended to follow brown in both environments. The survivors in the two areas possessed shell colors that

	Brown	Pink	Green
Beech-woodland	Common	Common	Rare
Meadows	Rare	Rare	Common
Deciduous woodland — Spring — Summer	Common Rare	Common Rare	Rare Common

FIGURE 7.6 Snail Selection The shell of the snail *Cepaea* occurs in three color phases: brown, pink, and green. In different habitats—beech woodlands, meadows—different-colored shells are common or rare. In deciduous woodlands, the frequency of the color phases changes from spring to summer.

blended them in—brown shells against brown leaf litter and green against grass. These were the survivors. The snails that possessed camouflage color traits tended to escape the notice of prowling thrushes. But what would you expect to find at the respective anvil rocks? At these rocks were the shell debris of those that perished. Where brown snails in brown leaf-litter survived, the shards of green snail shells predominated at anvil rocks. And where green snails survived in green meadows, the brown snail shell bits predominated at anvil rocks. Here, within the large area of England, the specific microhabitats could present quite different selective forces on the different shell colors.

Cain and Sheppard did more. Using similar approaches, they studied *Cepaea* in deciduous woodlands, early in the spring when the forest floor was still composed of brown leaf debris from the winter, and in the summer after the trees had leafed out in green growth. They found brown snails common in late spring (and green snails rare), but by summer this had reversed: green snails were common (and brown rare). The forest went from brown to green. As the habitat changed with the season, the selective advantage of the two color phases changed as well—brown enjoying the benefits of its camouflage early, but finding this disadvantageous as the forest greened. Green snails were conspicuous to thrushes in brown, late-spring forests, but blended in more as the forest greened. Here, with snails, as with peppered moths, selection pressures are demonstrated to act on color. The direction of selection depends upon the background of the habitat, which may change on an annual cycle from brown to green.

In these snails, there is variation in shell color. This variation is heritable. And amongst the snails, their survival is differential, based on shell color—some survive, others do not. The habitat sets the stage. Green shell against green meadows tends to survive; brown against brown does as well. But conspicuous colors, against an environment that sets them off, invite the attention of predators—they are spotted, and they perish.

Snails to Snakes

The water snakes around Lake Erie in the northeastern United States are an example of natural selection in action and also illustrate the local consequences of snake movements within this species. The water snake *Nerodia sipedon* is nonvenomous, although adults are quite aggressive and easily provoked to strike out at threats that come too close. It lives on land but, an accomplished swimmer, it dips into water to feed on fish and amphibians, and then returns to land to bask on warm soil and rocks. Throughout most of its range, which includes much of eastern North America, it is regularly dark-banded with alternating blotches on a lighter background. Such a color pattern provides camouflage protection against the dark, heavily vegetated habitats it occupies from visual predators—those that hunt by sight. However, within this species there are small enclaves with high variability in this banded color pattern. These occur exclusively on islands in western Lake Erie. On these islands, the banded individuals occur, but so do large numbers of completely unbanded individuals, leaving them with only the light background color.

This curious occurrence of numerous unbanded, white water snakes on the islands attracted the attention of field biologists. They began by scoring color from A (white, unbanded) to D (dark, banded) with intermediates between (figure 7.7a). They then gathered pregnant females from one of the islands and scored the live offspring. Most were C or D (dark, banded). They then carefully collected adults on the same island and scored them. Contrary to the color proportions in newborns, the adults were A to B (mostly white). A reasonable explanation for this newborn-to-adult difference could be that as banded snakes grow, they lose their banding. This proved not to be the case. Young snakes raised in the laboratory did not change color pattern as they grew into adults. If they were banded as young, they stayed banded as adults. What accounted for the differences from newborn to adult was not a loss of banding with

age, but instead the differential elimination of C and D color patterns. What was happening?

The islands, unlike the mainland, have lots of exposed, light-colored limestone, especially along the shorelines. Here, these island snakes forage for food, but are out in the open, with less vegetation for cover. Visual predators, such as gulls, raptors, and herons, are less likely to detect light-colored individuals against the light-colored rocks, giving these snakes an adaptive advantage over the more conspicuous, dark-banded individuals. The result is the differential elimination of color types, removing mostly dark-banded individuals from the island population, lost to predation (figure 7.7b). This differential predation falls mostly on the defenseless young. By the time surviving snakes become adults, they may be 3 to 4 feet in length, with an aggressive personality that can deliver an intimidating bite to any would-be predator, permitting the adult snakes to hold their own—they have literally outgrown their predators. But small, young snakes are easily dispatched if spotted. So if light, unbanded is favored over the banded color phase, why has natural selection not eliminated all banded snakes from the islands?

The answer is migration. As banded snakes, and their genes, are lost to predation on the islands, occasional banded immigrants from the mainland reach these islands and return some of the genes for banded color lost to predators. Migrations of individuals into the population can influence the composition of the phenotypes in the population, by replenishing genes lost to natural selection. But on these island snakes, we can also see the effects of natural selection on color phases, mediated through the community of visual predators, but leading to differential survival of young snakes based on their favorable (unbanded) or disadvantageous (banded) characteristics. We can also see in this example, as in that of the snails before, that the particular features of the microhabitat (islands versus mainland) can influence the results of natural selection.

TYPES OF NATURAL SELECTION

Any trait tends to vary—color, size, tolerance for heat, milk output, oil content, and so forth. In a group, there will be individuals with the trait at the extremes (at its "tails"), but most lie somewhere in between. For example, arranged and stacked by height, the cadets shown in figure 7.8a are from short to tall, with most falling in the middle. This distribution is described as a bell-shaped

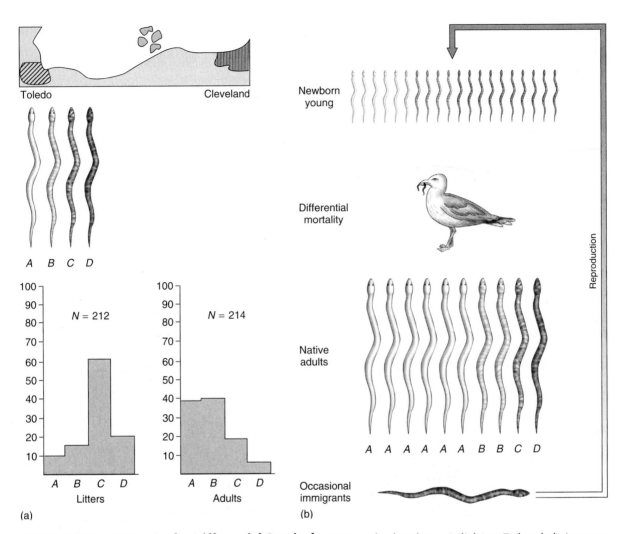

FIGURE 7.7   Water Snake Differential Survival   (a) Scored color phases A (light) to D (banded). In young born on the islands, most are C or D (dark and banded). But by the time the snakes reach adult stage, most are A or B (light colored). (b) On the islands, predaceous gulls feed on young snakes, usually spotting and eating the more conspicuous banded snakes, producing differential survival of mostly unbanded snakes. Occasional immigrants from the mainland return some of the genes for banded color.

curve (or bell curve) (figure 7.8b). Selection pressures acting on a population can affect variable characters in three ways as its focus is upon three different parts of the bell-shaped curve: stabilizing, directional, and disruptive selection (figure 7.9). Another, special case, of selection pressure is sexual selection.

Stabilizing Selection

In stabilizing selection, sometimes called "normalizing selection," the extremes are disadvantageous and eliminated, leaving the intermediate phenotypes favored and preserved by comparison (figure 7.9a). Applied

from generation to generation, variation in the trait is reduced, the range narrows, and the population becomes more uniform. For example, in humans, infants with intermediate weight at birth enjoy the highest survival rate; similarly, in chickens and ducks, eggs with intermediate weight benefit with the highest hatching success. Here, selection acts against the very small and very large. Extremes tend to be eliminated, intermediates favored. Most selection is probably like this on a day-to-day basis. Populations have been shaped to current environments. Any extremes tend to depart from the favored middle. This is sometimes referred to as the "if it ain't broke don't fix it" type of selection. At the

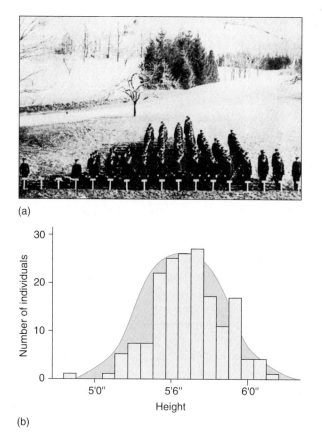

(a)

(b)

FIGURE 7.8   Bell-Shaped Curve   Arranged by height, these cadets show variation from short to tall, with most falling somewhere in between. This variation, common to most traits, is shaped like a bell; hence the name.

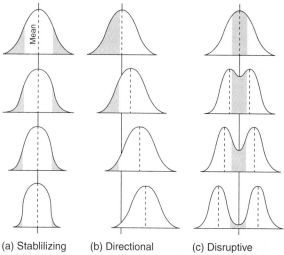

(a) Stablilizing   (b) Directional   (c) Disruptive

FIGURE 7.9   Types of Selection—Stabilizing, Directional, and Disruptive   The bell-shaped curve represents the distribution of a character in a population. The shading indicates where in that variation selection acts to eliminate individuals. (a) In stabilizing selection, extremes are eliminated, leading to a narrowing of the variation. (b) In directional selection, one extreme is eliminated, shifting the curve. (c) In disruptive selection, individuals with intermediate variation are eliminated, producing two bell-shaped curves at the extremes.

moment, things are working, and changes are likely to produce only maladaptive outliers. The population is settled in and suited to local conditions.

Directional Selection

When one extreme of a trait is disadvantageous, natural selection acts against it, favoring the rest of the individuals (figure 7.9b). Applied generation to generation, the average of the population shifts over time, moving away from the disadvantageous extreme. Such selection, which acts against one tail of the bell curve, is directional selection. We have seen examples already in artificial and natural selection. The dairy farmer who eliminates the low-milk-producing cows exerts artificial selection hoping to boost the herd and increase the average milk production. The water snakes, on the islands, show another example wherein the banded snakes at one extreme are selected against. The popula-

tion mean shifted as more light-colored, camouflaged snakes survived.

Disruptive Selection

Where intermediate phenotypes are at a disadvantage, selection is disruptive (figure 7.9c). Individuals with traits in the middle tend to be eliminated, dividing the population between the extremes. The result is to produce mostly individuals with characters at the two extremes of the bell-curve distribution. This condition is referred to as **polymorphism**, meaning that in the same species, two or more conspicuous and distinctive forms occur in abundance. The artificial selection for high- and low-oil corn is one example. The peppered moth may be another, wherein melanic and light phases predominate, but few intermediates survive.

Sexual Selection

Many worry, as did Darwin, about a type of selection that seems to fall outside of conventional selection—namely, outside the view that new adaptations arise to fit the organism to the assortment of challenges from its habitat. In this context, differences among species are related to the differences in habitats. But why then, in the same species, and living presumably similar ecological lifestyles, would you find major differences among individuals? In particular, why do we find major morphological differences between the sexes within the same species? Of course, copulatory organs differ between the sexes, but males and females also differ in ancillary features not primarily related to reproductive organs. Such distinguishing supplemental features differing between the sexes are *secondary sexual characteristics*. Male lions have a mane; females lack them. Male moose and other deer grow annual antlers; females do not. Adult male sea lions are large; females small. Peacocks sport bright colors and huge ornamented tails; peahens (females) lack such accoutrements. From beetles to bison, males and females within the same species differ morphologically, and exhibit different secondary sexual characteristics, a condition termed **sexual dimorphism** (figure 7.10).

At its most basic, evolutionary success comes down to reproductive success. Simple survival alone brings no evolutionary benefits if the organism fails to reproduce and thereby perpetuate its characteristics into future generations. To reproduce, an organism must obviously survive at least through a successful breeding effort. No matter how strong in build, how victorious in conflicts, how accomplished a predator, or how resilient against the elements, if individuals fail to reproduce and pass on their genes, biologically they are failures, losers in the evolutionary race. No matter the type of selection, evolutionary success is ultimately measured by reproductive success.

But the presence of sexual dimorphism testifies to a distinctive or special kind of selection in action, one unrelated to adaptation to environment. It is not competition for food or shelter that produces survival of the few, loss of others. Instead, members of one sex compete for the opportunity for preferential mating with members of the other sex. The advantage some individuals have over other members of their sex provides them with increased reproductive success. Those members of a sex (usually males) who have the greatest physical prowess or flashiest ornamentation enjoy, on average, greater success in attracting a mate and thereby benefit from reproductive success. This form of selection, termed **sexual selection**, places individuals directly in competition with members of their own sex ("the power to conquer," as Darwin put it) for success in attracting mates ("the power to charm").

During courtship, especially within vertebrates, males usually are in competition with each other for

FIGURE 7.10   Sexual Dimorphism   The large male California sea lion is distinctive from the surrounding, smaller females.

priority position amongst other mature males and for the attention of the females. On the other hand, females are particular when picking a mate. Apparently, females are impressed by masculine features. They tend, therefore, to choose males of larger size or strength, or those with more elaborate displays. Consequently, males with these traits fare better, breed more often, improve their reproductive success, and pass on their enhanced and distinctive secondary sexual characteristics to offspring.

Certainly, there are exceptions within animals. Many invertebrate animals simply cast their eggs and sperm, without fanfare or ceremony, into the aquatic environment. There, outside the body, *external fertilization* occurs. Sea urchins, sea stars, and sea anemones, just to name a few, reproduce by means of such external fertilization. There is little, if any, competition between mature males, and significant sexual dimorphism between the sexes is essentially absent. On the other hand, some animals with external fertilization may show extensive sexual dimorphism. Male fish often come decorated with bright colors, enhanced fins, and perform elaborate courtship displays to seduce a female. Other animals, especially those living on land, commonly employ *internal fertilization*, wherein the sperm are delivered into the female's reproductive tract and fertilization soon follows within her body.

Sexual selection occurs in plants, but in a limited way. Male plants do not directly display to each other to decide dominance or to female plants to attract their amorous attention. Instead, in plants, competition between males often occurs between their sperm, packaged in pollen. When pollen grains alight on the female carpel, their competing pollen tubes speed down its style, racing to the egg. The "winner" is the first to arrive, fertilize the egg, and exclude late-arriving, competing sperm.

Female flowers can reject pollen if the male producing it is genetically similar. Pollen may not germinate, growth of the pollen tube may be arrested, or plant embryos may abort. Such rebuff of the genetically similar male sperm by the female reduces incidences of self-pollination and favors out-crossing—interbreeding between relatively unrelated individuals. For many plants, as with eukaryotic organisms generally, out-crossing increases genetic variability and, hence, increases the chances of producing progeny with a wide range of characters to meet a wide range of habitats.

Sperm competition occurs in animals as well, again in relation to the mating system. In some animals, mating is *promiscuous* with no lasting or strong pair-bonds between males and females. Where bonds develop, at least for the length of a breeding season, the mating system may be *monogamous* (one male to one female) or *polygamous* (one to several of the opposite sex). Specifically, polygamous mating systems are usually *polygynous* (one male to several females), but occasionally the reverse occurs, *polyandrous* (one female to several males). When a female multiply mates (e.g., polyandrous, sometimes promiscuous), she carries away the sperm from several males. These sperm now vie for access to her eggs (fertilization) and the ultimate prize in evolution, reproductive success. Surprising new research turns up bizarre and brutal sperm tactics. In fruit flies, sperm from the second male to mate can often dislodge the stored sperm from the first; the fluid accompanying the sperm can poison stored sperm from earlier matings. In some species, the male has disproportionately large testes, producing a large volume of sperm to swamp the competition.

In many spiders and in the praying mantis, mating males often end up being part of the female's fresh dinner. For example, a male spider, upon successful approach to the female followed by onset of sperm delivery, may turn so as to present his body to the female. The female then eats the male, who consequently furnishes calories to the eggs he fertilized. By such suicidal behavior, the male also preoccupies the female and prolongs copulation while he continues to deliver sperm until the dinner is over.

In the Wild

Does the lavish opulence of males have consequences in the wild? So far, it seems to. For example, in the spring, male red-winged blackbirds arrive on reeds or cattails surrounding ponds. Here, they stake out and vigorously defend territories that will soon help them attract one or several females. Other males, trespassing on the territory of an established male, are assaulted by vocal and visual displays, including song calls accented by the presentation of colorful red and yellow feather patches, or epaulets, on an otherwise black body (figure 7.11). If these epaulets are blackened (experimentally painted over), this resident male, when returned to his territory, soon finds himself visited by increased numbers of interlopers—gate-crashing male competitors looking for a lush location of their own to attract a female. Conversely, epaulets of a dead, mounted specimen of a red-winged blackbird can be altered from black to doubly red. When placed within view of a living resident male, this bird mannequin receives attacks in proportion to its colors—few when painted black, to many when painted doubly red. Colorful epaulets make

FIGURE 7.11   Red-Winged Blackbird, Male
Note the epaulets—colored feather patches on its shoulders—that it uses in territorial and mating displays.

Each female flits about, visiting displaying males on their territories, then chooses one with which to pair. Together, the mated pair builds a mud nest and raises a clutch (group) of eggs. The female incubates the eggs, but both parents pull duty in feeding the chicks.

Barn swallows are sexually dimorphic. Males are bedecked in more vivid colors, but the most obvious difference is that males carry the longer tail feathers, called "streamers." Here was a wonderful natural situation inviting an experimental test. To test the effects of tail length on breeding success, tails of some males were shortened and, in another group, the tails were lengthened. Short-tailed males were produced by removing the feather's midsection and regluing the tip; long-tailed males were produced by inserting the removed feather section from the first group into the middle of the tail feather. (Separate control groups of swallows were also tested for handling and gluing effects—none were evident.) These groups with short and with long tails were then released, and their subsequent success in attracting females was observed (figure 7.12). It took almost four times as long for a short-tailed male to attract and pair with a female than a long-tailed male. As a consequence, long-tailed males started their first clutch sooner, completed it sooner, and therefore more often had time to start and finish a second clutch.

a difference. Males on their territories respond to the epaulet as if these color patches were a hostile gesture, a threat to the control of their real estate.

Barn swallows, insect-eating birds, provide another experimental example of secondary sexual characteristics with consequences in the wild. After wintering in Africa, barn swallows return northward to breed in Europe. Upon arrival in the spring, each male sets up a small territory and attempts to attract a willing female.

Consider This—

From Mate to Meal

Spiders typically spin nets of silk cast across insect flyways. When an unwary fly, bee, or moth collides with the sticky parts of the web, it struggles, buzzing away as it desperately beats its wings. This alerts the spider, which dashes out to wrap its meal in a cocoon of silk and then dines leisurely on its quarry.

Male spiders risk the same grim fate as insects—they start out as a mate but may end up as a meal. The male spider first spins web holders, loaded with sperm, carried on its short front legs. So equipped, he starts off in an amorous mood looking for females. Arriving at the edge of her web, the male may wave distinctive visual signals or strum on the web, taking care not to do so at the same frequency as the buzzing wings of an insect. If receptive, the female relaxes and the male approaches, tip-toe across the web. Reaching precariously under her abdomen, and exposing the rest of his body to her fangs, he inserts his sperm packets into a special receptacle. Job done, he now retreats—or, more accurately, sometimes he retreats. In some species, the male more commonly somersaults or turns to present himself to the female. Served up, he now becomes her meal. While he is being consumed, he continues to unload sperm into the female and occupy her attention away from other male competitors.

A similar fate often awaits a male praying mantis. While the male is coupled with the female, their copulatory organs locked, the female begins to dine on the male, eating off first his head; beheaded and all, the male continues to have sex with the female.

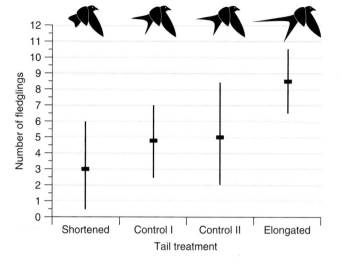

FIGURE 7.12   Barn Swallows   To test the effects of male tail length, four groups of males received four treatments. "Shortened" tailed males were produced by cutting out the middle of the tail and regluing the tip to the base. "Elongated" tailed males were produced by cutting and gluing the removed section (from shortened tailed) to the middle of the tail. "Control I" males had tails cut but then reglued, thereby exposing them to the basic treatment but without changing tail length. "Control II" males received no tail clipping or gluing. The solid boxes represent the average number of young each group of males produced during a summer; the vertical lines express the range of variation of the results (standard deviation). The control groups did not differ significantly from each other, but both differed from shortened males, which produced significantly fewer offspring, and from elongated males, which produced significantly more offspring. (After Møller, 1988)

This meant that during one breeding season, they got more of their genotype into the next generation (two clutches) than did their short-tailed competitors (one clutch). Sexual selection favored longer tail streamers, but there was more in this for long-tailed males than just increased reproductive output; namely, hanky-panky—sexual dalliance.

The polite term for this is *extra-pair copulations*, and in swallows this happens a lot, especially for males with a big, long tail to display. Scientists observing these swallows noticed that males occasionally solicited copulations with females other than their pair-mates. These extra-pair females were more likely to be tempted and succumb to these advances if their own pair-partner had a short tail, and males were more likely to succeed in their seductive advances if they had long tails to display. In this operatic mating system, long tails bring reproductive advantages. Sexual selection is happening.

Sexual selection is distinctive with distinctive consequences. Secondary sexual characters, driven by sexual selection, may attract mates, but at first they appear to lack adaptive benefits and may, in fact, hinder survival in natural environments. These secondary sexual characteristics are simply bigger, or brighter, or bolder. For a female seduced by such displays, it is probably not an aesthetic decision on her part. The research is still at an early stage, but so far it supports the view that sexual advertisements are health certificates. Only high-quality males with a sound physiology and robust health can produce such luxurious ornaments. Such exaggerated male features are outward testimony to inward quality. For example, diseased male birds produce ragged and

rumpled feathers, indirectly betraying poor prospects and inadvertently signaling a female to choose another.

A mature male lion grows a large, shaggy mane cloaking his head and neck. The mane is a signal advertising the male's condition, which females use to choose mates and other males use to assess rivals (figure 7.13). The mane, like a large wool coat, increases the heat-load on a lion, which must be robust to withstand the stress. Studies of African lions confirm that better-nourished and injury-free males sport a mane larger and darker than less-hearty challengers. The mane is an honest indicator of male quality. Rival males tend to avoid challenges with such mane-enhanced males, and females tend to choose them as mates.

Another view holds that cumbersome secondary sexual features test male stamina and quality. For example, in animals such as the peacock, the bright body plumage and huge, brilliantly colored tail bring it no advantages outside the breeding season. In fact, such bright colors draw the attention of predators and the tail encumbers it during flight or when fleeing. But a peacock that survives these impediments imposed by burdensome ornamentations the rest of the year is a peacock that must possess other offsetting, beneficial characteristics. A female choosing such an ornamented male is indirectly choosing these beneficial traits, and thereby endowing her offspring with them as well.

Female Choice

Take care here how to think about this idea of "female choice." Conscious decision is not implied—deliberate female decision is not required to understand the

(a)

(b)

FIGURE 7.13   Sexual Dimorphism in Lions   (a) Young male, with mane. (b) Female.

adaptive benefits of female choice. Females do not caucus—score males, 1 to 10, that strut before them—then make a premeditated decision. Even in humans, psychologists are discovering that women react to some unconscious cues, cues they are unaware of, when responding favorably, or not, to men. In animals, female choice should be seen simply as an innate, evolved response to an indicative secondary sexual cue presented by the male.

Sexual selection has consequences for the ways individuals are organized into societies or live solo lives. Certainly, within most vertebrates, males and females make different investments of time and energy in their offspring. In most vertebrate species, the male contributes only a cheap and inexpensive sperm; the female contributes an egg yolked with rich nutrients and may produce young that, upon hatching or birth, demand her effort and attention, without the aid of the errant male. Usually, but not always, it is the female who contributes most—yolk, incubation, birth, maternal care, protection. Consider, for example, a female grizzly bear. Af-

ter mating, she alone supports the growing young in her uterus through a placenta. After giving birth, she cares for and nurtures the young from her mammary glands. Threats to the young bear's safety are met by her protective action. She mates once, and having done so, is committed to this large investment, so she must initially choose wisely and well. But the male grizzly, after mating, departs for a more or less solitary life. It is to his advantage to mate with as many females as possible.

Some social systems have evolved in which the male contributes in a large and direct way to offspring survival. In many birds, males participate equally with the female in incubating eggs and later in bringing food to the nest for young. In primate societies and in many carnivores, such as wolves, societies are stratified, with a dominant top male and subordinates below in the hierarchy. These males may hold territories and the supportive resources contained within, and defend their immediate group from predators. In a variety of intriguing ways, complex social systems bring males and females together. Nevertheless, one consequence of sexual selection is the enhancement of different roles for males and females.

Sex Roles

Where sexual selection is important, males and females often meet different selective pressures within a reproductive context. The selection pressures on males are different from the selection pressures on females, simply because they are males or are females. Males and females evolve different morphological features that enhance these different sex roles on their way to successful reproduction, and sexual dimorphism is the result. But there is also an evolved social dichotomy— different social roles for males and females. Males and females live different lifestyles, meet different selection pressures, and evolve characters to suit them within the expectations of these different reproductive roles.

Sexual selection is still recognizable as a special kind of natural selection. Competition still occurs for preferential "access" to scarce resources (members of the opposite sex). Those competitors bearing traits that improve their mating chances fare better and, on average, pass on their genes to subsequent generations.

OVERVIEW

Natural selection happens. We see it in classic examples—peppered moths, snails, snakes. These and more recent studies verify that a natural culling of phenotypes is occurring and that this culling is not random, but discerning and natural. Natural selection of organisms is occurring against the environmental background of their own habitat. Thus, like a jury, natural selection does not summon up the cases. It acts on those brought before it.

Natural selection is similarly constrained. If there is no variation, there is no choice. Variation is the raw material of evolution. Without it, there is no change and consequently no evolution—natural or artificial. Tulips, when first brought to Europe, came in many colors, even striped, but never black. Using this natural variation, breeders separated, cultivated, and grew up these variations. All shades of tulips can be purchased today, but not black. Such a variant has never shown up, so none has been cultivated.

Evolution moves slowly by human standards. But artificial selection provides many examples of accelerated change—what could occur if the culling process

were intense and well focused. It shows what could occur quickly if the available variation in a population were surveyed, individuals with particular traits preserved, and breeding confined to these variants. Artificial selection favors individuals with traits that serve human purposes. Natural selection is a weeding-out guided by biological processes without human intents deliberately directing outcomes.

Natural selection comes in three types, with three different consequences. Stabilizing selection eliminates the extremes of a trait that are in disfavor. Maintained from generation to generation, the overall variation narrows and stabilizes around the mean of the population. Directional selection occurs when one of the extremes is disadvantageous; individuals with this extreme character tend to be lost from the population. Continued from generation to generation, the average moves away from this tail of the bell-shaped curve in the direction of the curve that does not receive such unfavorable treatment by selection. Disruptive selection acts on the middle of the bell curve, eliminating individuals with intermediate characteristics, favoring those at the extremes.

Sexual selection is a special case. Individuals still compete and face the evolutionary consequences of failure to do so successfully. Sexual selection may be stabilizing, directional, or disruptive. The difference is that the competition is based on sex, and the biological success is not measured by environmental success, but by mating success. Obviously, to mate successfully, individuals must survive and enjoy prior environmental successes in meeting the threats that challenge in their habitats. But one of the most challenging moments in the life cycle of many animals occurs during courtship and reproduction. An individual's major competitor at that moment is a member of their own species—in fact, a member of their own sex. Competition arises as a condition of the different sex roles females and males play in contributing to the production of offspring. Sexual dimorphism between the sexes is often the consequence.

Descent with modification is driven by two processes. One is relative reproductive success—*fitness*. The other is its prerequisite, survival success— the struggle for continuance in a particular habitat. When we look at the part of the evolutionary process concerned with survival, we are examining the ways natural selection acts to cull individuals from a population relative to immediate environmental demands. This helps us understand the source of adaptations.

But of course, any such survivor must meet the ultimate test, reproductive success. If mates are scarce or prove difficult to "charm," then natural selection favors secondary sexual characteristics that enhance reproductive accomplishments.

Some purists argue that we should speak primarily of reproductive success (fitness) when measuring evolutionary outcomes. This certainly makes some sense because it pinpoints this critical and deciding factor in determining evolutionary success—namely, reproductive achievement. Reproductive success is a result enjoyed by all evolutionary winners. But many features that foster an organism's reproductive success are often features that are in conflict with its survival. A secondary sexual characteristic (such as the male lion's mane) would be one, but not the only, example. Further, a focus on reproductive outcomes is a view too narrow because it underestimates the elegance and richness of this remarkable natural process we call evolution. Out of such a process come the splendor and lavishness of the biological diversity that enriches our world. Of course, these remarkable and diverse adaptations with which organisms come equipped are measured finally by the relative reproductive success they secure, but if we looked only at procreative outcomes, then we would miss the evolutionary processes that precede reproductive efforts. We would miss one of the important characteristics of evolution: the interaction of organisms with their environments and, in turn, the source of the variety of solutions to survival, the prerequisite to their reproduction. In short, we would slight the part of the evolutionary process that produces adaptations.

Our examination of selection should give emphasis to the conditions that make it happen. The weeding-out process acts within the context of the specific habitat where the individuals make their home. The island snakes live in a *microhabitat*, a restricted habitat different in its details from the snakes' mainland habitat. Consequently, the results of selection are different on the island. The snails of the fields and woodlands of England also meet their fate against different microhabitats, but microhabitats that change seasonally. Those with green shells fare better living in green meadows than in brown leaf-litter of woodland forests. As the seasons change, the background changes, and selection pressures on snail shell colors therefore change during the year. You cannot consider a particular characteristic in isolation from its environment and determine whether or not it is a favorable trait. The favorable or unfavorable consequences of a characteristic can be determined only against the environment in which it serves.

The point is as obvious as this: Feathered wings on birds in air can be advantageous. But feathered wings on fish in water would encumber and be disadvantageous. We cannot say ahead of time that wings (or any trait) are intrinsically adaptive. It depends. It depends on the context, the particular habitat in which the feature serves, and the lifestyle of the organism itself.

These examples of natural selection also reveal the particular component in the environment that does the culling. This specific environmental element that causes the elimination of unsuited individuals is termed the **selective agent**. The selective agent places demands on the individual organisms with which the organisms must cope in order to continue to survive and produce offspring. A selective agent may be either biotic or abiotic. A *biotic selective agent* is another organism—prey, predator, competitor, mutualist, or mate—that immediately influences survival or reproductive effort. We have covered examples: insect-eating birds were the selective agents on the survival of individual peppered moths; snail-eating thrushes were selective agents on snails; gulls, herons, and other visual predators acted as the selective agents on island water snakes. In sexual selection, it is often the female that acts as the selective agent, as she chooses a particular male with which to mate. An *abiotic selective agent* is an environmental demand on individuals arising from conditions of climate, physical topography, or habitat harshness. The bitter, cold climate of the Arctic acts harshly on individuals without the physiological, behavioral, or morphological means to cope. Storms may blow migrating birds, tardy in their seasonal departure, out to sea where they perish.

The importance of recognizing the selective agent in evolutionary studies is that we can identify the particular factor in the environment that actually causes differential survival of individuals within a population. In conversation you will often hear people talk about evolution acting to better the "species." Even otherwise-careful nature programs on television sometimes declare how a particular feature is beneficial to the "species." As we have seen in all our examples in this chapter, nothing of the sort occurs. Natural selection acts on the phenotype of an *individual*. A species is just the leftover result of persisting individuals who have endured and survived the environment of selective agents.

Selected References

Brown, C. R., and M. B. Brown. 1998. Intense natural selection on body size and wing and tail asymmetry in cliff swallows during severe weather. *Evolution* 52:1461–75.

Dugatkin, L. A., and J.-G. J. Godin. 1998. How females choose their mates. *Sci. Amer.* 278:56–61.

Møller, A. 1988. Female choice selects for male sexual tail ornaments in the monogamous swallow. *Nature* 332:640–42.

Møller, A. 1994. *Sexual selection and the barn swallow*. Oxford: Oxford University Press.

Williams, G. C. 1992. *Natural selection: Domains, levels, and challenges*. Oxford: Oxford University Press.

Learning Online

Testing and Study Resources

Visit the textbook's website at **www.mhhe.com/ evolution** (click on the book's title) to take advantage of practice quizzing, study/writing tips, timely news articles, and additional URLs for research on the topics in this chapter.

8

"Name the greatest of all the inventors. Accident."

Mark Twain, humorist

"Evolution is not 'of a very mystical nature.' It depends on accidents."

J. B. S. Haldane, geneticist

Variation: Spice of Life

INTRODUCTION

Henry Ford, the automobile tycoon of the early twentieth century, devised and perfected a way to produce cars cheaply so they were affordable to more than just the wealthy. The car has been a common contrivance in our culture ever since. His first Model T was affordable, dependable (for the day), and in great demand. He brought the car to the people. But there was a hitch: Henry Ford, with tongue in cheek, boasted that you could buy his inexpensive Model T in any color you wished—just so long as it was black. At the time, black was the only color available, a carryover from the horse-and-buggy days. Later the color choices would increase, become two-tone, and even striped. But when they were first produced, Model T's were only black. But of course one color is no choice at all.

Variety is also the spice of life. If all organisms within a species were identical, similar to all-black cars, there would be no "choice." Survivors of a later generation would be identical to their ancestors of an earlier generation. Species would be fixed; evolutionary change would not occur. No difference, no evolution. Differential survival depends on differential features. As Darwin saw, evolution needs choices, but where does variation come from?

MIXING IT UP

As it turns out, much genetic variation arises during normal sexual reproduction. For example, opportunities for new genetic variation occur during meiosis, when chromosomes are being prepared for packaging into gametes, and later during fertilization when egg and sperm fuse.

Recombination

Chromosomes come paired, one from the female and the other from the male. These chromosome pairs are homologous chromosomes, and they carry similar genes. During meiosis, homologous chromosomes pair up (synapse), and matched sections often exchange between the paired chromosomes. Genes reside on comparable sites—loci—on homologous chromo-somes. This allele exchange between homologous chromosomes is termed **crossing-over**. When cross-ing over occurs (figure 8.1), matched segments of ho-mologous genes are exchanged. However, the particular alleles present at these loci of exchange may differ. Thus, the exchange mixes up combinations of alleles on sections of the chromosome. This makes for new allele combinations, thus for new variation in ga-metes, and thus for the expression of new variations of traits in offspring.

Sex

In diploid organisms, sexual reproduction is a roll of the genetic dice, resulting in offspring that are a mix of pa-ternal and maternal genomes, about half from each. Offspring are not exact genetic copies of their father or mother, but a combination of both. Closely related, cer-tainly, to their parents, but offspring are different, in possession of their own genetic identity.

Even bacteria find a way to mix new DNA into their genomes. Recall that bacteria increase in number by simple division—mitotic division. No sex. Result-ing "progeny" are essentially exact copies of the orig-inal bacterium, forming clones or populations of identical progeny. But some genes jump directly from one bacterium to another as transmissible, short seg-ments of DNA called **plasmids**. Plasmids can be one to many genes in length and even move between dif-ferent species of bacteria. They leave the main DNA as little circles of genes and pass into suitable adjacent bacteria where they are incorporated into the DNA of the host bacterium. By such means, new genes are in-tegrated into the DNA of bacteria, adding variety to clones. Medically, this method of gene transmission is important because by plasmid transfer, virulence and even resistance to antibiotics spreads through patho-genic bacteria.

MUTATIONS

Early Work

Early in the twentieth century, the rediscovery of Mendel and the added discoveries of chromosomes clarified the basis of heredity. But for all the clarity brought to the mechanism of inheritance, the new sci-ence of genetics did not speak to the question of where genetic variation came from in the first place. Mendel's genetics solved one conceptual obstacle for natural se-lection: the mechanism of inheritance, the basis for transmission generation-to-generation of favorable traits—no blending of traits, but instead particulate in-heritance. However, it did not identify the source of new traits. That concept came from a happy accident turned into a useful idea by an observant Dutch botanist, Hugo DeVries, late in the nineteenth century.

DeVries worked with plant species, typically with the evening primrose. One day, when passing by an abandoned field, he noticed, growing wild, a form of the primrose unknown to him before. In follow-up ex-periments, both in the primrose and other plant species,

FIGURE 8.1   Crossing-Over During meio-sis, chromosomes duplicate and homologous pairs synapse. Chromatids exchange homologous sec-tions carrying alleles, producing recombinant daughter chromosomes with a different combina-tion of alleles.

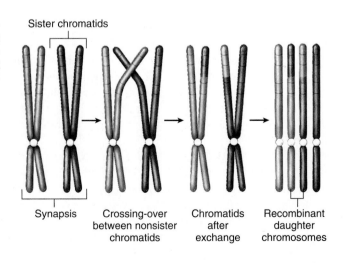

Sister chromatids

Synapsis   Crossing-over between nonsister chromatids   Chromatids after exchange   Recombinant daughter chromosomes

he noted that varieties occasionally occurred suddenly within a new generation of plants. These new characters appeared with no history in any previous generation. He called these abrupt hereditary changes **mutations**, and the organisms carrying them *mutants*.

Over the years 1901–1903, DeVries expounded the "Mutation Theory," the view that new species arose suddenly because of the abrupt appearance of distinct, large variations (mutations). The mutation theory drew many scientific disciples early in the twentieth century and was seen by some as an alternative to Darwin's ideas. Unfortunately, some zealots pushed this idea too far. They stood evolution on its head, arguing that abrupt, inheritable changes in an organism (mutations) were solely responsible for the appearance of new species (see chapter 9, polyploidy). In essence, gene changes drove evolution (DeVries), rather than simply producing variety upon which natural selection operated (Darwin).

Follow-up genetic research in the twentieth century confirmed the presence of mutations in the genetic material, although most of DeVries's supposed examples in the primrose turned out not to be alterations in individual genes. Instead, most of his examples resulted from new combinations of existing genes, such as those created by crossing-over. Nevertheless, the idea of gene mutations stuck and was later firmly established by careful genetic research. Mutations in the genome are now the recognized ultimate source of new genetic variation, the raw material upon which natural selection works.

Mistakes Happen

When cells divide, DNA is duplicated; repeated cell division leads to repeated duplication of DNA. A human being might have more than 40,000 useful genes, many more than this in some plants. During DNA duplication, mistakes happen, producing gene mutations. Mutation rates differ considerably from gene to gene, but generally only about one mutation occurs in a gene per 100,000 cell divisions. The gametes—eggs and sperm—hold genetic material that passes to future generations. The DNA in a single human gamete, if laid out end to end, would reach up to 6 feet in length. To fit into the gamete, the DNA is coiled, condensed, and packaged into the individual chromosomes.

The gametes, and their genomes, are the *germ cell* line. A mutation here is passed from generation to generation through the gametes. On average, about four such mutations occur per person, per generation. One in a million gametes may carry a mutation. All other cells of the organism, other than the gametes, are the *somatic cells* (*soma*, Gr., body)—up to 100 trillion in the human body. A mutation here stays within the particular individual, but may be deadly for the individual. Some types of cancers result from such somatic mutations, wherein mutations disrupt normal, controlled somatic cell division. The cell goes on a dividing rampage, dividing repeatedly without restraint. These cells pile up, forming a tumor.

Mutations affecting the germ cells are varied. Some act on single genes, others act on gene complexes, and some affect whole sets of chromosomes. Most are spontaneous errors in DNA synthesis and division, spoken of as "random" errors. Occasionally, radiation or chemicals act as **mutagens** that directly damage DNA and produce mutations.

Point Mutations

Point mutations in a gene result from one or a few alterations in the base-pair coding sequence of the gene's DNA, thus producing a new allele of the gene. At their simplest, genes are functionally integrated sections of DNA related to a particular trait. Recall from chapter 4 that the coding language of DNA involves codons—repeating triplets of bases within the long string of bases that make up the DNA. Three chemical bases code for one amino acid; amino acids chained together form proteins. By changing one or several of the bases, the affected codon changes; thereby, the particular amino acid specified by that codon may change; and therefore the sequence of amino acids assembled into a protein may change. Changing one of its bases can produce a new gene. This occurs most commonly by insertion, deletion, or substitution of bases (figure 8.2). Like crowding in line, *insertion* adds a nucleotide base into the normal row, bumping up all of the following bases in sequence by one. Because the decoding of DNA starts at one end and progresses sequentially in steps, three by three, to the other end, insertion can change not only its codon, but also all downline from it. *Deletion* does the opposite—it eliminates a base, but has the same effect of changing the reading of downline bases. *Substitution* results in just that—replacement of one nucleotide base by another, changing only one codon.

Sickle-cell anemia, a heritable blood disease, is an example of substitution. A defective hemoglobin

MUTATION	EXAMPLE RESULT
No Mutation (A)(B)(C)	Normal B protein is produced by the *B* gene.
Point Mutation Base substitution Substitution of one or a few bases (A)(B)(C)	B protein is inactive because changed amino acid disrupts function.
Insertion Addition of one or a few bases (A)(▮)(C)	B protein is inactive because inserted material disrupts proper shape.
Deletion Loss of one or a few bases (A)(C)	B protein is inactive because portion of protein is missing.
Changes in Gene Position Transposition (A)(C)(B)	*B* gene or B protein may be regulated differently because of change in gene position.
Chromosomal rearrangement (A)(C) (B)	*B* gene may be inactivated or regulated differently in its new location on chromosome.

FIGURE 8.2   Mutations in the Genome   Point mutations can occur through substitutions (change in bases), insertion (introduction of bases), or deletion (loss of bases) within the DNA. Major transposition of DNA segments can produce chromosomal inversion. Segments of DNA can be rearranged to new locations and even to other chromosomes, producing chromosomal translocation.

molecule is the result. Hemoglobin, contained in red blood cells, transports oxygen to active tissues. The large, defective hemoglobin molecule causes red blood cells to crumple—become sickle-shaped, making it difficult for them to speed through blood vessels. The misshapen red blood cells collect at critical spots, forming clots and causing pain; the warped cells have a shortened life span, which depletes their numbers and produces anemia (low numbers of red blood cells). The defective hemoglobin molecule is

made up of 574 amino acids, only one of which is improperly coded—valine in place of glutamic acid (figure 8.3). This error, in turn, can be traced back to a single base substitution in DNA. Such a single point mutation produces a new allele that manufactures a defective hemoglobin molecule, with a cascade of unfortunate consequences.

Gene Duplication

New alleles provide new ways to express old genes. Some mutations lead to new genes with new possibilities. One such method is gene duplication, which results from a mistake. As homologous chromosomes pair up during meiosis, crossing-over begins, but an

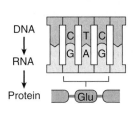
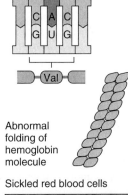
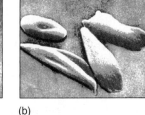

FIGURE 8.3   Sickle-Cell Disease   A single base change in DNA results in a change in one RNA codon, producing a protein with one substituted amino acid. But this single change is enough to disrupt the normal action of hemoglobin and, hence, of the red blood cells. (a) Normal. DNA codes via RNA for the amino acid, glutamic acid, one of many proteins in the hemoglobin molecule. (b) Sickle-cell disease. A single base change in DNA codes via RNA for a different amino acid, valine. But this replaces a critical amino acid that is important in proper folding of the hemoglobin molecule; the resultant defective hemoglobin molecule produces sickled red blood cells.

Consider This—

Sickle-Cell Anemia: Disease against Disease

In many ways, sickle-cell anemia is a very curious condition. The malformed red blood cells and anemia are most severe in individuals homozygous for the recessive mutant allele. What is curious is that even heterozygous people also get some sickling. Such an adverse allele clearly has an adverse effect on fitness and should have been eliminated. Yet it is the most common inherited blood disorder in the United States, affecting about 70,000 Americans annually; most are African Americans, where it strikes 1 in 500. This rate is unexpectedly high. Usually, the overwhelming majority of substitution changes create alleles with little advantageous effect. Amino acid sequences have been under selection for millions of years, so it is no surprise that a substitution today brings no new benefit. There must be some counterbalancing benefit

to the sickle-cell disease that accounts for its relatively high retention rate in humans.

As it turns out, people heterozygous for the gene—one copy of the normal allele and one copy of the mutant allele—enjoy some resistance to malaria. Apparently the infecting malaria parasite promotes sickling, prompting early elimination by the liver of the misshapen red blood cell and its resident parasite within. Where malaria is an environmental risk, the mutant allele is beneficial, partially balancing the disadvantages of the disease in oxygen transport. In such environments, the heterozygote enjoys benefits over both homozygote conditions: homozygous dominant (susceptible to malaria) or homozygous recessive (severe sickle-cell anemia). This is unusual, but the benefits of the heterozygote over homozygotes account for the unusually high incidence of the mutant.

error occurs wherein unequal crossover produces redundant sections of DNA on chromosomes (figure 8.4). The chromosome with the duplicated section of DNA now includes repeated copies of the gene (or genes) transferred during unequal crossover. Either gene can accumulate mutational changes without impairing normal function, because its twin gene is present to function normally. This seems to have occurred in normal human hemoglobin. The hemoglobin molecule is made up of six subunits, but only four are present at any particular stage in the life cycle of the individual. The four particular subgroups participating at any one time change during pregnancy, at birth, and within the young infant. The changing assortments of subunits reflect hemoglobin adjustments to different physiological demands of oxygen transport at these different ages. During vertebrate evolution, the gene presiding over hemoglobin manufacture apparently duplicated, and then duplicated again, giving us the six subgroups. The six are almost identical in their nucleotides, but provide differences in subunits that make up the hemoglobin molecule, and thereby fine-tune the ability of this composite molecule to

carry oxygen under different demands at different ages in the life of the human.

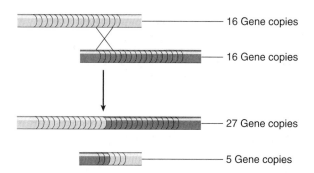

FIGURE 8.4   Unequal Crossing-Over   During meiosis, synapsed chromosomes occasionally pair out of register with each other, and crossover then occurs between nonhomologous sections. As a result, genes are duplicated on one chromosome, and deleted on the other. Unequal crossover events are thought to produce gene duplication in eukaryotic evolution, providing new possibilities for gene function.

Chromosomal Mutations

Large chunks of DNA can become flipped around within their chromosomes, reversing the sequence of genes within; this is *chromosomal inversion*. Sometimes pieces of chromosomes get transplanted to other chromosomes, a result known as *chromosomal translocation*. Both are quite disruptive, in particular because of the resultant difficulty in pairing homologous chromosomes during meiosis. But clumps of genes within an inversion sometimes act as a "supergene" and are inherited as a unit. Occasionally, such supergenes can bring major benefits. Within fruit flies, some inversions become more common in colder climates, and less so in warm climates, forming a gradient within populations from cold to warm regions. The fruit fly inversions contain specific combinations of alleles that function especially well in cold, wet weather, apparently favoring larger body size and acclimation to the temperature.

Another type of chromosomal mutation is **polyploidy**, wherein whole sets of chromosomes may be added. Commonly, the diploid ($2n$) number may double, forming a tetraploid ($4n$), and all double again forming an octoploid ($8n$). This is rare in animals, but common in plants. Polyploidy addresses a problem arising when part of a chromosome duplicates. Partial chromosomal duplications may survive in the individual plant where they first occur, but when gametes are formed, the duplicated chromosomal segment may lack a partner section of chromosome with which to pair during meiosis. This partner mismatch in turn produces defective gametes with an odd number or arrangement of chromosomes. Upon fertilization, this incomplete chromosome will not entirely restore the normal diploid number in the offspring, and consequently offspring are usually sterile. But by duplicating chromosomal number, polyploidy solves this problem of finding matching, homologous chromosomes. By first doubling, each chromosome automatically gains a homologous partner with which to pair during meiosis. For example, one of the most frequent errors leading to polyploidy occurs during meiosis where diploid ($2n$) gametes are formed. If the plant contains both male and female reproductive structures and self-fertilizes, then tetraploid ($4n$) offspring result. In turn, these new offspring ($4n$) form "haploid" gametes—now the norm for them, $2n$. When they mature, if they self-fertilize or cross with a similar

tetraploid sibling, then the chromosomal number is restored in their offspring—the new tetraploid number, $4n$. A breeding population of tetraploids is established. Polyploidy is a significant factor in plant evolution, especially in ferns and flowering plants wherein the majority of species are polyploid, evidence of past speciation via polyploid duplication of chromosomes.

It is not known why polyploidy should be so prevalent in plants generally, and in ferns and flowering plants in particular. New varieties and even species may form quickly, but of course each must meet the discipline of natural selection and find adaptive favor. From the introduction of this chapter, recall Hugo DeVries and the flower that inspired him, the evening primrose. The variety of primrose that he noticed in the abandoned field turns out to be a polyploid, or partial polyploid, of the primrose. Of course, he guessed or knew nothing of the sort at the time. It was a variety that appeared suddenly. A contemporary of his, William Bateson, would be first to notice another type of mutation, even more bizarre.

Hox Genes and Their Kingdoms

We owe the term *homeotic* to William Bateson (1861–1926) and to his interest in biological variation. He noted that normal body parts of animals and plants are often switched, transforming a part into the likeness of another, producing odd varieties. For example, he observed that, on occasion, the stamens of a flower transformed into petals. In 1894, he called such varieties *homeotic* (*homeo-*, same; *-otic*, condition) *mutants*. A more recent example comes from fruit flies. The repeating body segments of a normal fly are clumped into three body regions—head, thorax, and abdomen. The *head* includes eyes, mouth parts, and sensory antennae; the *thorax* has wings, legs, and haltere (balancing organ); the *abdomen* holds most of the body organs but lacks legs, wings, antennae or other appendages. Occasionally, in one generation, an abrupt, transforming mutation occurs. Close up, the homeotic mutant looks like it stepped out of a science fiction movie. A leg replaces the antennae on the head, or a second wing-bearing segment is added to the thorax, giving the mutant two pairs of wings. One body part is replaced by another.

Today we know that such major changes are due to **homeotic genes**—master gene switches that bring under their command legions of secondary genes responsible, in turn, for the formation of body parts. Although first worked out in arthropods (fruit flies in particular), similar homeotic genes have been found throughout the animal kingdom and even in plants and fungi (yeast). Although sometimes restricted to vertebrates, the term *Hox* **genes** is now more commonly used to embrace all these homeotic genes, wherever they occur. I will do so here. Before looking at the details of *Hox* gene action, we first need to understand the context in which they act.

Egg to Adult

The egg is one cell, the adult is billions of cells. To get from egg to adult, repeated cell division must occur, beginning with fertilization. Initially, division is restricted to cleaving the egg, but eventually proliferation of dividing cells contributes as well to growth in size of the embryo (figure 8.5). Each somatic cell formed by division contains an equivalent and full complement of DNA.

Because all cells have the same set of DNA instructions, any particular cell anywhere in the embryo could form muscle or nerve, or contribute to an arm or leg. But these cells and parts cannot appear randomly, or the embryo will be a scramble of bits and pieces in odd places. Arms must develop in the front, hindlimbs at the back; eyes must be on the head, and in fact the head must be on the front end, and so on. Placement and appearance of body parts must sprout in the embryo in the right positions. Organization is required. This organization begins by establishing basic body symmetry—front to back, top to bottom. Formally, a **polarity** is established in the young embryo wherein anterior and posterior ends (front and back) and dorsal and ventral (top and bottom) regions are delineated. Usually this is done through chemical gradients, where distinct chemicals are concentrated at one region and decrease toward the other, as for example, from front to back. Such chemical gradients, along with other chemical information, provide *positional information* within the embryo. The chemicals act as guideposts directing the subsequent positioning and placement of parts. By setting up this axis early, it is in place as a blueprint or chemical scaffolding to guide ensuing placement and building of body parts. In some animals, *Hox* genes actually turn on to set up body polarity; in others, polarity is established in the unfertilized egg. By whichever means, positional information is set up early, ready to direct placement of subsequent embryonic body parts and events.

Shaping Up: Positions and Parts

With the body polarity in place, the embryo can now be built, and most of the *Hox* genes work in this embryonic environment. Positional information within the embryo and environmental cues working through chemical intermediaries activate *Hox* genes, which in turn activate large banks of structural genes. *Hox* genes are *regulatory genes* that manage parts of the genetic program controlling structural genes; *structural genes* actually make products involved in building the phenotype. Particular *Hox* genes determine where paired wings form, where legs develop, or how a flower's parts are arranged. *Hox* genes are called master control genes because they may regulate 100 or more structural genes. Consequently, even a small change in one *Hox* gene can magnify into huge effects through the downstream structural genes over which it presides. There is amazing molecular similarity in *Hox* genes throughout the animal kingdom, further testimony here at the molecular level to the underlying evolutionary continuity between groups.

Hox genes are found in clusters with their loci lined up on chromosomes. The *Hox* genes in the cluster are in the same front-to-back order as the body part they affect (figure 8.6). A small change in one *Hox* gene in a cluster can produce large changes in the body region over which it presides, adding segments, or legs, or wings, or removing them.

Evolutionary Changes

Research continues. Many answers await research outcomes. But some promising correlations between *Hox* gene changes and major evolutionary events are apparent (figure 8.7). Major changes between major animal phyla are correlated with duplications in *Hox* genes or an increase in the number of *Hox* genes (figure 8.7a). The number of body regions over which a *Hox* gene presides may expand, thereby adding segments, or change the character of typical segments (figure 8.7 b, c). Through mutations that change downstream gene action, parts on segments are added or eliminated (figure 8.7 d).

Hox genes are elegant and complex. They are highly conserved anatomically (nucleotide sequences) and uniform in their expression (regulatory genes). What seems to have evolved is how they are activated and how downstream target genes respond. Research is turning up a more complex story. Apparently, some

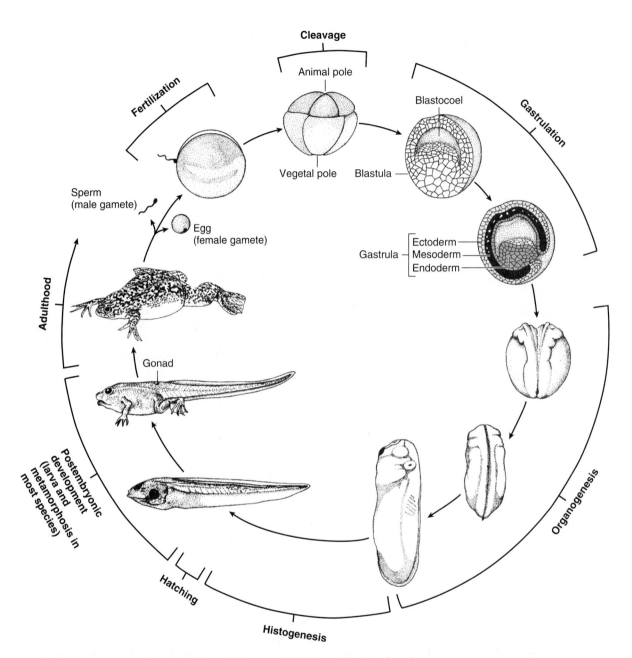

FIGURE 8.5   From Single Cell to Millions of Cells—Life Cycle of a Frog   A sperm fertilizes the single-celled egg, and cell division (cleavage) begins, leading to a multicellular blastula with a fluid-filled core (blastocoel). Major rearrangements (gastrulation) of formative cellular layers (ectoderm, mesoderm, endoderm) lead next to an embryonic stage wherein these formative embryonic cells become arranged into organs (organogenesis) and specific tissues (histogenesis). Upon hatching, the larva feeds and grows further, eventually undergoing a major anatomical change (metamorphosis), becoming a juvenile and then an adult frog, which reproduces to repeat the cycle.

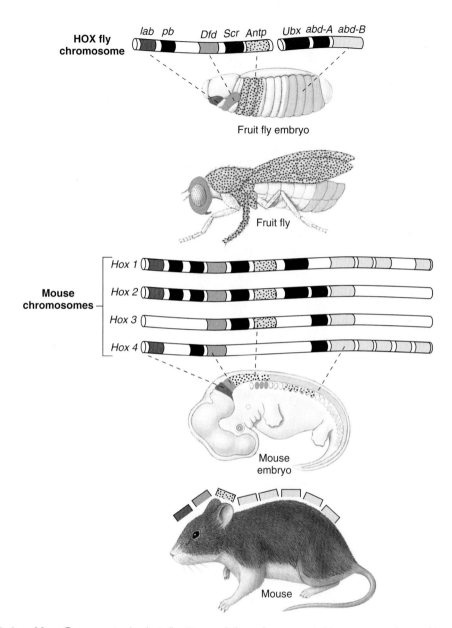

FIGURE 8.6 *Hox* **Genes**   In the fruit fly *(Drosophila melanogaster)*, *Hox* genes are located in clusters on a single chromosome, the HOX fly chromosome. In the mouse *(Mus musculus)*, similar genes are located on four chromosomes. In the fly and mouse, these genes control the development of front-to-back parts of the body.

Hox genes are turned on and off repeatedly during embryonic development, responding to the changing chemical and anatomical conditions within the developing embryo. Not only do *Hox* genes simultaneously turn on legions of structural genes, but some can directly and selectively control single, individual downstream genes as well. *Hox* genes turned on at one stage in embryonic development may be turned on again later but produce a different effect. *Hox* genes and their triggers may remain more or less the same, but

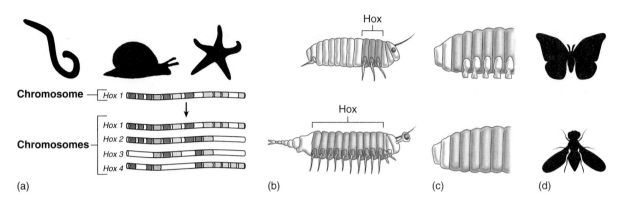

FIGURE 8.7   Evolutionary Changes Via *Hox* Genes   Several major changes are thought to be based on changes in *Hox* genes and in their pathways of control of structural genes. These include changes in the number of *Hox* genes producing phyla-level changes (a), broad changes of *Hox* expression over body regions (b), local changes of *Hox* expression (c), and changes in regulation of downstream genes or in function, here changing the second-segment wings of a moth or butterfly into the haltere of flies (d).   (After Gellon and McGinnis, 1998)

downstream tissues respond differently. Within flies, the pair of halteres, riding on the thoracic segment behind the single pair of wings, is apparently a modification of the wings that occupied that position in ancestors (figure 8.7 d).

OVERVIEW

Even if big mutations occur, the discipline of natural selection still rules. Gene mutations set the boundaries of variation, but natural selection culls out the unworkable, the maladaptive, the disadvantageous. Variation and selection are partners producing descent with adaptive modifications. Variation produces novelties that must meet the demands of the environment. A mutation producing featherless, or nearly featherless, birds may or may not be advantageous. If occurring in a bird that depends on flight to escape predators, then the featherless trait is highly disadvantageous and individuals carrying the mutation are likely lost from the population. On the other hand, if the featherless trait occurs in a bird on a remote island that is predator-free, then it may be advantageous and spread within the population. Variation arises from the basics of normal meiosis, and from the variation that comes from mixing of new combinations within the process of sexual reproduction.

Variation also arises from mutations within the genome. With millions of years to experiment, to shape, and to refine organisms, it is not surprising that new

mutations usually are deleterious. They disrupt a fine-tuned and harmonious working system. But if large-scale morphological changes are to occur, then major gene changes occur from time to time. The different groups of animals are quite diverse. Yet underlying the diversity is a shared, ancient structural blueprint based on genetic modifications at points, in chromosomes, and within master genes. The great structural similarity, at the level of the genes, testifies again to the solid evolutionary relationship among all animals. But this common genetic blueprint also suggests the genetic basis of major adaptive changes. Both large and small, variation produces the opportunity for evolutionary descent with modification. Phenotypic changes, based on underlying heritable genetic variation, may collect within a population and lead to divergence of groups and formation of new species. This is the process of speciation, which we will turn to in chapter 9.

Selected References

Cameron, R. A., K. J. Peterson, and E. H. Davidson. 1998. Developmental gene regulation and the evolution of large animal body plans. *Amer. Zool.* 38:609–20.

Gellon, G., and W. McGinnis. 1998. Shaping animal body plans in development and evolution by modulation of Hox expression patterns. *BioEssays* 20:116–25.

Kalthoff, K. 2001. *Analysis of biological development.* New York: McGraw-Hill.

Learning Online

Testing and Study Resources

Visit the textbook's website at **www.mhhe.com/ evolution** (click on the book's title) to take advantage of practice quizzing, study/writing tips, timely news articles, and additional URLs for research on the topics in this chapter.

9

"Paleontologists, conditioned to vertical thinking, considered speciation to be the change of a phyletic lineage over time. But since there is a steady extinction of species, where do the new species come from? This has been the problem from Lamarck and Lyell [geologist] on. The answer which Darwin found was that species not only evolve in time but also multiply."

Ernst Mayr, evolutionary biologist

Speciation

INTRODUCTION

It seems odd at first that in his book *The Origin of Species* Darwin devoted almost no space to the actual steps leading to the origin of species. By implication, natural selection favored individuals with superior characteristics against the environmental backdrop in which they served. The consequence—descent with modification—led to change through time, producing new species in the process. We can understand Darwin's light touch on the actual steps leading to new species if we recall his overriding preoccupation with the faulty ideas of geologic time and inheritance that stood in his way. Further, little help came from the discipline of population biology—the study of how individuals within a local community interact reproductively—because population biology was mostly a product of the twentieth century, when the genetic composition of a community became better studied and understood.

Today, the term **speciation** refers to the process by which new species arise. As to the term *species*, it is useful to think of this as a concept with various definitions serving various purposes. The *species concept* expresses one's view of the role of a species in nature or the simple operational method used to delineate it as a matter of convenience.

The species concept helps us recognize evolutionary events in several ways. An inventory of the many different kinds of plants and animals documents the great diversity of life and implies that a diversifying force or phenomenon is responsible. This leads us into a search for evolutionary mechanisms that might produce just such *numbers* of species. But it is not just the presence of species that betrays a diversifying process. The pattern of speciation implies something about the *pattern* of natural selection. For example, there are generally more species in tropical regions than in temperate regions. Does this imply that natural selection acts differently in one region than in the other? Species—their numbers and pattern of distribution—are an outward manifestation of the underlying processes that produce them. Consequently, the definition of species we adopt will affect our view of species in nature and of the evolutionary issues inviting our attention.

SPECIES DEFINITIONS

The definition used to define a species depends upon the problems faced in identifying a species and the use one wishes to make of the term *species*. No single definition exists that applies with equal utility to both sexually and asexually reproducing individuals. One of the most important definitions, applied to sexually reproducing individuals, views a species as a reproductive community.

Biological Species

A **biological species** is defined as a reproductively isolated community in which all individuals potentially or actually interbreed amongst themselves but are genetically isolated from other groups. Such a species is a self-contained genetic community. This definition includes two important qualifications: The members *potentially or actually* interbreed. This ensures that members distantly placed geographically are included. Today, for example, the geographic range for cougars (mountain lions) is much restricted. But at one time cougars ranged from Alaska, through North America, Central America, and to the southernmost reaches of South America. It was quite unlikely that an amorous male cougar in Alaska would meet and mate with a receptive female in Argentina. But *potentially* they could successfully breed, if they met, and so they are considered to be members of the same species by this definition. Note also the criterion of *interbreeding*, central to this definition. When the ability to interbreed is used as the criterion of membership in a species, we avoid the use of arbitrary or subjective judgments to demarcate one species from another. Individuals themselves supply the answer: If individuals can potentially interbreed, then they define themselves as members of the same reproductive community (species).

The advantage of the concept biological species is that it defines a species on the basis of criteria important to their evolution: reproductive isolation, a measure of the amount of exchange (or not) of genetic variation. Arbitrary decisions are kept to a minimum, as the breeding behaviors of the members of a species self-define the boundaries of their own species. Occasionally, quite different species do successfully interbreed, such as wolves and dogs. But even if exceptions are allowed, the criterion of interbreeding is also the limitation of the concept. Rarely are time and resources available to study each species' interbreeding in detail. This would be time-consuming and usually impractical to complete. Consequently, in practice most species are defined by other criteria.

Morphospecies

Individuals that look alike (similar phenotypes) are placed together as a **morphospecies**, or morphological species. Contemporary species exhibit "natural" breaks in anatomical appearance (figure 9.1a). We can easily note the differences among a white-tailed deer, an elk, and a moose. Most species are defined on the basis of phenotypic similarities and distinguished from other species on the basis of their collective differences.

The advantage of this concept is that it can be applied more easily than one based on reproductive habits alone. Most sexually reproducing species are recognized today by the diagnostic criterion of morphological similarity. But the practical application of this concept requires some arbitrary decisions, especially if the individuals are similar in appearance.

Paleospecies

Suppose we had a relatively complete series of fossils, such that there were no "natural" breaks in the appearance between extinct and extant (living) individuals, and extant forms were connected with a complete and unbroken series of intermediate forms back to a common ancestor (figure 9.1b). How then would we identify species boundaries? Paleontologists, in particular, face this problem. One solution is to use the continuous sequence of change itself as the basis for defining a species. Specifically, a chronological series of similar forms is a **paleospecies** (chronospecies).

Agamospecies

Certainly, these definitions do not exhaust the ways to define a species and the ways to use a particular species concept in biology. Many scientists today prefer to measure species by the percentage of genetic relatedness rather than by shared phenotypic similarity. This is mostly an operational convenience used to delimit species **taxa** (sing., **taxon**), a related group of organisms. The biological species concept applies only to sexually reproducing organisms, but many plants and animals reproduce only asexually. Some plants, for example, propagate through vegetative reproduction. Such asexually reproducing organisms may constitute an **agamospecies** based on the amount of

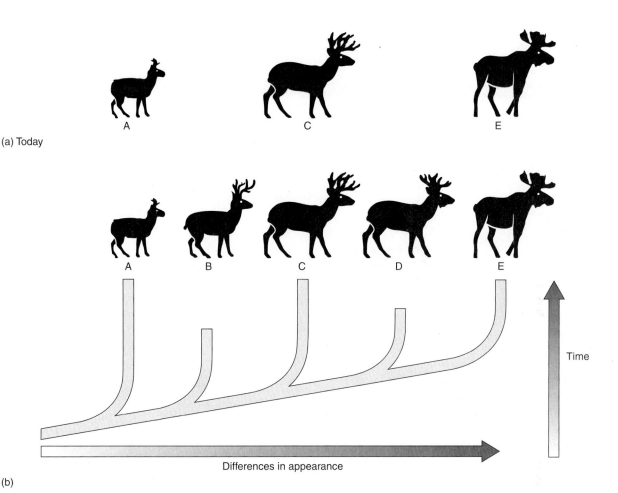

(a) Today

(b)

FIGURE 9.1   Defining Species   (a) Morphospecies. Viewed today, at one moment in time, species A, C, and E are clearly distinct, demarcated by current natural discontinuities between them. (b) Paleospecies (chronospecies). Viewed historically, through time, discovered fossil intermediates (B and D) fill in the missing gaps above, giving us a more or less continuous series with no obvious discontinuities between them.

variation and gaps in the variation of phenotypic features, including similarities in molecular traits.

For our purposes, we shall use the biological species concept, but be prepared to see other biologists employing alternative concepts in other contexts. With these ideas in mind, let's get back to our main purpose—namely, identifying how new species arise.

THE PROCESS OF SPECIES FORMATION

New species are not inevitable. There is nothing in natural selection itself that automatically churns out new species, like some internal engine spontaneously throw-

ing out new species to populate the world. Evolution does not "cause" speciation; external conditions must be right for new species to form. Four steps characterize most processes of speciation wherein a single ancestral species gives rise to new descendant species (figure 9.2).

Four Steps to Speciation

1. *Single population.* The ancestral species is a single reproductive community—a single species. All members actually or potentially interbreed and are genetically isolated from other species.

2. *Barrier develops.* Subgroups of the ancestral species become divided, separated from each other

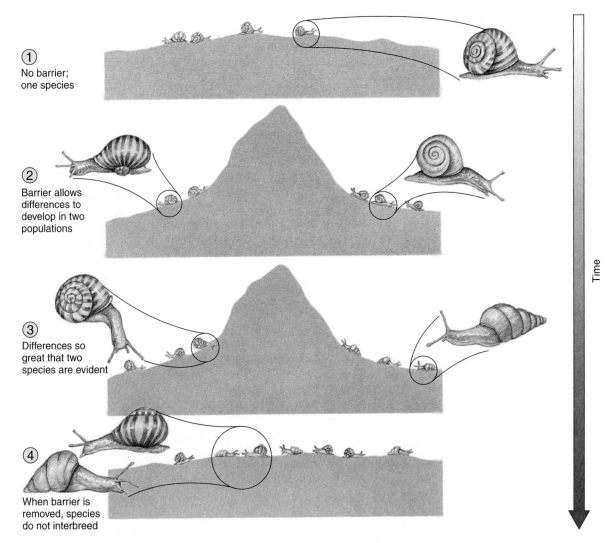

① No barrier; one species

② Barrier allows differences to develop in two populations

③ Differences so great that two species are evident

④ When barrier is removed, species do not interbreed

Time

FIGURE 9.2   Speciation   Four steps lead to speciation: First, a single species is an interbreeding reproductive community. Second, a barrier develops, dividing the species. Third, separated into different habitats, the divided populations become differentiated through the accumulation of differences. Fourth, so different have the separate populations become, that when the barrier disappears and they overlap again, interbreeding does not occur.

by a barrier. It is perhaps easiest to think of this barrier as a physical barrier, such as a mountain range or wide body of water. However, other types of barriers can effectively divide a species into separate, ecologically isolated populations. For example, some members of a species may prefer different regions of the forest canopy and lose frequent ecological contact with other members of the species living on the forest floor. The separated populations are said to be **allopatric**, meaning that their ranges do not overlap.

At this early point, the species is simply divided. Members still can interbreed if brought together. At this point in the speciation process, there is no particular advantage or disadvantage to interbreeding between the separated populations. Natural selection does not now see a new species in the offing and compel the individuals into greater and greater genetic isolation until the populations become reproductively incompatible. An external barrier simply divides the populations, isolating them from one another. One way to speak of this is to say that **gene flow** between the populations is interrupted. Variation arising within one population is not shared with the other via interbreeding.

3. *Differentiation.* Now divided, each population experiences a slightly different environment that delivers its own demanding suite of selective pressures. Different features will be successful in each location. Through successive generations, the accumulation of the character changes driven by local conditions leads to genetic distinctiveness of each separate population. Traits that served well in the ancestral group may now find disfavor in the newly divided population that is adapting to the different environment. Each population undergoes independent evolution, and if the barrier persists and enough time passes, the separate populations become genetically distinct. Differentiation has occurred.

4. *Barrier disappears.* Whatever the barrier, it now disappears. The mountain range is weathered away; the lake dries up; separate stands of forests merge. Whatever divided the populations now withers away or vanishes, and the populations come into contact. The populations are now said to be **sympatric**—overlapping with each other geographically. But by now they have diverged so substantially that they are genetically incompatible, unable to produce viable offspring even if the occasion arises. Where there was once a single species, now there are two.

Isolation and Diversification

Geographic isolation is usually a prerequisite to speciation, but not always. Geographic isolation results from division—a population is split into separate groups by some barrier, as described earlier—or from isolation arising through dispersal, the colonization of a distant island or the invasion of an uninhibited region. Strong winds, unseasonable weather, or other chance events may scatter individuals to distant habitats and separate them from the source population from whence they derive. Darwin saw this on the Galápagos Islands. In separated populations, gene flow is reduced and local conditions prevail. Speciation that arises from geographic isolation, whether by division or dispersal, is called *allopatric speciation.*

However, geographic isolation does not always precede speciation. Speciation that arises without geographic isolation is called *sympatric speciation.* This is less common than allopatric speciation. One of the most common mechanisms of sympatric speciation (speciation in place) is through polyploidy in plants. A hybrid cross between two species may be sterile because the sets of chromosomes, derived from female and male of different species, do not find and therefore pair with one another (see chapter 8). But if a chromosome anomaly occurs, and the chromosome number of such a hybrid doubles, then this doubling automatically produces partners, effectively homologous chromosomes that can pair and produce viable gametes. This polyploidy—doubled chromosome numbers—permits the hybrid to breed with itself, but instantly prevents backcrosses with either parent; the parents have fewer chromosomes and lack the ability to properly pair with the hybrid. Instantly, a new species arises in the same geographic region as its "parents." About half of all flowering plants, including familiar wheat, cotton, tobacco, bananas, and potatoes, have a polyploid episode in their evolutionary history.

Accentuated Reproductive and Ecological Isolation

When divided earlier by a barrier, individuals in each population became genetically distinct. This genetic differentiation may have been accompanied by the evolution of distinctive courtship behaviors, habitat preferences, dietary fondnesses, and other traits specific to their environment. These differences can be carried forward as the barrier disappears and become the basis for continued separation from the other population. Each population is now its own species, genetically distinct and reproductively incompatible with the other. But mistakes can occur. Individuals that attempt to interbreed or actually interbreed with the other species incur a detrimental cost. They spend much time and energy preparing for offspring; if births ensue, parents expend costly care on hybrid offspring that may do poorly. Such interbreeding individuals would be selected against, and individuals that practice reproductive isolation would be favored. The result is the *active* selection now for **reproductive isolating mechanisms**—behaviors, physiology, and morphological features that prevent such wasteful interbreeding. The result would be enhanced **reproductive isolation**, now expanded to accompany the genetic isolation that evolved earlier.

Besides their genetic and reproductive distinctiveness, members of a species form an ecological unit, sharing similar ecological requirements. Once the barrier to interaction falls, the previously divided populations—now different species—come into competition for very similar resources. This favors the evolution of ecological separation, the development of phenotypic features that permit the species to avoid direct competition with one another. The result is that the species develop enhanced **ecological isolation** in addition to their reproductive and genetic isolation.

Note that the most important of the four stages is differentiation. The barrier brings about the conditions, but it is the differentiation of the separated populations that makes speciation happen. While the populations are separate, they are exposed to different selective forces. Distinctive features, genetically based, collect. Thus, by the time the barrier falls, the groups are already distinctive and genetically incompatible. But if they do attempt to interbreed, then now the interbreeding would be biologically disadvantageous. Selection now favors behaviors and other traits that enhance reproductive isolation.

REPRODUCTIVE ISOLATING MECHANISMS

Reproductive isolating mechanisms (RIMs) are obstacles to interbreeding between genetically distinct species. By such devices, members of the same species tend to interbreed only with each other, avoid breeding with members of other species, and thereby avoid producing hybrid offspring. Hybrids often carry an assortment of various traits from their parents, but consequently may not be well suited to the environments of either parent. Their fitness is low, and parents that attempt or succeed at interbreeding to produce such hybrids incur unfavorable reproductive costs. Examples of such reproductive isolating mechanisms can be grouped into those in effect before and those acting after fertilization of the egg. The fertilized egg is termed a zygote, hence reproductive isolating may occur before or after fertilization—prezygotic and postzygotic mechanisms (figure 9.3).

Prezygotic Mechanisms

If species occur in different *geographic* locations, they are physically separated from interbreeding. Hybridization may be avoided through *ecological* devices. Two species may be sympatric but follow different ecological patterns. The cottonmouth snake frequents fresh water, feeds on fishes and frogs, and is aggressive in temperament. The closely related copperhead snake tends to be terrestrial, also feeds on frogs but includes a higher proportion of rodents in its diet, and is more retiring. Different lifestyles reduce the chances of meeting and interbreeding. *Behavioral* differences, especially during courtship, can prevent hybridization. The various species of shorebirds, for example, tend to breed during the same times of the year, but actual mating is preceded by elaborate, species-specific courtship displays. A male may get whipped into a heated froth and display to a female. But if she is of another species, then his ardent display delivers no recognized erotic encouragement, fails to provide the behavioral signals needed to bring her into a similar state of amorous readiness, and interbreeding is avoided. *Seasonal* or *temporal* differences may reduce chances of interbreeding. Sagebrush commonly blooms and produces pollen in the spring; rabbitbrush flowers in late summer. Or consider green lacewings, a group of insects common in most of North America. Adults are green-bodied with transparent (lacy) wings and feed on other insects and on nectar, pollen, and other such foods. Their larvae, in particular, are large, voracious predators that stalk aphids and sometimes mites, rushing forward with great pincer jaws to dispatch such prey, earning lacewings the name "aphid lions." These larvae are such effective predators, that lacewing eggs are commercially sold and deliberately introduced into domestic gardens where, upon hatching, they become welcome biological controls on insect pests.

Two closely related species of green lacewings, one pale and the other dark, hybridize in the laboratory but apparently do not do so in nature. The absence of hybridization in the wild is due to seasonal differences in breeding—their mating seasons are offset and do not overlap—and to differences in location. The pale-green lacewing frequents deciduous foliage; it breeds in late winter and again in summer, with a break in between. The dark green lacewing is restricted to evergreen trees where its dark color provides camouflage; it breeds once in early spring.

Mechanical incompatibility is another isolating device. The interlocking copulatory equipment of some male and female insects acts like a lock and key, preventing hybridization between individuals of different species. Prevention of *gamete fusion*—eggs and sperm do not fuse and fertilize—offers another point during reproduction to deter production of hybrids.

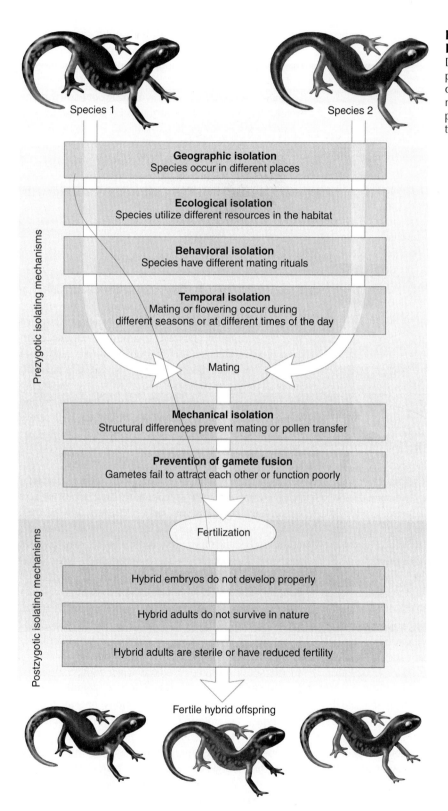

FIGURE 9.3 Reproductive Isolating Mechanisms (RIMs) Different mechanisms prevent reproduction between individuals of different species. These mechanisms may occur premating or postmating, as illustrated here with two species of salamander.

Species 1

Species 2

Prezygotic isolating mechanisms

Geographic isolation
Species occur in different places

Ecological isolation
Species utilize different resources in the habitat

Behavioral isolation
Species have different mating rituals

Temporal isolation
Mating or flowering occur during different seasons or at different times of the day

Mating

Mechanical isolation
Structural differences prevent mating or pollen transfer

Prevention of gamete fusion
Gametes fail to attract each other or function poorly

Fertilization

Postzygotic isolating mechanisms

Hybrid embryos do not develop properly

Hybrid adults do not survive in nature

Hybrid adults are sterile or have reduced fertility

Fertile hybrid offspring

Postzygotic Mechanisms

Occasionally, breeding between members of different species does occur but yields no reproductive benefits. Consequently, reproduction returns no procreative advantage, and the time and energy spent on courtship and gametes is squandered.

Further investment in an unpromising mating is often terminated by mechanisms that prevent continued reproductive development. The sperm, if successfully transferred to the female, might not fertilize the egg, or the zygote might soon die. Sometimes the crossbreeding produces a zygote that develops to term and is born, but this hybrid is not viable or may itself be sterile. A familiar example is the mule, the product of a cross between two species, a horse (female) and a donkey (male). (The reverse cross, horse [male] × donkey [female] is called a hinny—also sterile, and slightly smaller because the donkey mother is small.) For humans, the hybrid mule is a favored beast of burden, but it is sterile and can breed no offspring. To get another mule (or hinny), breeders must again start with a horse and a donkey.

Natural Selection and RIMs

Natural selection is the cause of divergence and differentiation, but may not be the direct cause of RIMs, whose presence tends to be *passive* and inadvertent. During speciation, members in the separated populations evolve different traits under the guidance of natural selection, adapting them to different ecological demands. Their separate differentiations produce ecological divergence. But once natural selection causes survival traits in these populations to diverge significantly, reproductive isolation is usually a side effect. If populations have been reproductively separated for a long time, then the accumulation of separate changes in the populations can make them incompatible once they again become sympatric. Natural selection has had no hand in actively and directly favoring RIMs. In these populations, the genetic and then reproductive isolation is just a by-product of their separate histories.

Occasionally, natural selection can be *active* and directly favor RIMs, especially after the two genetically distinct populations become sympatric. If populations come into contact before reproductive isolation is complete, then natural selection might be involved at this late stage in favoring the reinforcement or even the appearance of RIMs. If the species interbreed, adaptively inferior and less-fit hybrids may result, a quite unfavorable consequence. If reproduction fails, the time and energy invested in these compromised offspring are wasted. A whole breeding season may be lost in this futile effort. Further, if the populations, when previously allopatric, develop specific adaptations to particular habitats, interbreeding when sympatric could disrupt the favorable complex of local traits. Hybrids would be a mix of features, rather than the best of one or the best of the other. Here, selection would favor prezygotic mechanisms that prevent wasted expenditure of energy and effort on nonproductive breeding activity. This would also prevent the infusion of unfavorable traits into well-adapted local populations.

PATTERNS OF SPECIATION

The diversity of traits and the patterns of speciation in organisms are outward manifestations of the unseen processes responsible for the diversity and patterns of species formation. There is no built-in engine compelling natural selection to produce more and more species. Yet there are lots of species, millions and millions of them, distributed unevenly and in surprising configurations geographically. If we can understand these patterns then we can understand the processes.

Clines

A **cline** is a gradient or gradual change in a character within a species over a geographic area. Color or size can change progressively over the range of a species. Such character gradients, if genetically based, invite our attention because they may represent an early stage in the speciation process—namely, the stage when local populations start to exhibit adaptive adjustments to local conditions. One example is the graded tolerance of frog tadpoles to water temperatures from northern to southern populations (figure 9.4). In the north, tadpoles of the leopard frog (*Rana pipiens*) survive low temperatures but perish when temperatures reach about 27 ° C (80 ° F). In the south, tadpoles tolerate water temperatures up to about 34 ° C (92 ° F) but perish at temperatures below about 12 ° C (54 ° F). Populations intermediate in geography are intermediate in their tolerances. The north-to-south shift in this trait makes sense, as local environmental temperatures change from north (cold) to south (hot). Other environmental factors such as day length, seasons, humidity, foods, and predators change as well, so other traits might be expected to show clinal variation as well, reflecting local adaptations to local conditions through the range of the frog.

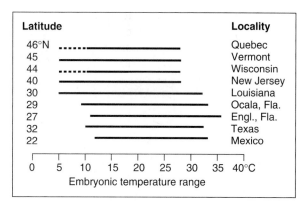

FIGURE 9.4   Clinal Variation   In the leopard frog *(Rana pipiens)*, tadpoles exhibit a range of temperature tolerances, generally enduring colder temperatures in higher (northern) latitudes and warm temperatures at lower (southern) latitudes.   (From J. Moore)

Even within this single species, populations tend to become more and more genetically distinct the farther apart they are along this north–south gradient. The biologist John Moore explored this experimentally (figure 9.5). He mated females from the north with males collected from populations progressively southward into Mexico. He scored genetic compatibility by the number of surviving hybrid embryos (tadpoles) from each of these crosses. Hybrid survival decreased, especially along the eastern part of the frog's range, the farther south the males were collected. Consequently, even within leopard frogs, there is some genetic separation between populations at the extremes of its range.

In fact, it turns out that the frogs were ahead of the scientists. When crosses between populations exhibited defective embryos, biologists suspected that the clinal variation had begun to lead to differentiation along a geographical gradient. This seems to have occurred (figure 9.6). Across their range, leopard frogs are externally similar in appearance. But within this overall range, four species are now recognized on the basis of distinctive mating calls. Apparently these mating calls are sufficient to reduce the incidence of interbreeding, and thereby avoid the embryonic defects this hybridization would produce.

Clines show us that within the full range of a species, there is a surprising degree of adjustment of local populations to local conditions. Some of this can be physiological. As the organism grows, it adjusts to the environment—for example, growing a thick coat of fur in cold winters or shedding hair to thin the coat in hot summers. But for many local populations, much of the

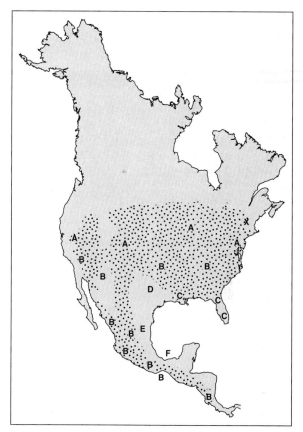

FIGURE 9.5   Reproductive Success   In the leopard frog *(Rana pipiens)*, eggs from females in the north were fertilized with sperm from males progressively farther to the south. The degree of embryo or tadpole abnormalities was scored, from *A* (normal young) through progressively more abnormalities to *F* (high death rate). This study was done by J. Moore in 1949. Today, this study and others prompt biologists to actually divide leopard frogs into subspecies or even different species.   (From J. Moore)

accommodation to the local environment is built in, part of the genotype and ready to go. Within a species, such a local population that is genetically adapted to a particular local environment is called an *ecotype*. Ecotypes, within a species, do not overlap geographically. Sometimes ecotypes exhibit distinctive external features, such as color or pattern or anatomy, and are thereby visually detected. Such a visually identifiable ecotype is termed a *subspecies*, a distinctive allopatric subgroup of the species. Frequently, such subgroups are not distinctively marked for visual identification with the naked eye, because their accommodation to the local environment is an internal adaptation, a hardiness to the local habitat. A cline is composed of such a geographic series of ecotypes, or subspecies.

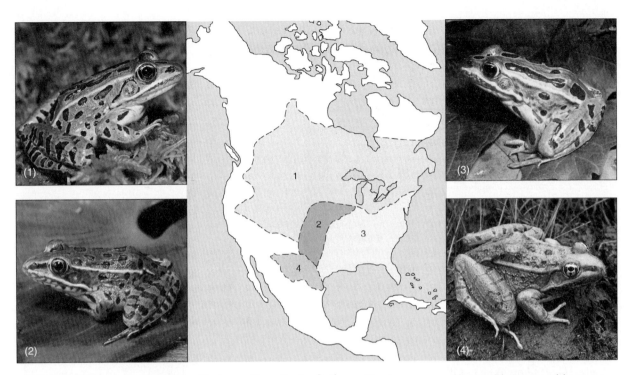

FIGURE 9.6   Leopard Frogs Today—Zygotic Isolation   The four groups of leopard frogs resemble one another closely in their external appearance. But early tests of interbreeding produced defective embryos in some combinations, leading biologists to suspect that these might be different subspecies or even different species. Research on males' mating calls indicates that the various groups differ substantially, and that such prezygotic behavior separates and reproductively isolates members of each group, producing four species: (1) *Rana pipiens*; (2) *Rana blairi*; (3) *Rana utricularia*; (4) *Rana berlandieri*.

Yarrow

If selection acts to adapt local populations to local conditions, then the local organisms should do best in that area and less well outside the local habitat. A study to test this hypothesis was done on yarrow. The yarrow (genus *Achillea*) is a member of the sunflower family. These perennial plants grow from sea level to high montane meadows at timberline. Using transplant studies, vegetative cuttings from the same plant can be tested in other environments outside its immediate habitat, and the cuttings' response to life in these transplanted locations scored. The study took a 200-mile west-to-east transect across the yarrow's range in California that ran from sea level to timberline in the Sierra Nevada mountains, crossing the lower coastal range of mountains and the San Joaquin Valley in between. Natural variation in yarrow height ranges from 6 feet (San Joaquin Valley) to a few inches (timberline) (figure 9.7a). The biologists used three different field growing stations within the yarrow's range at three different sites/elevations, west to east: Stanford (sea level), Mather (4,600 feet), and Timberline (10,000 feet). Biologists collected cuttings from four populations along this transect and planted the cuttings in these three field growing plots.

The growth responses of transplanted yarrow are shown diagrammatically in figure 9.7b. Individuals from sea-level populations did progressively less well at higher elevations, perishing when planted at timberline. Individuals from timberline populations survived plantings at lower elevations, but remained small, never attaining the height of low-elevation individuals. Within the extreme ranges, west to east, ecotypes have developed within yarrows. Certainly physiological processes still affected growth, such as adjustments to temperature, rates of photosynthesis, and moisture. But each of the populations adapted to local conditions, with consequences for the genotype. In particular, hardy phenotypes for montane environments have been selected at timberline, and specific phenotypes have been selected for individuals at sea level. The result is genetic distinctiveness—ecotypes—within local populations that result from the local selective pressures.

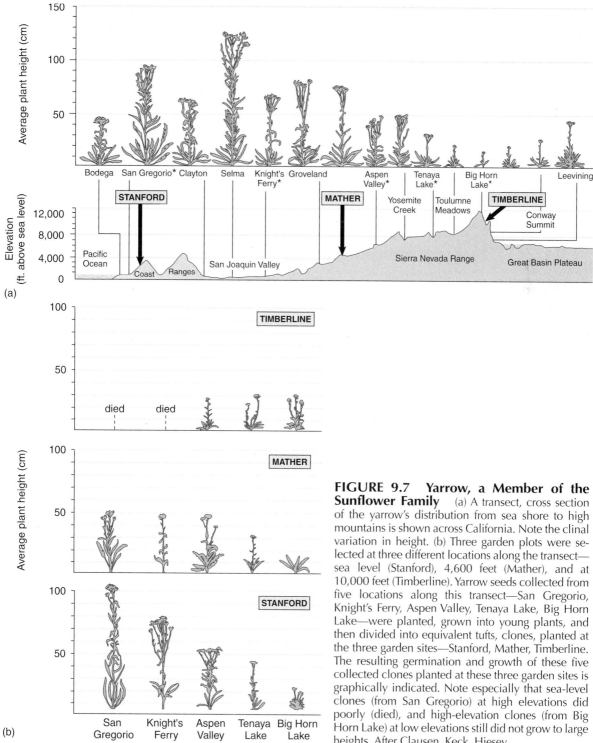

FIGURE 9.7 Yarrow, a Member of the Sunflower Family (a) A transect, cross section of the yarrow's distribution from sea shore to high mountains is shown across California. Note the clinal variation in height. (b) Three garden plots were selected at three different locations along the transect—sea level (Stanford), 4,600 feet (Mather), and at 10,000 feet (Timberline). Yarrow seeds collected from five locations along this transect—San Gregorio, Knight's Ferry, Aspen Valley, Tenaya Lake, Big Horn Lake—were planted, grown into young plants, and then divided into equivalent tufts, clones, planted at the three garden sites—Stanford, Mather, Timberline. The resulting germination and growth of these five collected clones planted at these three garden sites is graphically indicated. Note especially that sea-level clones (from San Gregorio) at high elevations did poorly (died), and high-elevation clones (from Big Horn Lake) at low elevations still did not grow to large heights. After Clausen, Keck, Hiesey.

Ring Species

An odd thing happens when a species ends up chasing its tail. Occasionally a clinal variation within a species does not occur linearly (from north to south, for instance) but bends around a geographic barrier and the extremes of the species meet. Such a species is referred to as a **ring species (Rassenkreis)** because it encircles the geographic barrier but is not actually divided in two by the barrier. Each local population within the species interbreeds with adjacent populations. However, and this is what is odd, the populations at the extremes that meet are reproductively incompatible (figure 9.8).

The various times of glacial advance during the Pleistocene cornered many plants and animals in **refugia** (sing., *refugium*)—small, persistent pockets of available habitat. This happened to herring gulls, which subsequently dispersed out of these pockets to take up residence in habitats that opened as melting glaciers retreated. Today, herring gulls, or really a complex of related gulls, form a ring species living in open waters encircling the uninhabitable Arctic. They stretch, through adjacently interbreeding populations, from Western Europe, across North America, and from Alaska to Siberia; but when they again reach Western Europe, this end of the

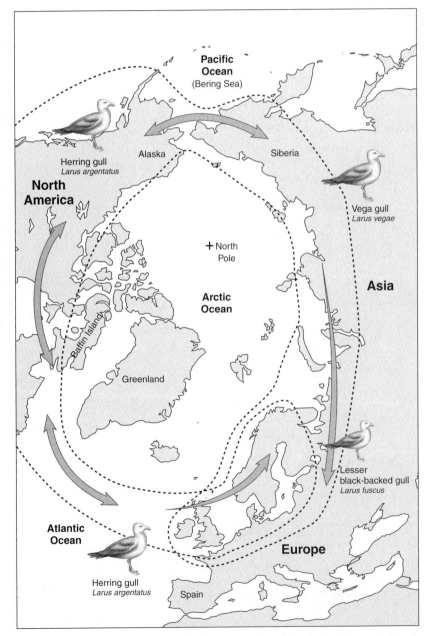

FIGURE 9.8 Ring Species—Herring Gulls As glaciers retreated, herring gulls (*Larus argentatus*) were released out of a north Pacific refugia spreading one way across North America and into western Europe; and spreading in the other direction across Alaska into Siberia. From Siberia, as the herring gull now extended its range further across Asia, it tended to differentiate, producing a subspecies (or species by some ornithologists) such as the vega gull (*Larus vegae*) and farther west the lesser black-backed gull (*Larus fuscus*). Eventually its current circumpolar distribution became established (dashed lines). Adjacent subspecies interbreed (solid arrows), but where the ends of the circular range of the herring gull meet and overlap in Europe, there is very little interbreeding. (Simplified originally from Mayr, 1963)

circle has become differentiated into a smaller gull that picks up some distinguishing darker markings, now recognized as the lesser black-backed gull. Where the herring gulls and lesser black-backed gulls meet along the coasts of Europe, they remain reproductively isolated, practicing differences in ecology and behavior so that they rarely hybridize in this region of overlap (figure 9.8).

The salamander ensatina (*Ensatina eschscholtzii*) has a range from Canada to southern California, where the dry Central Valley divides the species into coastal and inland arms. Populations along the coast tend to be

uniform in color; inland populations tend to be patterned. Where the terminal ends of these two arms meet, south of the Central Valley, the salamanders are so differentiated that little, if any, significant interbreeding occurs. Yet both types are members of the same species, but they are recognized as a ring of ecotypes or subspecies (figure 9.9).

Parallelism and Convergence

The previous examples illustrate that different habitats produce different selective pressures and thereby different populations—ecotypes—within the same species. The reverse is also true. The same habitats produce similar selective pressures and, thereby, organisms with similar general characters. Where closely related species take on similar traits, the pattern is termed **parallelism**; where the species are distantly related but enter similar habitats, the superficial resemblance that may result is called **convergence**.

The marsupial (pouched) mammals of Australia superficially resemble many placental mammals found elsewhere (figure 9.10). Placentals such as the mole, anteater, mouse, lemur, flying squirrel, ocelot, and wolf have marsupial counterparts. Yet all of these Australian counterparts spring from a different mammalian group, the marsupials—which converged on similar designs in the pursuit of similar lifestyles.

The "true" cactus occurs in the New World, coming equipped with sharp thorns that discourage bites by large herbivores. But in Africa, other desert plants belonging to different families also have evolved convergent forms with long spines or short thorns and leafless stems (figure 9.11).

The occurrence of similar traits in different species is real enough, accounting for homoplasy (see chapter 6) and illustrating the consequence of similar selective forces upon different organisms. But when comparing organisms, the distinction between the two types, parallelism and convergence, is often arbitrary. As a convenience, we may just speak of convergent evolution to notice these similarities between distant groups.

Latitudinal Gradients of Species Diversity

Within major taxonomic groups, species are not evenly distributed across a geographic gradient. For example, fewer mammalian species occur in far northern latitudes but numbers increase toward lower

Consider This—

Flaming Retreats

For thousands of years, maybe more, the Great Plains of North America were subjected to fires at the end of scorching summers. Grasses and other low-growing plants flowered in the spring, set seed, then dried up in the heat of the summer sun. Lightning strikes set these tinder-dry plants ablaze, and great swaths of prairie burned each year. Although the plants burned, many of their fire-adapted seeds survived to sprout the next spring and replenish the prairie community. In an odd and incidental way, frequent summer fires swept away larger competing plant species, producing a fire ecology to which the grasses and prairie wildflowers were suited.

With the arrival of Europeans, most of the Great Plains was turned to farming, displacing the prairie plants with fields of wheat, corn, and other crops. The prairie grasses and wildflowers retreated into refugia, which were unsuitable for farming and therefore left to the specialized fire-adapted flora. One of these retreats was along railroad tracks, where an occasional spark from the train wheels set off fires in the brush adjacent to the tracks. This was just the sort of environment to which the prairie flora were adapted, and there they survived. Efforts to restore prairie followed. Today there are set-aside areas, often subjected to controlled burns, that provide deliberate refugia for the prairie flora.

FIGURE 9.9 Ring Species—Salamanders The ensatina salamander (*Ensatina eschscholtzii*) occurs from Canada to Southern California with interbreeding between adjacent populations through this range. In California's Central Valley—a dry, hot, lowland area—the species is divided into a coastal arm and inland arm. However, where these two arms of the species meet again in Southern California, interbreeding does not occur.

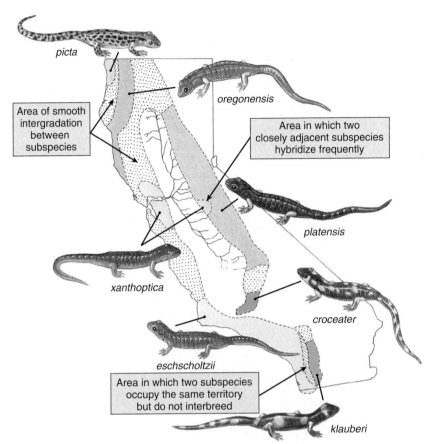

picta

oregonensis

Area of smooth intergradation between subspecies

Area in which two closely adjacent subspecies hybridize frequently

platensis

xanthoptica

croceater

eschscholtzii

Area in which two subspecies occupy the same territory but do not interbreed

klauberi

latitudes (figure 9.12). Notice that the numbers of species increase progressively along the latitudinal gradient from far northern regions toward the tropics. Alaska has 20 to 30 mammalian species; the lower continental United States has just under 100; in Mexico and Central America, there are well over 100. Not all groups show this temperate-to-tropic gradient in species numbers, but birds, bats, and many reptile species do. Why should this gradient exist in species numbers? Various suggestions have been offered. For example:

Evolutionary time. The northern parts of temperate regions have been free of permanent ice sheets since about 8,000 years ago, when retreating sheets of glacial ice opened up the temperate landscape. This has not been, by this view, enough time for speciation to fill up the recently exposed countryside with new species. By implication, the area today is unsaturated with species.

Climate stability. The swings in winter to summer climate extremes mean that a species must have a broad range of tolerances, to survive. Thus, few species are supportable in temperate regions.

Productivity. Northern areas, because of the tilt of the Earth on its axis, receive less sunlight annually, especially during winter months. This means less energy for photosynthesis and plant productivity and, indirectly, less food for herbivores and, in turn, for carnivores. Fewer resources translate into fewer species. In tropical regions, productivity is high; in temperate regions, by comparison, productivity is low. The differences translate into high and into low numbers of species, respectively.

Other ideas center on arguments that competition, predation, and the basic complexity of the forests place limits on the number of temperate species or favor diversification of tropical species. The issue is still

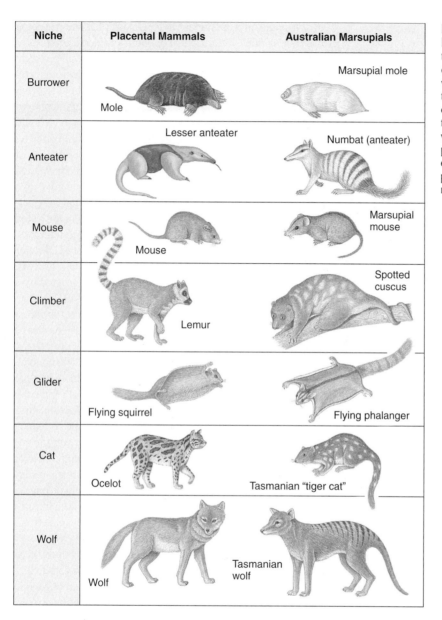

Niche	Placental Mammals	Australian Marsupials
Burrower	Mole	Marsupial mole
Anteater	Lesser anteater	Numbat (anteater)
Mouse	Mouse	Marsupial mouse
Climber	Lemur	Spotted cuscus
Glider	Flying squirrel	Flying phalanger
Cat	Ocelot	Tasmanian "tiger cat"
Wolf	Wolf	Tasmanian wolf

FIGURE 9.10 Convergent Evolution—Mammals Australian marsupials resemble placental mammals in the rest of the world. Within the relative isolation of Australia, the marsupials entered habitats similar to those of their placental counterparts elsewhere. Under similar selective pressures, Australian marsupials evolved features similar to those of placental mammals, but upon a marsupial theme.

Euphorbiaceae
Africa

Asclepiadaceae
Africa

Cactaceae
New World

FIGURE 9.11 Convergent Evolution—Cacti Different families of desert plants have evolved similar adaptations to the desert's dry, hot conditions—namely, succulent shoots with spines. Two such plant species are shown from Africa. The third, from the New World, is the endemic member of the true cactus family (Cactaceae).

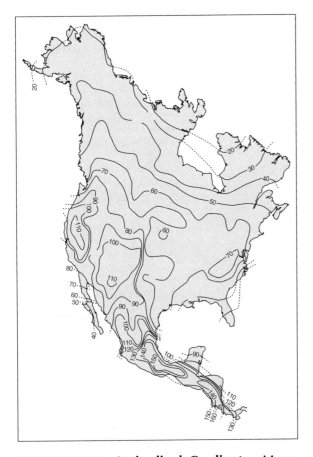

FIGURE 9.12   Latitudinal Gradients—Mammals   The numbers of mammal species, from high latitudes (north) to low latitudes (south), are shown along the lines. Note the general increase in the number of species from north to south across the various latitudes.

open, and we will not try to settle it here. But for us, species gradients hold a significance. The presence of latitudinal gradients implies an underlying process—some fundamental biological phenomenon, an unknown engine of species diversity, working in a progressive way from temperate regions to the tropics. And the sheer numbers of species, per se, provide an outward manifestation of this underlying phenomenon. Although not providing the answer directly, the graded changes of species abundances, temperate to tropical, are the external symptom inviting our attention to diagnose the underlying natural cause.

OVERVIEW

If you stop to think about it, it is curious that distinct "species" occur at all. Why aren't organisms simply dis-

tributed in continuous and uninterrupted interbreeding populations? Why would continuity of selection pressures not result in continuity of interbreeding populations? In fact, clinal variation seems to satisfy just such an expectation. Even more curious is that speciation should appear in patterns of abundance, such as latitudinal gradients. All is curious because speciation is not inevitable, nor should it be seen as an evolutionary "goal." Perhaps parceling up continuous populations into species is a consequence of patchy and parceled habitats. Perhaps the speciation process occasionally is self-propelled, driving its own completion. Perhaps the answers will be found in genetics. Clearly, speciation is still a subject with unanswered questions. Some things, however, we do understand.

When it occurs, speciation is a product of circumstance, not of active natural selection for the process. Gene flow between parts of a species is interrupted by a barrier. The separated parts of the species continue to evolve, but because they are separated and meet different environments, the subsequent evolutionary trajectories of each is different. By the time the initial barrier disappears, the separated populations have already differentiated. They have become genetically distinct and incompatible with each other, so that interbreeding in areas of overlap is disadvantageous. Now enters natural selection, in a larger way, by favoring reproductive isolation to reduce the disadvantages of ineffective breeding among the genetically distinct populations. Individuals attempting to or actually producing hybrids with members of other species incur a high reproductive cost, compared to those that do not. Relative to other individuals, those producing hybrids have a lower fitness. Distinct, but closely related and similarly designed, the two new species likely vie for access to similar resources, placing them in competition with each other. Ecological isolation is now favored. Speciation therefore is characterized by attainment of genetic isolation. Reproductive and ecological isolation follow, guided by natural selection.

Large populations hold large variation, shared through gene flow. This provides much upon which natural selection can act. But if the population is too large and overwhelming, then small, favorable variation is swamped through interbreeding before it finds adaptive favor. Not enough of the variants reach critical numbers. The isolating barrier may change this. The isolating barrier not only separates parts of the population from each other, but it also reduces the size of the population receiving the influx of swamping genetic variation. We need not, and should not, think of this barrier as dividing one population into just two. Instead, the barrier more often produces many fragmented and isolated groups from the original single population. These isolated groups are smaller, favorable characteristics may collect without being swamped, and new favorable characteristics come to predominate.

Clines and ring species illustrate that within a species, local populations adapt to local conditions. Thus, within a species range, selection pressure can be different enough to develop locally specialized subspecies. The conventional view is that this local differentiation sets the stage for, but does not yet complete, speciation.

Selected References

Barton, N. H. 2001. Speciation. *Trends Ecol. Evol.* 16:325.

Cain, A. J. 1971. *Animal species and their evolution.* London: Hutchinson University Library.

Schilthuizen, M. 2001. *Frogs, flies, and dandelions: The making of species.* Oxford: Oxford University Press.

Learning Online

Testing and Study Resources

Visit the textbook's website at **www.mhhe.com/evolution** (click on the book's title) to take advantage of practice quizzing, study/writing tips, timely news articles, and additional URLs for research on the topics in this chapter.

10

"It is, however, the specialized links and networks of interaction among species that have produced much of the diversity of life and the organization of communities. It is the evolution of these linked lives that has produced Darwin's entangled bank."

John N. Thompson, evolutionary biologist

Co-Evolution

INTRODUCTION

Co-evolution is an evolutionary handshake or a punch in the chops, depending upon who wins and who loses. It involves the joint evolution of two or more species as a consequence of their ecological interaction. Although genes are not exchanged between the interacting species, each species is partially dependent upon or threatened by the other, thereby producing reciprocal selective pressures. We have seen examples already. The competition between two ecologically similar species may drive them apart, producing specialists, a consequence of their co-evolutionary interaction. A back-and-forth, a give-and-take, produces ever-changing consequences for each partner in a co-evolutionary relationship. Sometimes co-evolution is beneficial to both, other times it is detrimental to one species. Predation produces reciprocal selective pressures between predator and prey. As rabbits evolve characters that improve their effective escape from wolves, wolves must evolve to keep up; as wolves improve, this in turn favors rabbits that improve. The same is true for parasites and their hosts, as well as herbivores and plants.

SYMBIOSIS—GOOD, BAD, AND UGLY

Symbiosis means living together, an ecological coupling of two or more species in a co-evolutionary relationship. The degree of profit or loss from symbiotic interactions can be generally categorized. If both partners benefit, the relationship is **mutualism**. Honeybees visit flowers where they feed on nectar and collect pollen for transport back to their hive to feed their brood; the flowers dust the honeybees with pollen, turning them into inadvertent carriers of the pollen on their subsequent foraging trips to other flowers. Both honeybee and flower benefit. If one species benefits and the other is at least not harmed, then the relationship is **commensalism**. Large grazing herbivores flush insects from the grass they eat. Insect-eating birds often follow such herbivores, waiting to catch the spooked insects when they fly up from hiding places in the grass. The herbivores lose little (and gain little), but the birds benefit. When one species loses and the other gains, we recognize the relationship more specifically as **predation** (predator and prey), **parasitism** (parasite and host), or **herbivory** (herbivore and plant).

Consider This—

Arms Race

The back-and-forth, the give-and-take, of co-evolutionary relationships produces reciprocal changes in ecologically linked species. Where the interaction is antagonistic, as in predator–prey interactions, the successful advance of one species must be met by the adaptive response of the other. This escalating interaction is referred to as an "arms race," analogous to contending nations wherein the buildup of threatening armaments in one nation is met by the buildup of armaments in the other nation, then back and forth with increasing threats and counterthreats. In the biological world, venomous snakes inject toxins into their prey to quickly dispatch it and thereby reduce retaliation from the prey (usually rodents) that can inflict serious bites with sharp front teeth. Rattlesnakes strike and quickly release envenomated prey, leaving the dispatching of the rodent to the injected toxins. But many populations of ground squirrels susceptible to rattlesnake attack have evolved a resistance to rattlesnake toxins that gives them some protection from the snake venom. In turn, rattlesnakes have evolved more potent venoms, and the ground squirrels seem in turn to have evolved even more effective resistance. Metaphorically, this is an arms race between predator and prey, rattlesnake and ground squirrel.

PLANT–ANIMAL CO-EVOLUTION

When a herbivore approaches, plants can't duck for cover or dash for safety. They stand and take it. Plants are fixed in place. Their defenses reflect the fact that plants are rooted to a location.

Spines

Bison, deer, and pronghorns on the North American continent and wildebeest and antelope on the African savannahs pose a threat to plants. Spines (thin, needlelike projections) on plants are a defensive weapon (figure 10.1). When the grazing herbivore bumps its tender nose into the spines, it gets unpleasantly poked. The herbivore retreats and more likely moves on to other plants with less daunting defenses. Cactus and cactus-like plants are equipped with such spines, and enjoy their benefits in thwarting or deflecting herbivore attacks.

Chemical Warfare

Some herbivores have tough tongues and steely mouths and can eat even plants with spines. Small insect herbivores (phytophagous insects) evade the spines by simply navigating around them to safely reach the fleshy plant tissue below the bristling sharp points of the spines. Plants under such attack often produce chemicals that are unpleasant in taste or even toxic in effect; these are termed *secondary chemical compounds*. They are "secondary" in that they are additional chemicals,

FIGURE 10.1   Cactus, with Projecting Thorns
Prickly pear.

not part of the assortment of primary plant chemicals used directly to support growth and maintenance metabolism. Secondary chemicals may be derivatives of primary compounds, but their central function is to provide the plant with chemical weapons to frustrate natural enemies.

The nicotine in tobacco plants affords some protection against herbivorous insects. When chewed on by insects, tobacco leaves release the nicotine into the insects, disrupting their metabolism and probably carrying a foul taste—a sensible encouragement to give up the habit. Seeds of some morning glories include d-lysergic acid, close chemical cousin to LSD, with hallucinogenic properties, which is not likely a coincidence. Some biologists propose that any herbivore munching on morning glories enjoys altered states of perception but becomes inattentive to its own survival challenges, more likely falls victim to a predator, and is eliminated from the community of herbivores threatening the plants. Members of the mustard family include mustard, radish, horseradish, cabbage, and watercress, all of which contain mustard oil. We tolerate and enjoy the pungent aromas and tastes of spicy foods, but to insects these sharp flavors signal tissue toxins and tend to be passed over. Some plants, such as milkweeds and dogbane, produce milky sap containing cardiac glycosides, named for their potentially lethal effects on the heart of vertebrates but equally toxic to many insects.

Plants have been in chemical conflict with herbivores, insects in particular, for millions of years. As quickly, in evolutionary time, as plants evolve chemical defenses against insects, insects evolve resistance to chemical toxins. We should not be surprised, then, when humans, late participants in this chemical warfare against insects, are quickly defeated by insects that evolve resistances to human-engineered chemical pesticides. Insects have been waging this old battle of chemical warfare against plants for millions of years. Humans, with their pesticides, are just the most recent organism, with its chemicals, to enter the skirmish.

Mutualisms

Interactions between two species are often favorable to both. Aphids have long mouth parts that, like straws, allow them to tap into sap beneath the thick bark of shrubs and trees to drink up the sugary fluids (carbohydrates). What they do not digest, they excrete as a droplet of water containing undigested sugars, known as "honeydew." This is flicked off their abdomens and drifts to the ground as a sticky droplet. Some species of ants collect this excreted honeydew, using its sugars as a resource. They may do so before the aphids flick it to the ground, and directly solicit the aphids to release the drop of honeydew (figure 10.2). And in some interactions between ants and aphids, the ants may actively protect the aphids from their predators, carnivorous insects, and transport aphid eggs. Both benefit. Aphids are the ants' "cattle," turning a resource not easily avail-

FIGURE 10.2   Ants and Aphids—Mutualism
These ants tend their "herd" of aphids, which in turn secrete fluids rich in sugars drunk up by the ants.

able to them (tree sap) into an accessible resource (honeydew). Aphids benefit from the protection afforded by the attending ants.

Ants and Fungus

Attini ants, denizens of the New World tropics, culture and then eat the fungus (genus *Leucocoprini*) in safe, underground "gardens" within their burrows. The fungus is grown on a mixture of excrement fertilizer, from the ants, and fresh plant material gathered aboveground by the ants. Many species of Attini ants, sometimes known as "leaf-cutter ants," forage for fresh leaves and flowers and raid crops by cutting and carrying a piece of the plant underground to enrich their gardens. Once underground, the ants chew up the leaf pieces into even smaller pieces (1–2 mm), munching them into a pulp, and insert the pulp into the garden plots. The ants then pluck tufts of the fungus from elsewhere in the nest and plant these in newly prepared gardens. Within about a day, the fungus has overgrown the new garden, a "crop" the ants then harvest to meet their nutritional needs.

Overall, both ants and fungus benefit. The ants feed upon these fungus gardens and keep replenishing them. For the ants, it is a sustainable crop. For the fungus, it is care and nurturing. If left unattended, the fungus is quickly overrun by competing species of fungi. Therefore, the fungus, although harvested, benefits from the attention and cultivation of the ants. Specifically, ants chew up the antifungal defenses of the plant (such as its waxy coating), and the fungus degrades the anti-ant defenses (insecticides) of the plant, such as its secondary chemical compounds, so these insecticides are not passed to the fungus and, in turn, consumed by the ants.

But as any experienced gardener could predict, this chummy world of ant farmer and fungus garden invites the invasion of "weeds." The weed is a virulent parasitic fungus (genus *Escovopsis*). Unchecked, this parasitic fungus overgrows and destroys the fungal garden, leaving the ants without a food supply. In response, ants carry on their bodies a bacterium (genus *Streptomyces*) that produces antibiotics specifically targeted to suppress the growth of the parasite and keep it in check.

Ants and Acacia

Acacia trees are a widespread genus with species distributed in tropical regions around the world. A focused study in Mexico clarified the close, and mutually beneficial, current relationship between the local acacia tree (*Acacia cornigera*) and its associated colony of ants (*Pseudomyrmex ferruginea*). Along its branches, the acacia provides the ants with small pools of nectar ("nectaries") drunk up by the ants and Beltian bodies (packets of food) picked and fed to ant larvae (figure 10.3a); further, the inflated, hollow thorns of the acacia provide a home in which the ants live (figure 10.3b). The ants protect the acacia from browsing mammals, which the ants swarm over when they start chewing on the acacia branches. The ants also attack herbivorous insects that alight on the acacia. Further, the foraging ants eliminate adjacent, competing plant species by eating their leaves, defoliating them. The favorable effects of the ants on the acacia were demonstrated in the 1960s in a series of clever experiments.

The ecologist, Daniel Janzen, removed the ants from their associated acacia trees by clipping off branches that held the colony of ants or by temporarily eliminating the ants with insecticide spray. He found that healthy acacia trees, deprived of their ant colony, quickly became vulnerable to unchallenged attacks by phytophagous insects—insects equipped with mouth parts to chew up or suck fluids from plants. And without the ants to maul overlapping competing plants, the acacias were overgrown by competing shrubs and trees. Essentially, the ants provide a standing army for the acacias, and the acacias provide room and board for quartering the ant troops.

The association between ant and acacia is complex, but how might it have evolved? Certainly not overnight in one big evolutionary boom, and certainly not by conscious decision by ants and plants. If we envision a period in the past before each was in such a complex association, we can then hypothesize the small steps leading up to the intricate relationship we find today between ants and acacias. Perhaps the first step

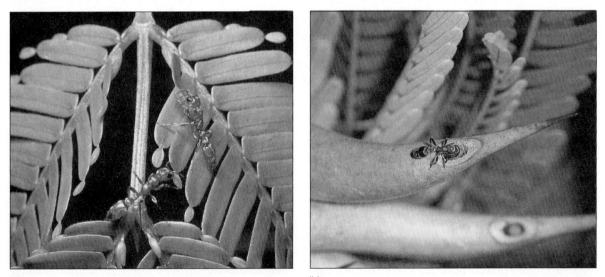

(a)   (b)

FIGURE 10.3   Ants and Acacias—Mutualism   (a) Ants feed off the Beltian bodies produced at the tips of leaves by the acacia tree and off nectaries along the stems. (b) Ants live in the hollow, swollen thorns of the acacia. The ants protect the acacia from phytophagous insects and from overgrowth of competing species of plants.

would be the *evolution of thorns* by the acacia. Thorns would be co-opted by ants in the distant evolutionary future, but their initial adaptive advantages were not in anticipation of this distant biological role. Instead, acacia trees with projecting spikes would enjoy the immediate advantages of dissuading the attacks of large and effective browsing mammalian herbivores. Next, the *facultative or casual use of these thorns* as nest sites by ants would bring both acacia and ants into opportunistic association. Ant colonies in residence that did not defoliate the acacia enjoyed the advantages of preserving their new nest sites. Feeding off-site from their host acacia would be selectively favored. As an added by-product and benefit, foraging off the host acacia that holds their nest sites *defoliates competitor plants* impinging on the host acacia. Even before taking up residence on the acacia, ants likely evolved the beneficial behavior of defending their nest sites, wherever they were, by swarming over and attacking animals that threatened. Such a defensive behavior would be carried with them as they established residence on the acacia. Consequently, any phytophagous insect that alighted on the acacia to dine would be treated similarly as a threat. Ants would swarm over and attack the insect, driving it off or turning it into a meal for themselves. The by-product of this attack would be to *fend off acacia enemies* as well, increasing the fitness of these protected trees. By now, the co-evolutionary relationship of ants and acacia is close and interdependent. At this point, any acacia that *produced supplementary food* would be at a selective advantage over acacias that did not. Ants would be more likely to take up residence on the acacia providing nectar "bribes" and Beltian bodies as food for their larvae. And such acacias would support a healthier ant colony that is able to beneficially reciprocate.

Overview: Ants and Acacia   Each step is a prerequisite to the next, but each step is not taken in anticipation of good things to come in the distant future. Each step is adaptive in its own right. But once in place, a new step changes the dynamics of ecological interaction between evolutionary partners. New selective pressures arise. Casual associations become selectively favored, leading to the preservation of individuals with genetic mutations enhancing the favorable benefits of the interaction. Reciprocal benefits "snowball," leading in small steps to intricate and complex co-evolved species. When we meet other complex associations or structures, we should remember ants and acacias. Later we will look to other

ways that complex features may arise relatively rapidly. But for now, note that we do not have to explain the appearance of complex features in terms of one huge, evolutionary boom. Small steps may build complex associations or structures. Evolution has the time.

Honeybees and Flowers

One major evolutionary event within plants occurred during the second half of the Mesozoic era: Angiosperms (flowering plants) debuted. The gymnosperms (conifers) preceded them. Gymnosperms produce pollen, packets of sperm, carried usually by the wind to female ovules where fertilization occurs. Instead of wind, flowering plants often employ animal vectors, directed transfer agents, to carry pollen more or less directly from male to female flowers. Butterflies and hummingbirds are examples of animal pollinators. Many insects are flower pollinators, but honeybees are perhaps the most recognized. Honeybees are just specialists, active most of the summer foraging for food supplied by flowers.

Flower colors delight the human eyes and their scents bring enjoyment to our world of smells. But in the world of biology, the colors and smells are quite practical. They both catch the attention of honeybees and, when fully developed, signal the availability of food and ripe pollen (figure 10.4). The flower supplies the visiting honeybee with nectar and pollen. Nectar feeds the visiting bee, which harvests pollen as food for other bees and for the young brood back at the hive. Honeybees have special bristles on their legs—"pollen baskets"—which are gradually filled with pollen during visits to many flowers in succession. Eventually pollen baskets form lumps of collected pollen that are carried back to the hive. While busily harvesting pollen, honeybees fortuitously serve the plants they visit. Dusted on or collected by the foraging honeybees, pollen accumulated at one plant is carried to another, where it might be brushed off to then fertilize eggs in the ovules of the flowers. Honeybees benefit from the resources gathered on visited flowers. Flowers benefit from the delivery service of their pollen by honeybees. The honeybee becomes an agent for the flower, spreading pollen—an inadvertent errand girl (foraging bees are genetically female, although sterile). The honeybee is not being a good Samaritan, doing the flower a favor. It is out to satisfy its own needs. But in the course of doing so, the honeybee delivers pollen. Both benefit, bee and plant. The honeybee gains a resource from the flower; the flower gains a traveling agent to deliver its pollen.

FIGURE 10.4   Flower of an Orchid   Note the distinguishing nectar guides, the spots near the center of the flower.

Hummingbirds and Flowers

Hummingbirds feed on insects. But they are outfitted with specialized long beaks and tongues to drink up nectar, along with a hovering flight to hold their aerial position while doing so. Overall, they are specialized to tap resources supplied by flowers but they also act as pollinators as they move from flower to flower. Unlike the community of insects available continuously throughout the summer, hummingbirds may visit one site only during a short block of time. Hummingbirds migrate on a seasonal basis, moving south into tropical and subtropical regions with the approach of winter, returning north as spring arrives. Angiosperms, which cater to hummingbirds as their pollinator, bloom during the spring as hummingbirds pass through on their way to northern breeding territories. Not only do such flow-

ers open in anticipation of hummingbird arrival, but many are tubular and narrow at their base, an anatomical design that ensures hummingbird visitation. This tends to prevent roving insects from pirating nectar, saves it for hummingbirds with long beaks and tongues, and rewards the bird for its visitation. But during its visitation, the hummingbird face and beak get dusted with pollen, carried to other flowers as it satisfies its own quest for high-energy nectar.

Practical and Functional

Bright flower colors and aromatic smells inspired poets to, well, wax poetic about how flowers titillate and delight human senses. A gift from the gods. As biologists first worked out the practical role of colors and scents, others simply dismissed the idea that insects could recognize and respond to color. An early experiment to address this view took advantage of butterflies. In an area frequented by butterflies, biologists placed around the grounds small, clear dishes of sugar water on colored backgrounds. They then scored the number of butterfly visitations to these colored sugar-water dishes. Butterflies could tell the difference. They clearly preferred two particular colors, yellow and purple. These turned out to be the flower colors they preferentially visited as well.

Flowers' colors not only attract insect pollinators; they also direct and funnel insects to the pollen/nectar. Some flowers look like targets, with their nectar at the bull's-eye; some flower colors identify the landing areas from whence to start the search for food within the flower (figure 10.4). Such color markings are *nectar guides*—visual direction signs for the arriving pollinators, pointing them to the nectar. To our eyes, some flowers seem dull and drab in ordinary light. Our eyes do not see in the ultraviolet range of light, but the eyes of some insects do. "Drab" flowers might exhibit ultraviolet "color" nectar guides that are visible to insects but not to ourselves.

Commensalism

Some pollinators get little benefit from their service to the plants they visit. For example, "looking-glass" orchids, named after their shapely flower, are visited not by bees, but by wasps. At first, that puzzled biologists. Wasps are usually scavengers and predators. They may occasionally steal nectar, but they are more likely to pounce on a fly, bite off its wings, and carry its hapless body back to their brood. For wasps, unlike honeybees, pollen is a nuisance, not a resource to be harvested. For flowers, wasps would seem a poor pollinator. In fact, looking-glass orchids do not even produce nectar, yet wasps visit, and hence the puzzle.

What biologists eventually discovered is that wasps visit these orchids not to find food but to satisfy the prurient interests of male wasps—for sex! These flowers imitate female wasps poised and ready to mate (figure 10.5). The flower emits a fragrance that is apparently seductive to male wasps; the flower itself is well-formed and shapely, fashioned into the voluptuous body shape of a female wasp, and poised in an inviting posture. The hopeful male wasp flies in and alights, but for all of his amorous expectations he only gets bonked on the head with pollen, and flies off, pollen dusted, in search of more satisfying relationships. Spotting another "female" orchid flower, and apparently short of memory or following determined instincts, he alights again, delivers pollen, but gets bonked again with more pollen that dusts his body. It is not known for how long or for how many times a male wasp continues his unrequited quest, but looking-glass orchids are a very successful group of plants.

Skunk Cabbage

The skunk cabbage is named for the unpleasant odor, to human senses, it gives off when flowering. Humans of-

FIGURE 10.5 Orchid Flower that Mimics Specific Female Wasps The flower's mimicry attracts male wasps, which arrive attempting to mate. Instead, the males only get doused with pollen, which they carry to the next expected amorous rendezvous.

ten describe this plant as smelling like rotting meat or dung. But to the sensitivities of many insects, these welcome odors signal useful resources. Flies and scavenging beetles seek out decaying meat or dung, respectively, where they lay their eggs. The eggs hatch and then feed as larvae (grubs) in such resources. To these insects, smells we find repugnant signal the promise of a nearby inviting site to lay their eggs. The flower of the skunk cabbage is wrapped in an enveloping leaf. When the flower ripens, it warms, up to 40 °C, volatilizing the chemicals within that drift out into the air to attract the insects. The insects follow the scents upwind to the flower, climb into the flower, get doused with pollen as they scurry about the flower's innards, but eventually weary of their search and fly off. Eventually the attractive smell of rotten meat or dung from another flower brings them in again, where they deliver the pollen previously collected, as they continue their search for a suitable egg-site.

Overview: Commensalism Certainly, the looking-glass orchid benefits from the amorous travels of the male wasp, but the wasp receives no biological return. Similarly, the skunk cabbage benefits from insect pollinators, but the insects find no beneficial resource. In fact, these wasps and other insects in such examples of co-evolution lose time and energy in hopeless pursuits. Consequently, some biologists argue that these are not examples of commensalism (benefit/no harm), but of behavioral parasitism (benefit/harm)—only the plants gain in the interaction, the insects lose. How we classify such examples is less important than noticing the basis for current co-evolutionary relationships, so we might understand how they evolved.

Initially, ancestors to skunk cabbages may have depended on all available insect pollinators. But skunk cabbages are physiologically adapted for life in wet and soggy places, dim light, and shade. Any ancestor that produced an odor attractive to insect denizens of similar habits would increase the likelihood that local insects would investigate. Doused with its pollen during the insect's investigation, the plant loads its pollen on a traveling insect, and thereby increases its fitness by increasing the chances its pollen will be delivered to and fertilize eggs in another plant. Subsequent evolution would enhance the tight co-evolutionary relationship, producing specializations of the plant (heat, volatile odors, morphology of the flower) that further improved its success. The odor need not appeal to human senses. The skunk cabbage only needs to catch the attention of insects drawn to the odor—fetid meat or smelly dung—a built-in attraction to some flies and beetles, in which it is a sensory delight already in place as part of their own reproductive quest.

PROTECTIVE COLORATION AND SHAPE

Colors and shapes also help many organisms address the threats they face from enemies.

Camouflage

Sometimes it pays to hide. Organisms that conceal themselves from enemies may do so with **camouflage** or **cryptic** coloration and shape (figure 10.6). For example, in addition to sharp thorns and chemicals, plants hide behind colors and shapes that enable them to be overlooked by browsing herbivores. Rounded "stone plants" live in arid and open habitat, swollen with water to see them through dry seasons. These lush plants are also favored foods for herbivores, if discovered. But they are the shape of inedible stones, and silver skinned—easily overlooked by browsing herbivores venturing through.

Many animals, like plants, do not get up and run away. During daylight hours treehoppers pose as rose thorns. The twig caterpillar looks like an uninteresting dry stem (figure 10.7). Even if a browsing bird bumps against it, it only falls to the ground and continues the deception, remaining motionless, like a broken bit of stem. Many tropical katydids have wings that resemble leaves—even to the veins and blemishes—nothing of apparent interest to insectivorous birds. Most vulnerable animals secrete themselves away from view, especially during daylight hours. But a species of tropical frog is out in plain view, on top of leaves, during daytime. With legs tucked under its body and its splattered color pattern, it looks like just bird droppings—nothing of interest to its visual predators. When it's safe, it becomes active, moving about in search of its own quarry.

Warning Coloration (Aposematic)

Sometimes it pays to advertise. Coloration that publicizes an organism's presence is **warning coloration** or **aposematic coloration**. If an animal possesses features that can hurt its predators, then it has a defense. The poison arrow frogs of tropical South America are brightly colored, advertising the toxic skin secretions they produce. They are named for a use to which aboriginal peoples of the area put them. Caught by hand (the toxins must be ingested or enter the bloodstream, not just touched, to be dangerous), the frogs are stressed by holding them over a hot fire, which provokes release of their skin toxins. The skin toxins are dripped onto darts or arrow tips. The darts or arrows are shot from

FIGURE 10.6   Camouflage (Cryptic) Coloration and Shape   This dwarf seahorse (center) is camouflaged within the branches of this colonial sea fan on a reef off the Solomon Islands.

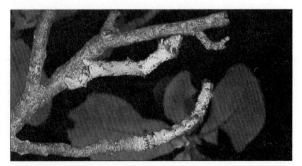

FIGURE 10.7
Inchworm Caterpillar, Resembling a Twig

blowguns or bows at monkeys and other animals high in the branches. When the point penetrates the prey, the toxins spread and the wounded animal falls to the ground, where it is collected by the hunters and returned to camp as food. The cooking denatures—inactivates—the frog toxins.

For most enemies, other than humans, the frog's bright coloration advertises the harm it can inflict if eaten. A predator that attacks and bites the frog finds the toxins distasteful and nauseating, and tends to quickly release the frog. But in the encounter, the frog might be injured, even if it is eventually rejected. Any conspicuous trait, such as coloration, that makes this frog stand out as being noxious would be favored. A knowledgeable predator, one that has previously encountered such a frog and come away with a bad experience, would now tend to notice the identifying features and avoid the frog on a second meeting. By identifying the traits that make the animal conspicuous, the predator avoids the prey without first biting into it. Thus, the frog's warning coloration, which makes it distinctive and advertises its dangers, is of selective advantage.

Sometimes animals do both—hide *and* advertise. The screech owl, when approached at a distance, tucks itself up against the tree, closes its headlight-like eyes, and tries to become inconspicuous. But if you close in on the owl to become an approaching threat, then the owl puffs up, spreads its wings, displays its bright eyes, and tries to look formidable. Venomous rattlesnakes will do much the same. If a human threat is at some distance, the snake is likely to remain silent, not move or rattle its tail. But if you approach closer, it turns into a flurry of display. The snake rattles, may hiss, and assumes a dangerous strike posture. All of this is its warning behavior.

Sometimes it pays to startle. If you are a small, insectivorous bird gliding through the forest in search of a meal, you may not be expecting the appearance of a threat to you. You spot a likely butterfly meal on the branch ahead and start for it, when suddenly two large eyespots flash in its place before you (figure 10.8a). The surprise and uncertainty give you pause. Perhaps it is a predator, a threat to you (figure 10.8b). In the instant you hesitate, the butterfly makes its escape. Eyespots occur in moths, on caterpillars, and even on the rumps of frogs. They are thought to play a similar biological role—namely, to startle the predator long enough to make an escape or unsettle it enough to discourage it from completing its attack.

MIMICRY

Sometimes it pays to impersonate. The superficial resemblance between two or more organisms that results from a co-evolutionary relationship is termed **mimicry**.

(a)

(b)

FIGURE 10.8   Startle Response, Eyespots on Butterfly   (a) When discovered in its cryptic disguise, this butterfly suddenly flashes eyespots on the underside of its wings, startling the predator, and giving it an extra moment to make its escape. (b) The eyespots are thought to confuse the insect-eating bird with its own predators, such as an owl, causing the bird to pause.

Energetic biologists have identified many variants. In a general sense, most examples of camouflage might also be considered cases of mimicry. The looking-glass orchids mimic female wasps, the twig caterpillar mimics a dried stem, the katydid mimics leaves, and so forth. But the term *mimicry* is usually applied in a strict sense to two general categories of mimicry: Batesian mimicry and Müllerian mimicry, each named for the first person to publicly recognize the type (figure 10.9).

Batesian Mimicry

In **Batesian mimicry,** one species is dangerous or distasteful and can back up its boastful warning coloration with unpleasant consequences for would-be predators. This species is the *model*. The other look-alike species is the *mimic*. The mimic is brightly colored and shaped like the dangerous model, but is an imposter, because it lacks the ability to inflict unpleasant consequences on a predator.

Danaus plexippus   *Limenitis archippus*

(a) **Batesian mimicry:** Monarch *(Danaus)* is poisonous; viceroy *(Limenitis)* is palatable mimic

Heliconius erato   *Heliconius melpomene*

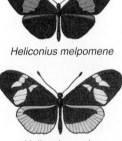

Heliconius sapho   *Heliconius cydno*

(b) **Müllerian mimicry:** two pairs of mimics; all are distasteful

FIGURE 10.9   Mimicry   (a) Batesian mimicry between a toxic monarch (model) and a harmless viceroy (mimic), left and right, respectively. (b) Müllerian mimicry. Both ecologically sympatric pairs are distasteful, and both have warning coloration.

The insect-eating blue jay is faced with a problem. The two species of butterfly upon which it might feed are superficially similar in color and pattern, but one is distasteful and the other a tasty meal (figure 10.9a). The monarch butterfly contains sequestered in its tissues noxious chemicals (cardiac glycosides). If eaten by the bird, these chemicals are released, causing nausea and vomiting. The bird who has once tasted a monarch will later avoid other similarly colored butterflies. The viceroy butterfly usually contains no such noxious chemicals and, if eaten, is enjoyed with no unpleasant side effects. But the community of insect-eating birds that have had a bad experience with the brightly colored monarch tend to avoid similar color patterns such as that of the viceroy. The viceroy (mimic) benefits from predators' aversion to the unpleasant monarch (model).

The unpalatable chemicals are not manufactured by the monarch but instead are picked up during their larval stage. Like most butterflies, the life history of the monarch includes eggs that hatch into a caterpillar or larva. The caterpillar is a feeding machine, dining voraciously on the leaves of the milkweed plant. When grown, the caterpillar sheds its skin and becomes a chrysalis, wherein it undergoes a metamorphosis, or transformation, into the winged adult, which eventually emerges and becomes the reproductive adult butterfly (figure 10.10). As part of its own chemical defense, the milkweed produces the noxious chemicals cardiac glycosides. Monarch caterpillars are one of the few caterpillars that feed unaffected on milkweed. They have evolved a protective mechanism whereby the toxic plant chemicals are safely sequestered in concentrated fat bodies, away from the caterpillar's general body tissues. These chemicals are passed through the chrysalis to the adult butterfly, now inflicting its noxious effects on birds that eat the adult butterfly.

Birds' learned aversion to monarch toxins can be demonstrated in the laboratory (figure 10.11). Biologists raised one group of monarch caterpillars on milkweed plants; these metamorphosed into bright adults that carried the toxins. They raised a second group of monarch caterpillars on plants without toxins; these metamorphosed into identically colored bright adults that did not carry the toxins. Fed a nontoxic monarch, the blue jay rips off its wings and gobbles it down. Next, when the unsuspecting blue jay is fed a monarch with toxin, the bird similarly rips off the wings and eats the body, but almost immediately experiences gastric distress. It vomits up the unpalatable butterfly. Now when the blue jay

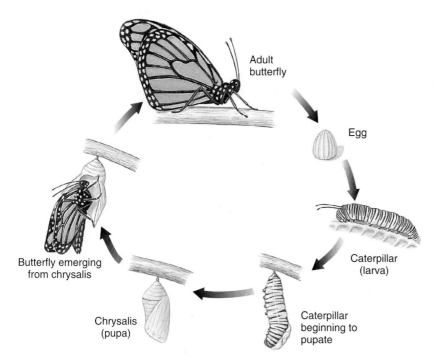

FIGURE 10.10   Life Cycle of Monarch Butterfly   Adult monarch butterflies are protected from birds and other predators by the toxins in their tissues. These toxins are incorporated from the milkweeds they feed on as larvae (caterpillars).

Adult butterfly

Egg

Caterpillar (larva)

Butterfly emerging from chrysalis

Caterpillar beginning to pupate

Chrysalis (pupa)

is again offered a nontoxic monarch, it refuses and avoids the offered meal. The blue jay has developed a learned aversion to the monarch color and pattern.

In the wild, we find complex, co-evolutionary relationships among many species—milkweed, viceroy, monarch, and the community of insectivorous birds. How might it have evolved? As we did with the ants and acacia, we might usefully go back in time and propose a plausible series of steps from the early (simple) condition to the current (complex) condition of Batesian mimicry. Certainly one of the early steps would be the *evolution of toxin in the tissues of milkweed plants*. The milkweed is doing no favors for monarch butterflies, but rather enjoying the adaptive advantages it receives from these toxins, which discourage plant-eating insects. Next, the *monarch evolves a tolerance for the milkweed toxins*; the adaptive advantage is that the caterpillar can exploit a resource unavailable to competing species. As the sequestered toxins are passed through the chrysalis to the adult, the *adult acquires toxins noxious to its predators* and is likely to be avoided by its predators. At this point, any individual with a distinctive color, easily recognizable by visual predators, would be favored because it avoids being mistaken for a tasty butterfly. This would favor the *evo-*

lution of warning coloration by further reducing the chances of a mistaken attack. An experienced predator, one with an unpleasant encounter with a bright-colored monarch, would tend to avoid any other species of butterfly that can be mistaken for the monarch. For the

(a)   (b)

FIGURE 10.11   Blue Jay Learning Aversion to Distasteful Monarch   (a) This hand-reared blue jay, having never eaten a monarch, rips off the wings and gobbles down the body. (b) The toxins quickly make the blue jay sick, and it spits up the monarch. Thereafter, even if presented a monarch lacking such toxins, the blue jay refuses it.

tasty viceroy, this favors the evolution of visual traits mimicking the unpalatable monarch. The *closer in appearance of viceroy (mimic) to monarch (model), the more the viceroy benefits from predator avoidance.*

In Batesian mimicry, the mimic benefits but the model may actually find its survival threatened, especially if the harmless mimic becomes too common. If the mimic becomes common, the visual message can be reversed and the predators can be taught to seek out the butterfly rather than avoid it. Brightly colored, the mimic butterfly stands out to visual predators. If eaten and enjoyed (rather than sickening the predator), then the rewarded predator more likely searches for other similarly colored, bright and tasty butterflies. This increases the risk that the model, too, will be attacked. From the model's standpoint, it is advantageous to not be confused with the tasty mimic.

Other cases of Batesian mimicry have been proposed. One is between the venomous and dangerous rattlesnake (model) and the nonvenomous gopher snake, or bull snake (mimic). When discovered and cornered, the rattlesnake rears back in a defensive posture from which it threatens to strike and deliver its venom; it shakes its specialized tail vigorously, producing a rattling sound of the loosely interlocking segments; and it may hiss. When the nonvenomous gopher snake is threatened, it will similarly assume a defensive posture, also spreading its head into a triangle shape. It has no tail rattle, but it can vibrate its tail; if it does this against dried leaves, the vibrations produce a buzzing sound. It also hisses loudly. Taken together, the overall gestalt mimics the rattlesnake.

A similar case of supposed Batesian mimicry occurs between a distant relative of cobras—the highly venomous coral snakes (model)—and some brightly colored species of the nonvenomous kingsnakes (mimic). Both are brightly colored with rings of black, yellow, and red around their bodies. In North America, coral and kingsnakes can be distinguished with this effective, if somewhat cruel, poetic advice: "Red and yellow, kill the fellow," meaning that if the red and yellow bands touch, the snake is a deadly coral snake; "Red and black, friend of jack," meaning that if the red and black bands touch, the snake is a harmless kingsnake.

Others have questioned this case for Batesian mimicry, suggesting that the bright bands of the coral snake might not be warning colors at all but cryptic colors. Although predaceous birds see in color, most predatory mammals do not. Foxes, wolves, badgers, raccoons, weasels, and coyotes see in shades of gray. When such marauding mammalian predators turn over

rocks and rubble in search of prey, the coral snake they discover looks black and white banded. In motion, these bands would pass the eyes of a predator at a high "flicker frequency" and be seen, not as bands, but as a blended gray (figure 10.12). Video and traditional film projection take advantage of this effect in humans. Although each frame of a motion picture film we watch in a theater is displayed for a split second as a single, distinct image, the rate of projection is about 30 individual frames per second. This exceeds our flicker frequency—our ability to resolve these as separate images—and these individual images blend into a "motion" picture sequence. A motionless, banded snake approached by a mammalian predator may startle quickly into motion, but as it does, its image, to the eye of the predator, would change from banded to blended gray. From the standpoint of the predator, the banded snake it sought to grab would disappear in front of it. Or at the least, the predator would experience momentary confusion, giving the snake a chance to escape. If the banding of coral and kingsnakes is seen similarly as shades of white and black, then the banding may be cryptic and not aposematic.

For still different reasons, others challenge the existence of Batesian mimicry in snakes, because, they claim, a venomous snake biting a naïve predator would kill it, allowing no opportunity for the predator to carry away the lesson learned. However, motmots, tropical birds related to kingfishers, feed on lizards and small snakes. Motmots are likely to encounter lethal prey, coral snakes sporting brightly colored red, yellow, and black bands that advertise their danger. It is not surprising that motmots have evolved innate avoidance of coral snakes but not of other snakes. Young motmots can be hand-reared from hatching so they are deprived of any social experience with coral snakes or contact with adult motmots that exhibit the avoidance behavior. The hand-reared birds have no opportunity to learn avoidance, either from a bad encounter with a snake or by watching adults. Nevertheless, hand-reared motmots avoid coral snakes upon their first encounter. They even avoid thin wood dowels, snake size, painted with adjacent, repeating yellow and red bands. However, other colors painted into stripes have no similar effect on young, inexperienced motmots, which instead usually investigate such models. Learning is not necessary, as motmot avoidance of coral snakes seems to be a genetically based, innate behavior the birds possess from birth.

In South America, where predation on snakes can be especially intense, brightly banded species of ven-

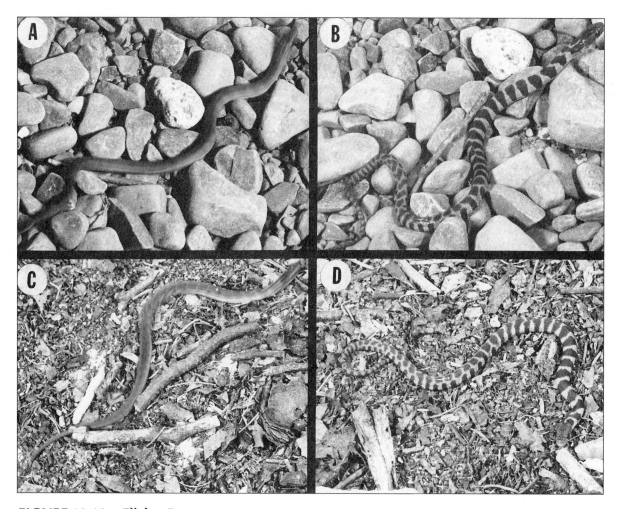

FIGURE 10.12   Flicker Frequency   A newborn watersnake shown crawling (a, c) and motionless (b, d). In motion, the snake's banding pattern looks evenly gray, as it would when exceeding the flicker frequency of a predator. (From Pough 1976.)

omous coral snakes also occur. Here, sympatric (overlapping) with them, is a species of nonvenomous snake whose coloration pattern is similar. A poor but painful defensive strike, survived by the predator, could reinforce its avoidance of coral snakes. Or innate avoidance may serve from the time the predator is born. These do seem to be convincing examples of Batesian mimicry: a brightly colored, dangerous coral snake; a brightly colored, nondangerous mimic.

Müllerian Mimicry

In **Müllerian mimicry,** both species come to superficially resemble each other, but both species are un-

palatable or dangerous to their predators. Unpalatable butterfly species may come to resemble each other in sporting similar warning coloration. In a sense, each unpalatable species is the model for the other. Unlike Batesian mimicry, where the model is at increasing risk with increasing abundance of the mimic, in Müllerian mimicry the co-evolutionary message is not contradictory. The predator community is reinforced in its aversion whichever species it attacks, because both are unpalatable. Whichever species the predator eats, it comes away with a bad experience. The more the better. The larger the number of individuals with similar warning coloration that can back it up, the more reinforcement occurs (figure 10.9b).

Other Types of Mimicry

The careful study of mimicry in and between plants and animals has led to the identification of many specific types of mimicry. Most are simply derivatives of basic camouflage or warning coloration, so only one other type of mimicry will be mentioned—*aggressive mimicry*. In aggressive mimicry, the prey do not mimic something predators avoid or overlook; instead, the predator does the mimicking. It is a wolf in sheep's clothing. Some predaceous fireflies flash at frequencies that mimic the flashing of females of other species. When the amorous male of this other species arrives to investigate, he is grabbed and eaten. The angler fish is dressed in leafy cover so it looks like a harmless algae-covered rock. It waves a lure in front of its mouth, something attractive to small fish that come to investigate. When in reach, the angler fish turns the inquisitive fish into a meal.

The wrasses (pronounced "rasses") are a family of marine fishes that are common in densely populated coral reefs. Wrasses can be found in warm seas, but most occur in tropical waters where these bright colored, carnivorous fishes feed on tiny crustaceans, snails, worms, and other small invertebrates. Some wrasse species specialize as *cleaner fish*. Cleaner fish feed on the ectoparasites and bits of dead skin on other fish, removing them, hence their name. Both the cleaner fish and the fishes it services benefit—an example of mutualism. The cleaner fish gains a meal; its customers gain a cleaning (figure 10.13).

The customers served by the cleaner fish can be some of the most formidable predators on the reef. Mistakes in identification can be costly. What has evolved is a reciprocal system. Cleaner fish are brightly and distinctively colored. When a fish arrives to be cleaned, the cleaner fish runs through an elaborate display, flashing these colors. The customer relaxes—some biologists describe it as almost going into a trance—it opens its operculum, the protective bony cover over the fleshy gills, and this allows the cleaner fish to dart in and about, picking off parasites along with diseased and damaged tissue. Job done, the cleaner fish dashes off to other customers, and the spiffed-up fish returns to its predaceous ways in the reef.

Each cleaner fish occupies a little region on the reef. Surrounding fish visit its "spa" to be serviced. A team of divers following one cleaner fish counted more than 300 fish customers visiting in a 6-hour period. If the cleaner fish is removed from the reef, the health of other fish around its spa declines. Unfortunately, into this chummy community comes an imposter, an aggressive mimic. These fishes, about the same size and color as the cleaner fish, belong to a different family of fishes altogether. When an unsuspecting customer arrives, they display their colors, and the impostor recip-

FIGURE 10.13 Cleaner Fish and its Customer The cleaner fish is the small, striped fish working around the relaxed mouth of the larger fish, gathering up bits of debris.

rocates, mimicking the displays of a proper cleaner fish. The unwary customer relaxes. But instead of picking off parasites, this aggressive mimic dashes in to take bites out of the fleshy gills and chunks of living flesh from fins and tail. This aggressive mimic is a predator, masquerading in "sheep's clothing."

OVERVIEW

Linkage and Liaisons

Ecological couplings can lock species into interdependent co-evolution. Long-term options may be lost to short-term benefits of a symbiotic relationship. On the island of Mauritius, off the coast of Africa, biologists discovered an odd situation with the *Calvaria* tree and the extinct dodo. Many old trees were present, but young sprouts were few. The old trees produced seeds with thick protective coats, but few seemed able to germinate easily from within the sturdy seed. Mauritius was the island home of the flightless dodo bird until the seventeenth century, when the species became extinct (figure 10.14). The bird had been hunted by sailors, supplying fresh meat after a long sea voyage. The population steadily dwindled until the last dodo was killed in 1681—one of the first modern species extinctions at the hands of humans.

The dodo was a seed-eating bird and probably dined on the seeds of many trees, including the *Calvaria*. As in most such birds, a muscular part of the stomach, the gizzard, works small pieces of swallowed gravel against ingested seeds to grind them down, abrading and breaking through the protective coat around the young plant tissue within. Apparently, the

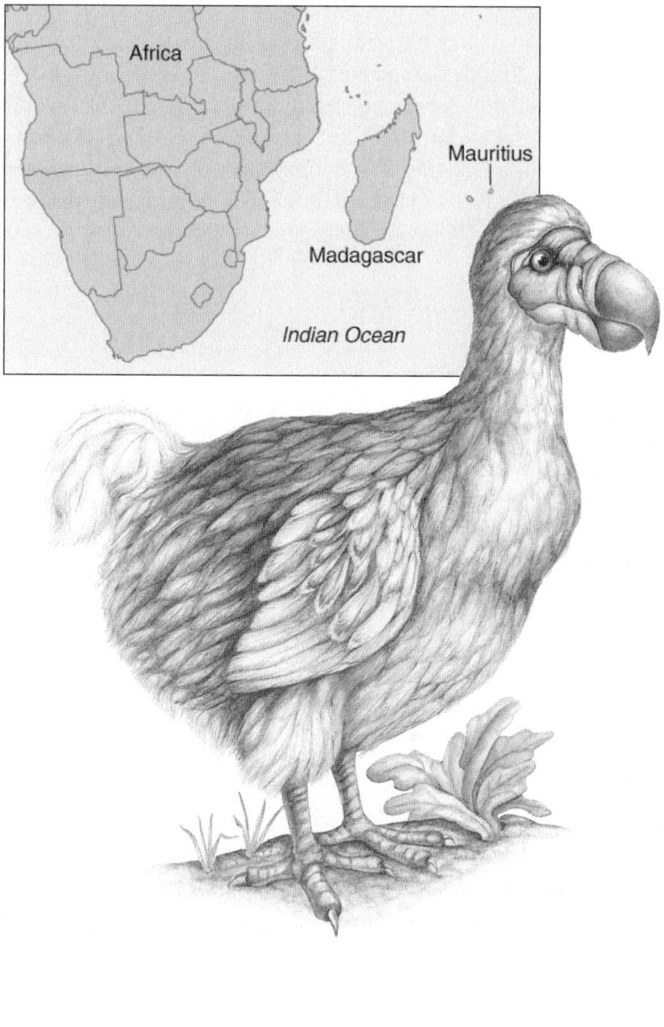

FIGURE 10.14 Dodo Bird (Extinct)
The flightless dodo lived on the island of Mauritius off the coast of Africa until the last one was killed in 1681. It fed on plants and seeds, including the seeds of the *Calvaria* tree. These seeds had evolved thick coats to survive the passage through the grinding gizzard of the dodo. With the extinction of the dodo, these seeds no longer made an abrading trip through a bird's digestive tract, the coat remained thick, and the young tree embryo could not so easily germinate.

Calvaria had evolved an especially thick coat that protected the seed during its passage through the grinding gut of the dodo. Passed out in the feces, the thinned but intact protective coat could be broken through by the young plant within and germinate. But in the absence of the dodo, the seeds did not pass through the digestive tract, no abrading occurred, the seed coat remained too thick to permit frequent germination, and the young tree embryos usually died, imprisoned within their own seeds.

With the extinction of the dodo, the *Calvaria* lost its linked co-evolutionary partner, and its long-term fate was in doubt. Biologists took some of the seeds and fed them to turkeys, surrogate dodos, who, like the dodos, are equipped with a grinding gizzard. Some of the seeds safely passed through the turkeys, were collected, planted, and germinated. They may have been some of the first *Calvaria* to do so since the dodo's extinction 300 years before.

Pollinators and their flowers are also intertwined ecologically. Hummingbirds gain a nectar meal, and flowers gain a pollinator. The spring migration of hummingbirds overlaps with the flowering of the plants they visit. But if the plant changes flowering time, it risks losing its pollinator; if the hummingbird changes course or migration time, it may meet no opened flowers at crucial legs along its journey. Both are linked in their evolutionary destinies.

Remodeling

Previously existing natural features become the basis for co-evolutionary relationships. Aphids exude honeydew with or without ants present. But ants, foraging on plants, come across this discarded waste and start a mutualistic interaction with the aphids. Fish had parasites before cleaner fish took up the business of picking them off. Before taking up the cleaning business, ancestors of cleaner fish probably fed on small food and just extended this to the surface of other fish. The skunk cabbage draws in pollinating flies, producing an odor that already attracted flies, the smell of decay. It is thought that a behavioral change precedes an evolutionary change. The change in behavior sets the stage. Any genetic mutation that enhances this beneficial behavior would be favored and establish a genetic basis for what behavior began. Evolution proceeds by remodeling, not by major new construction. An ancestral feature or association becomes refined.

Complex

In our visit with the subject of co-evolution, we met complex associations. But complexity need not intimidate nor cause us to abandon what we have learned of natural selection so far. The mistake would be to try to explain complex associations of today as the results of an overnight evolutionary binge of yesterday. The fungus cultivation by Attini ants arose approximately 50 million years ago, through a series of small steps. Complex associations of today started as simple ecological partnerships of the past. Steps to the present may be small steps, each adaptive in its own right—not in anticipation of the bright future millions of years away, but adaptive for the immediate and favorable effects the change brings to those at the moment. But once a new association is in place, the dynamic of the ecological relationship changes, perhaps now setting the stage for the next evolutionary step. Critics of biological evolution argue that complexity alone speaks against the intervention of natural selection in living systems. "How," they ask, "might natural selection account for the complex associations we meet today?" The answer is: step by step, one step at a time. The mistake would be to try to account for intricate co-evolutionary associations in one big evolutionary boom. Instead, the steps add up. One successful change may make the next step advantageous, and more steps follow.

Selected References

Currie, C. R., J. A. Scott, R. C. Summerbell, and D. Malloch. 1999. Fungus-growing ants use antibiotic-producing bacteria to control garden parasites. *Nature* 398:701–4.

Mueller, U. G., T. R. Schultz, C. R. Currie, R. M. M. Adams, and D. Malloch. 2001. The origin of the Attini ant-fungus mutualism. *Quart. Rev. Biol.* 76:169–97.

Pough, F. H. 1976. Multiple cryptic effects of crossbanded and ringed patterns of snakes. *Copeia* 1976:834–36.

Schultz, T. R. 1999. Ants, plants and antibiotics. *Nature* 398:747–48.

Thompson, J. N. 1994. *The coevolutionary process.* Chicago: University of Chicago Press.

Wickler, W. 1974. *Mimicry in plants and animals. Reprint ed.* New York: McGraw-Hill.

Learning Online

Testing and Study Resources

Visit the textbook's website at **www.mhhe.com/ evolution** (click on the book's title) to take advantage of practice quizzing, study/writing tips, timely news articles, and additional URLs for research on the topics in this chapter.

11

"Natural history involves the study of not only anatomy, physiology, systematics, distribution, but also behavior and ecology."

Mary F. Willson, evolutionary biologist

Life History Strategies

INTRODUCTION

The life history of an animal reaches from conception to death, and along the way includes all the animal's characteristics—anatomical, behavioral, reproductive, and physiological attributes, from fertilized egg, to larva, to juvenile, to adult. To reproduce successfully, the individual must first be a successful juvenile, and before that, a successful larva, and before that, a successful egg. The life history of plants includes steps through a lifetime from gametophyte to sporophyte stages. Each stage in the life history of a plant or animal is under selective pressure to suit local environmental conditions. As a convenience, we sometimes speak of *life history strategies* to recognize the adaptive advantages of these patterns that characterize all stages of the life cycle from egg to adult. Certainly, plants and animals do not consciously reflect on their local environmental demands and then deliberately fashion the anatomical, behavioral, and reproductive features needed. As with other characteristics, life history features evolve under the guidance of natural selection, which culls varieties that do not meet the local adaptive requirements of habitat and life cycle stage.

LIFE HISTORY TRAITS

Lizards

The habitat of the side-blotched lizard, *Uta stansburiana*, stretches from central Washington State (United States) in the north to parts of Mexico in the south. Although all the populations throughout this range constitute one species, populations in the north and south differ in life history features (table 11.1), showing clinal variation in life history traits. Insects are the major food item in the diets of all populations. Individuals in the southern populations are larger, females in particular. Because mature females are larger, they lay more eggs than do females in northern populations. These differences correlate with differences in habitat. Southern lizards have a longer reproductive period, 5 to 6 months, compared to 2 to 4 months of favorable summer reproductive weather in the north. The

Table 11.1   Lizards

Life history features of the side-blotched lizard, *Uta stansburiana*. Notice the differences in body size especially in female size at maturity, and in the number of eggs produced during one breeding season between individuals at southern and northern extremes in this species.

	Southern	**Northern**
Food	Beetles, termites, ants, grasshoppers	Beetles, ants, crickets, grasshoppers
Body size	Large	Small
Female size	Large at sexual maturity	Small at sexual maturity
Eggs per season	More	Fewer

Source: From Pianka and Parker

longer season means that southern lizards are out foraging longer and grow larger in size. There is also indirect evidence, from scars and other markings on the body, that southern lizards experience more predation pressure than do lizards to the north.

These are the same species of side-blotched lizard, north to south, but their life histories are slightly different through this latitudinal gradient: there is a clinal variation in their life histories. These life history differences are correlated with differences in habitats, a response of lizard life histories to different selection pressures, producing clinal variation within the overall life history itself.

Guppies

The guppy, a common aquarium fish, is native to the small streams in Venezuela and the nearby island of Trinidad. Here, small populations can be found in relative isolation occupying freshwater pools separated by waterfalls and rapids. Over several decades, populations of guppies were studied through laboratory and field experiments by John Endler, David Reznick, and many colleagues. In multiple studies, these biologists documented adaptive changes in guppies' life history characteristics—the target of natural selection. We shall look at just a few examples.

Guppies in different pools can have quite different life history characteristics. In some pools, male guppies sport prominent secondary sexual characteristics: bright body colors and enlarged tail fins that trail like banners to exhibit flashy colors; in other pools, males are drab, lacking these particular secondary sexual characteristics (figure 11.1). Laboratory and field experiments show the reason why—predators. In the wild, guppies are preyed upon by fish predators, such as the large pike-cichlid and the smaller killifish. To test the effects of predators, researchers collected varieties of guppies from bright to drab in these pools. These were raised together in artificial streams in a greenhouse under simulated habitat conditions similar to the natural pools. Predators were added to some of these artificial streams; other streams were kept predator-free. After only about 10 generations, predator-free streams contained brightly colored male guppies; in streams where guppies were exposed to predators, males were drab (figure 11.2). A parallel study was conducted in the wild. Drab guppies were transplanted to predator-free pools. Returning after two years (about 6 to 8 generations), males in these predator-free pools were now colorful, just like those living in other predator-free pools (figure 11.1).

The bright, secondary sexual characteristics of male guppies certainly draw the intended favorable attention of females. Unfortunately for male guppies, these advertising colors also attract the attention of hungry predators. Bright males are picked off by predators more often than drab males, which consequently survive more frequently and enjoy a relatively higher reproductive success, passing along their drab features. Formally stated: If bright and drab guppies result from absence or presence of predators, respectively, then, in the absence of predators bright guppies should be

Consider This—

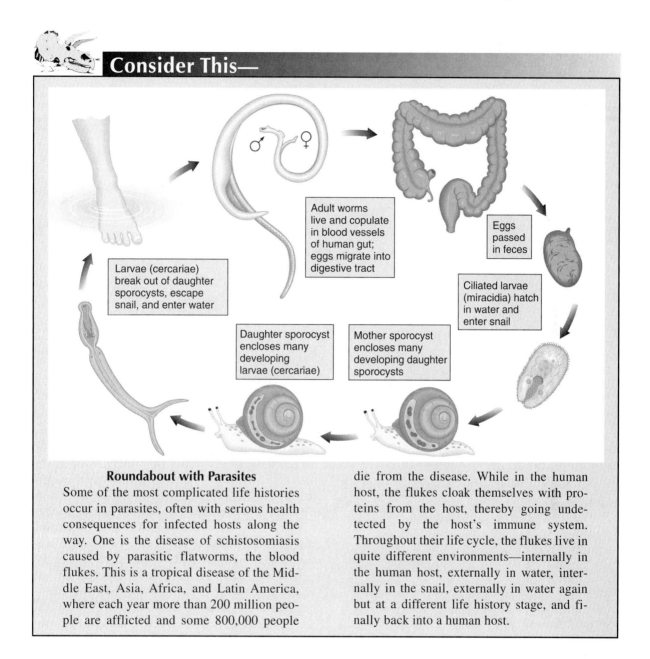

Adult worms live and copulate in blood vessels of human gut; eggs migrate into digestive tract

Eggs passed in feces

Larvae (cercariae) break out of daughter sporocysts, escape snail, and enter water

Ciliated larvae (miracidia) hatch in water and enter snail

Daughter sporocyst encloses many developing larvae (cercariae)

Mother sporocyst encloses many developing daughter sporocysts

Roundabout with Parasites

Some of the most complicated life histories occur in parasites, often with serious health consequences for infected hosts along the way. One is the disease of schistosomiasis caused by parasitic flatworms, the blood flukes. This is a tropical disease of the Middle East, Asia, Africa, and Latin America, where each year more than 200 million people are afflicted and some 800,000 people die from the disease. While in the human host, the flukes cloak themselves with proteins from the host, thereby going undetected by the host's immune system. Throughout their life cycle, the flukes live in quite different environments—internally in the human host, externally in water, internally in the snail, externally in water again but at a different life history stage, and finally back into a human host.

---

prevalent, and in presence of predators drab guppies should be prevalent. Both laboratory and field experiments support these predictions.

As an aside, notice that the bright guppy colors enhance reproductive success but decrease survival success when predators are present. Overall, fitness results from an overall trade-off between the advantageous (mating success) and disadvantageous consequences (attract predators) of bright, showy guppy color.

An assortment of other life history features correlate with predators as well. In some natural pools, guppies are large, reproduce later in life, but produce only small broods of young. In other pools, the guppies are small, reproduce at a young age, but produce large broods of young (table 11.2). One explanation for

FIGURE 11.1 Male Guppies, Friends and Foes Guppies occupy pools separated by waterfalls. In pools where predators are present, males are drab; where predators are absent, male guppies are brightly colored to attract females. Shown are a colorful and a drab guppy (*Poecilia reticulata*), and a predator, the pike cichlid (*Crenicichla altra*).

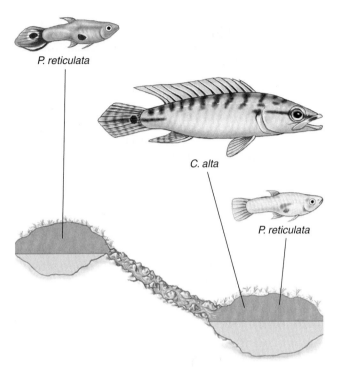

P. reticulata

C. alta

P. reticulata

these life history differences might be that guppies possess the ability to adjust their phenotype by simple physiological or behavioral alteration. This is termed **phenotypic plasticity**. By such a view, these life his-

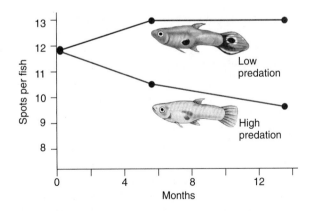

FIGURE 11.2 Male Guppies, Evolution of Color Varieties After several generations, guppies raised in low- and high-predation environments evolve different features. As measured by the number of bright spots, males become more brightly colored (low predation) or drab (high predation).

tory differences result from fine-tuning of the phenotype, not from different heritable, genetic differences. To test this, and eventually eliminate this possibility, researchers collected guppies with different life history traits from pools and raised them in the laboratory. The results were clear. By quantifying the similarities of relatives, scientists established that the differences in traits were heritable, not just physiological adjustments. The life history differences between populations (pools) are genetically based, and therefore the variation observed in the wild is likely the consequence of differences in fitness.

As with male color, predation is one factor contributing to the evolution of life history differences such as body size, reproductive timing, and brood size. To test this, guppies were transplanted from one pool to another, exposing them to different regimes of predators. Prior field observations noted that guppies are eaten by at least two different fish predators. One is the killifish, a small predator with a small mouth that preys mainly on small, young guppies. When killifish are present, most mature guppies are large. The other predator is the pike-cichlid, a large predator, that preys mainly on large guppies. When it is pres-

Table 11.2   Guppies

Life history features of guppies living in pools subject to predation by killifish and pike-cichlid predators. Transplantation of small guppies from pike-cichlid pools to killifish pools leads to evolution of larger-size guppies and other changes in their life history traits.

	Killifish	**Pike-Cichlid**	**Killifish (no guppies)**
GUPPIES Natural	Large Mature late Small brood	Small Mature early Large brood	
Transplant			Large Mature late Small brood

Note: Guppy sizes are exaggerated for emphasis

ent, most mature guppies are small (table 11.2). The transplanted guppies came from pools where the pike-cichlid was present, so the guppies were characteristically small early to reproduce, and produced large broods. These were transplanted into pools with killifish but no guppies were present prior to the transplant (table 11.2).

When the researchers returned (about 11 years later), the life history of the transplanted guppies had changed. Under predation from the killifish (prefers small guppies), the surviving adult guppies present were now large, they matured later, and produced small broods (table 11.2). The life history of the transplanted guppies had evolved, producing features that increased chances of survival from killifish predation and successful reproduction. The guppies that had escaped predation when they were still small had devoted available energy to rapid increase in size, and grew beyond the size prefered by their killifish predators. The consequence was to breed later but produce smaller broods.

How the trade-offs are made between body size, reproductive age, and brood size depends upon environmental conditions. With guppies, it depends partly on the predator and its preferences—a biotic factor.

With the side-blotched lizards, it depends partly on climate and growing season—abiotic factors. It is not easy to evaluate evolutionary consequences of life history strategies, but there is a tool at hand—time and energy budgets.

TIME AND ENERGY BUDGETS

There is no free lunch. A new adaptation brings a cost that must be covered from somewhere else. The reason is that energy is limited. An individual organism can gather up only so much of the available resources during its lifetime. The energy derived from these resources is limited, but the organism has many demands to meet. It must grow, contend with threats from predators, endure inclement weather, maintain and repair its tissues, reproduce, and meet many other environmental stresses. Consequently, the limited energy must be allocated judiciously among these essential, and often competing, survival and reproductive demands. Like a thrifty shopper on a budget, the energy available must be spent sparingly on many basic functions and not splurged on just one item. Time, too, is limited. There are only 24 hours in a day; a growing season lasts only so long. Both *time* and *energy* are in short supply, budgeted among basic functions. An organism's **time and energy budget** is limited. Not all demands can be fully met. Trade-offs must be made. One activity that receives more time and energy often must come at the expense of another activity. Let's look at some examples.

Abiotic Factors

The rock pipit is a small bird frequenting coastal regions of Britain and Europe. How it allocates time and energy to different tasks was studied by a biologist in the middle of the twentieth century. As rock pipits fed in the intertidal areas in and among rocks, searching for insects, J. Gibb recorded the amount of time they allocated to three activities—foraging, resting, and territorial defense (table 11.3). In one mild winter, pipits devoted about 72% of their time to foraging, 20% to resting, and 8% to territorial defense during daylight hours. In a harsh winter, these percentages changed, with foraging demanding more time (91%), time taken from the other activities: resting now received 8% of their time, and territorial defense 1%. In response to harsh environmental conditions, the birds adjust their activities, taking time from one to add to another. In harsh winters, insect food densities may be low. Moreover, rock pipits are endothermic, requiring more energy in cold weather. Overall, they must spend more time foraging and, consequently, less time in other activities. To meet different climatic demands, the rock pipit is likely making simple behavioral adjustments in its allotments of time and energy to various competing activities. However, in some organisms, differences in life history characteristics are genetically based.

The swift, a swallow-like bird, is a "determinate" layer. This means that the number of eggs a female lays is preset and fixed; she does not lay additional eggs if some are lost. In contrast, the familiar barnyard chicken is an "indeterminate" layer. When a laying hen returns to find her clutch size reduced (eggs harvested by the farmer), she lays more to replace the missing eggs. Up to a point, she will keep doing so, which is one reason the domestic chicken has been incorporated into farm production.

For the female swift, the fixed number of eggs laid is set by her genotype. Swifts are polymorphic; some females lay just two eggs, others just three eggs per clutch, but thereafter no replacement eggs. But, upon reflection, this is curious. Presumably, swifts laying three eggs instead of two would get more of their genotype into the next generation—about one-third more. Three-egg swifts should increase in frequency, replacing two-egg swifts, and come to characterize future populations. But this has not happened. This puzzle caught the attention of English ornithologist David Lack. To examine this, Lack unobtrusively located swift nests with two- and with three-egg clutches. He then followed parental success in raising the chicks from hatching to fledgling, the point at which they flew

Table 11.3   Rock Pipit

Time allocations of the rock pipit distributed among three activities—foraging, resting, territorial defense. Comparison between mild winters and harsh winters.

	Mild Winters		Harsh Winters	
	Hours	**% of day**	**Hours**	**% of day**
Foraging	6 1/2	72%	8 1/4	91%
Resting	1 3/4	20%	39/60	8%
Territorial Defense	3/4	8%	7/60	1%

Source: From Gibb 1956

Table 11.4   Swifts

Clutch size in swifts, 2 or 3 eggs, and average numbers successfully fledged. Mild and harsh years are compared.

	Clutch Size	Young Fledged
Mild Years	2	1.9
	3	2.3
Harsh Years	2	1.0
	3	0.9

Source: From Lack 1966

off on their own. In relatively mild breeding seasons, he found that, on average, more young were raised in three-clutch than in two-clutch nests, but in harsh years the reverse was true (table 11.4).

In harsh breeding seasons, food is less abundant and harder to find for the insect-eating swifts. Swifts producing three young face literally more mouths to feed, but have less food to go around. The time and energy of parental care is spread too thin, the three chicks suffer, and overall fewer young successfully fledge compared to those from smaller, more manageable clutches. By contrast, in good years there is enough food to go around for all the chicks in large clutches. Swifts investing time and energy into three young fared better. From year to year, conditions varied, favoring one or the other life history characteristic—clutch size—in different years.

Biotic Factors

Life history characteristics are also affected by biotic factors—competition and predation, for example. We saw this earlier in this chapter with guppies, whose life history features responded to predation or its absence. In particular, the reproductive strategy of an individual is often the target of adjustment.

Beans and Beetles

The Leguminosae are a family of bean plants. The Bruchids are a family of beetles, pea weevils. Both are residents of Central America. These beetles lay their eggs

in the developing pods of bean plants. When the eggs hatch, the larvae feed on the seeds. For the bean plants, each seed is a package of hereditary material producing offspring of the next generation. But the seeds must, of course, survive the beetle attack. One conceivable adaptive strategy of attacked plants might be to increase seed toxicity and thwart the beetles. Plants genetically disposed to make seeds with chemicals toxic to beetles survive, on average, better. The problem with such a strategy is that beetles evolve, too, producing detoxifying systems to neutralize seed toxins. It seems that this evolutionary route employing toxins was not taken by this family of beans. Instead, the beans under beetle attack evolved alternative life history features. In attacked plants, individual seeds are small, but they are produced in large numbers, yielding a high total output of seeds; chemical defense is reduced; vegetative growth is low. In plants free of attack, individual seeds are large, but fewer in total output; chemical defenses are in place; vegetative growth is high. For beans under attack, smaller seeds force the beetles to spend more of their time finding and processing them, reducing the foraging effectiveness of the beetles. By increasing the total output of seeds (total biomass), the likelihood increases that some seeds will escape the notice of foraging beetle larvae and survive. But the beans pay a price. The trade-off they make is lower growth and lower chemical protection from other threatening insects. However, when under beetle attack, this life history strategy returns overall greater fitness for these bean plants.

The evolutionary responses of the beans to beetle attack involve reallocation of time and energy. How do we represent this? Consider two genotypes with the same amount of time and energy, but different allocations. In one, much of the energy goes to growth and antipredator chemical defense (toxins); less goes to total reproduction (figure 11.3a). In the other, most energy goes to reproduction, drawing away from antipredator chemical defense and away from vegetative growth (figure 11.3b). Which of these two strategies returns the greatest survival benefits depends on how the gains and losses balance out. This, in turn, depends upon local environmental conditions. Where beetle attack is a problem, the beans allocating more energy to reproduction (small seeds, large total output) fare better, although at the cost of growth and chemical defense. Where plants are free of beetle attack, the beans allocating more energy to chemical defense and growth fare better,

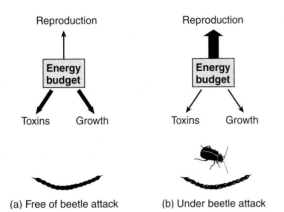

Reproduction

Energy budget

Toxins   Growth

(a) Free of beetle attack

Reproduction

Energy budget

Toxins   Growth

(b) Under beetle attack

FIGURE 11.3   Energy Budgets, Beans Free of and under Beetle Attack Life history strategies evolve under different environmental demands. This can be diagrammatically represented with alternative energy budget allocations. The size of the arrow represents the size of the energy investment. (a) Free of beetle attack, beans allocate more to *Toxins* and *Growth* than to *Reproduction*. (b) Under beetle attack, beans evolved a strategy of increased *Reproduction*, overwhelming beetles with a large output of seeds, but at the expense of *Toxin* production and vegetative *Growth*.

although at the cost of reproduction (large seeds, low total output).

The interplay of biotic factors can lead to adaptive changes in various features of an organism's life history. An organism cannot optimize and commit all of its energy to just one activity, or it would leave itself vulnerable to deficiencies elsewhere. The balance of trade-offs between competing demands determines an organism's success.

OVERVIEW

Successful reproduction is preceded by a prior lifetime of successful contention with the threats to survival. From egg to adult, an organism must possess the successive adaptations upon which its survival depends at each stage in its life history. A flowering plant putting forth its buds must, before that, have grown up to a mature individual, perhaps contending with stifling shade or threats from herbivores; before that, it must have germinated on welcoming ground in favorable weather; before that it must have been carried as a seed to a hospitable environment. At each stage—seed to seedling to

adult plant to flowering plant—the plant possesses adaptations serving it at each step along the way in its full life history. Failure at any step denies it reproductive success, an evolutionary failure.

Adjustments in a plant's or an animal's life history may result from phenotypic plasticity or from genetic changes. If variation is heritable (genetically based), then fitness amongst individuals varies and there will be differences, eventually, in relative reproductive success, with evolutionary consequences. Natural selection acting on heritable variation will cause these differences in reproductive outcomes. But as our examination of life history characteristics also illustrates, there may be multiple traits representing multiple or conflicting solutions to selection pressures.

The concept of time and energy budgets helps us think about how selection might strike a balance between contending traits in the same individual. For any organism, both time and energy are necessarily limited, because the hours of a day are limited and resources are limited. Compromises must be struck. Time and energy allotted to various metabolic and functional requirements must be apportioned in such a way that, overall, the organism meets the various demands of its environment so that it survives to reproduce. Plants must do many things well—grow and mature, produce defenses (structural or chemical), maintain tissues, and reproduce, to name just a few. But time and energy cannot be spent on just one activity, such as growth, or other activities will suffer. Such a plant may grow very tall, but succumb to predators for want of defenses or fail to reproduce for want of energy. Trade-offs must be made and compromises established between conflicting demands. Time and energy are apportioned between various activities. Organisms striking the best balance enjoy the relatively greater overall fitness. The challenges from predators, climate, competitors, and other environmental demands affect the overall distribution of time and energy among various activities and determine which trade-offs best serve the evolutionary success of the organism.

Selected References

Endler, J. A. 1980. Natural selection on color patterns in *Poecilia reticulata. Evolution* 34:76–91.

Janzen, D. 1969. Seed-eaters versus seed size, number, toxicity and dispersal. *Evolution* 23:1–27.

Lack, D. 1966. *Population studies of birds*. Oxford: Oxford University Press.

Reznick, D., R. J. Baxter, and J. A. Endler. 1994. Long-term studies of tropical stream fish communities—The use of field notes and museum collections to reconstruct communities of the past. *Amer. Zool.* 34:452–62.

Reznick, D., I. Mark, J. Butler, and H. Rodd. 2001. Life-history evolution in guppies. VII. The comparative ecology of high- and low-predation environments. *Amer. Nat.* 157:12–26.

Learning Online

Testing and Study Resources

Visit this textbook's website at **www.mhhe.com/evolution** (click on the book's title) to take advantage of practice quizzing, study/writing tips, timely news articles, and additional URLs for research on the topics in this chapter.

12

"It looks to the uneducated eye that species are *designed* to fit their environment, or that particular species as a whole are "trying" to adapt. The truth, of course, is very much otherwise. Individual organisms are simply pursuing their own procreative interests as best they can. As result, the species evolves to be better adapted. The appearance of higher level design or intentionality is an artifact."

Ronald N. Giere, philosopher of science

Life in Groups

INTRODUCTION

For many animals, life is lived in groups. In groups, potential mates are on hand; in groups, partners can be found to harvest a resource or stalk large prey; in groups, there is safety from threats. But in groups, there is a cost. The interests of the individual are often subordinated to those up the hierarchy. The efforts and risks taken by one individual can be exploited by the others. In groups there is give-and-take. For many animals, the costs of group life are too high, and life is lived mostly alone. The balance between solo and social life depends on the benefits of one or the other. Who wins and who loses in a group is not easy to determine. In fact, some biologists have argued that the group, itself, is the unit by which individual fitness is measured.

Alarm Calls

When danger threatens, an individual may fight or flee. Sometimes one or more members of the animal group may give off a call, usually a high-pitched yell or distinctive screech, and nearby members of the group scatter for cover or take wing in a burst of frantic flight. These vocal calls are **alarm calls** that alert others to the danger. From an evolutionary point of view, alarm calls are a puzzling behavior. If you have spotted an approaching predator, why call out and give away your position, thereby drawing particular attention to yourself?

Dangers emerge from many directions. For example, vervet monkeys of Africa produce alarm calls when approached by threatening, predaceous snakes, mammals, or large birds. Each type of predator poses a different type of threat, so different alarm calls are emitted. When a snake (e.g., python) is encountered, the alarm call is low-amplitude, alerting other monkeys nearby. They respond by looking at the ground, the most likely place to find a slow-approaching snake. When a stalking mammal (e.g., leopard) is discovered, the alarm call is a very loud, low-pitched series of chirps (figure 12.1a), and the monkeys scatter for a secure sanctuary in the trees. When a large bird (e.g., eagle) approaches from the air, the alarm call is short, loud, staccato grunts (figure 12.1b) signaling the approach of a menacing bird. The monkeys look up or just immediately beat a hasty retreat into dense, covering vegetation, making

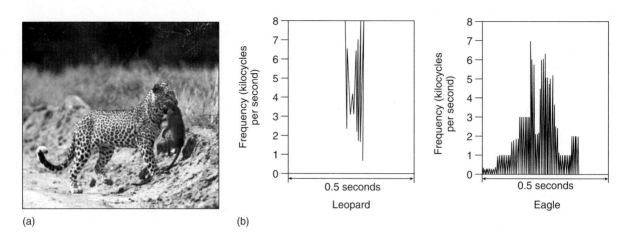

(a)   (b)

FIGURE 12.1   Alarm Calls   (a) Vervet monkeys fall prey to leopards, as well as to snakes and eagles. Each—snakes, leopards, and eagles— practice different predatory styles and arrive from different locations. Depending upon the predator, the monkeys emit different alarm calls that elicit different adaptive escape responses. (b) Shown are the abrupt chirp call (leopard) and staccato grunts of the threat-alarm bark (eagle).

it difficult for a diving eagle to maintain visual contact. The calls alone elicit the retreat response. If recorded tapes of the calls are played back, the vervet monkeys respond in these same specific and distinctive ways to each of the three types of alarm call.

Belding's ground squirrels of North America live in burrows excavated themselves. Here, they gather into large colonies and raise their families in these warrens tunneled into the earth. Predators threaten from the air, and from the ground—coyotes, weasels, badgers, to name a few. When a predator is detected, the spotter emits an alarm call—a whistle if the threat is by air, a series of short, sharp chirps if by land. When a hawk approaches, the first ground squirrel to spot the danger emits a whistle, and bedlam breaks out amongst the colony as all within earshot dash for safety into the nearest burrow. As a result, the hawk rarely secures a meal of ground squirrel. But if it does, the hawk more often catches a noncalling member of the colony, not the individual giving the alarm call. On the other hand, if a terrestrial predator approaches, the alarm call is given, all scatter for cover, again usually denying the ground predator a meal. But if the terrestrial predator is successful, the caller emitting the warning call is about twice as likely to be caught as a noncaller.

Here is the issue: Why would an individual ground squirrel emit an alarm call if that draws the predator's attention to itself and thereby increases the chances it will be caught and eaten? Alarm calls are common features in mammals and birds, especially if they are the subject of predator interest. Alarm calls are widespread and at

first seem to be disadvantageous to the individual signaling danger. If disadvantageous, our initial expectation is that natural selection should quickly eliminate conspicuous callers from the population, leaving the overlooked but consequently more successful noncallers. Clearly this has not happened. This paradox at first stumped evolutionary biologists.

One suggestion is that, in fact, individuals that give alarm calls benefit directly from the advantages of such a vocal warning. For example, alarm calls alert the predator to the fact it has been spotted. The calling individual signals that it is ready to make an effective escape, and the stalking predator, relying on stealth, moves on, having lost the decisive element of surprise. Or the alarm call sets colleagues all around into chaotic pandemonium, confusing the predator, deflecting it from the caller. The predator loses focus, and the caller escapes amongst the chaos. This seems to occur when ground squirrels respond to an aerial predator. Hawks, if discovered, have low predatory success, and their few captures are usually the noncallers. The caller appears to benefit from the mayhem produced by its alarm call. Or perhaps the caller gains by simply causing all neighbors to scatter to safety, denying the predator a meal, making the thwarted predator less likely to return in the future to this unproductive location.

But not all callers benefit. For ground squirrels, predators also threaten from the ground. When responding to terrestrial predators, the alarm caller actually may draw attention to itself, increase its exposure and vulnerability, and fall prey more often than its silent neighbors. Some biologists propose that in such circumstances the caller is acting "unselfishly." Such behavior is termed **altruism,** wherein an individual's trait or behavior (here, emitting alarm calls) *reduces* its own relative chances of successful reproduction (it more often gets caught and eaten), but this same trait or behavior *enhances* the relative chances of others in its group to survive and reproduce successfully.

Individual Selection and Group Selection

In the mid twentieth century, the idea of animal altruism gained followers as other examples of apparent unselfish behavior accumulated. In response to threats, male baboons step forward to intercept an approaching predator, thereby protecting females and young but increasing their own chances of meeting an untimely end. For some biologists, altruistic behavior was linked with the idea that group benefits outweigh individual advantages: Although, as with baboons, some individuals expose themselves to greater danger (males to predators), the group overall benefits. Some biologists saw selection acting at two levels: one was **individual selection,** acting on the particular phenotype of one organism, the other **group selection,** acting on favorable traits held in common by the group. Extending this, some envisioned individuals as engaged in "selfish" behavior (giving preference to their own individual advantage at the expense of the good of the group) but that group selection involved altruistic behaviors (individuals "gave up" their advantages for the good of the group).

Hang on, because these early conceptual issues became even more complicated. I promise we will sort this out shortly, but first let's finish this history—because, as we will see, this history and its misguided views are back to trouble us again today. Independently of these growing ideas about individual versus group selection, selfish versus altruistic behavior, a very influential book was published in the mid twentieth century by V. C. Wynne-Edwards entitled *Animal Dispersion in Relation to Social Behavior.* The book's primary argument was based on the observation that if we go out into nature, seldom do we see animal populations actually outstripping their resources. Certainly, animal populations have the potential for astronomical growth, but most populations seem to level off and hold their numbers in check at sustainable levels. Wynne-Edwards allowed for exceptions. Locusts and lemmings occasionally overproduce, but in most species he observed, food supply and population size seemed matched and in balance. The balancing

mechanism did not seem to be disease (he didn't see many diseased animals) or starvation (he didn't see many animals starving), so he reasoned it must be something else. He argued that we do see animals engaged in behaviors that limit overall group size. Birds on territories breed, the others don't, so recruitment of new individuals into the population is limited. Many animals have a social hierarchy, so that top males breed, lower-ranking males don't, and this limits overall population size. In other words, Wynne-Edwards argued, animals themselves sacrifice personal survival and fertility in order to help control population growth—group selection—based on altruistic behavior. Animals exercise the necessary personal restraint to keep group size in check and matched to food supplies.

Some generosity is in order when critiquing such a view. In its day, many of the remarkable and now long-term field studies of animals in the wild were not available. Today, even a casual observer of an adventure or nature television channel knows that many animals die of starvation; disease can decimate a population; and most animals die in their early years, sometimes in the first month of life. Culling is going on; the astronomical numbers of animals being produced are also being drastically reduced by harsh natural processes. Natural casualties, not unselfish behavior, bring numbers down to sustainable levels. So, the observations upon which group selection was based were overdrawn and, in some characterizations, were simply wrong. But the mechanism itself championed here—group selection—was flawed as well.

To understand, consider a hypothetical case, but one applicable to all instances of supposed group selection. Consider the "cheater"—the individual who exhibits "selfish" behavior, lets others do the dirty work, and consequently benefits from the "altruistic" sacrifices of comrades. Suppose that instead of joining other male baboons to defend from predators, the cheater holds back, hides in a tree, and avoids the danger other males meet. The males that step forward to defend the group exhibit "altruistic" behavior with group benefits; the cheater exhibits "selfish" behavior serving only himself. If the attacking leopard dispatches the defending males, then the cheater benefits. The other males are killed, he survives and thereby enjoys the subsequent reproductive success his dead companions do not. In stark, evolutionary terms, who has won? The cheater. Who has lost? The altruists. Group selection, as envisioned by Wynne-Edwards and others of his time, has no culling mechanism to select against the interests of the individual and allow group benefits to prevail.

Let's try to update this, bring it forward to our understanding of evolution today, and place it in a modern context. I will try to make good on my promise above to sort through this early confusion.

ALTRUISM VERSUS SELFISH BEHAVIOR

The terms in which the debate is cast are unfortunate terms. The terms are intuitive, certainly, but also misleading. They quickly stake out the differences in viewpoint. Group values (altruism) prevail or individual benefits (selfish) trump them. These terms are value loaded with human ethical judgments: altruism—good; selfish—bad. We are measuring animal behaviors by human ethical codes, rather than by the stark discipline of nature's own rules, which are indifferent to ethical standards. The lioness dashing across an African savannah after a prime wildebeest likely harbors no qualms about meat-eating, nor is she tempted to abandon her carnivorous ways and take up the life of a vegetarian. In contrast, humans do not act on instinct or cultural imperative alone. Religion and philosophical systems provide ethical standards to inspire action and against which human behavior is measured. And with human ethical standards come the judgmental terms of *good* and *bad*. Applied to animal actions, such value-burdened terms as *altruism* and *selfishness* encumber the ideas with self-righteous human motivations, not bare animal incentives.

In fact, even within human social commerce, these terms do not serve us well, and quickly fail as useful characterizations. Consider an individual who hands over money to a homeless person. Is this human giver altruistic or selfish? The saint would say altruistic—you give money to help another human being; the cynic would say selfish—you give money to make yourself feel pious. Almost any human action can be reduced to the self-serving. All that this says is that *altruistic* and *selfish* are ambiguous ethical terms that do not even serve us effectively in our own human society. They are even less appropriate when employed to describe animal actions.

To make these terms useful in biology, we need to define them more carefully, and free of pious human values. Altruistic behavior, biologically speaking, is characterized by loss of fitness by the giver to the benefit of neighbors. Selfish behavior is the opposite, gain for the giver at the expense of neighbors.

KIN SELECTION

When first envisioned, group selection assumed that altruistic traits are genetically based and thus transmissible to future generations. But it was also assumed, initially, that benefiting members of the group are not necessarily closely related. A breakthrough in sorting out the significance of individual and group selection came with the understanding that the degree of relatedness between altruist and beneficiary matters. Clearly, parents defending their vulnerable offspring have much invested, not the least of which is each parent's genetic endowment in each offspring: half from the father and half from the mother. But relatives also carry, in diminished proportions, shared genes. The concept of kin selection is intended to recognize this.

Inclusive Fitness

An individual benefits *directly* by helping its own offspring, and the individual benefits *indirectly* by helping relatives that carry proportionately some of the same genes, such as nieces, nephews, aunts, and uncles. Natural selection that favors actions benefiting offspring and relatives is **kin selection**. It is a kind of individual selection because individuals benefit—or, really, their particular genotype benefits—through kin benefits. The genetic contribution an individual makes to future generations includes the successful results of its own reproductive effort *and* the favorable effects that individual has on promoting its shared genotype in relatives. Together, the success, directly and indirectly, in promoting one's own genotype is **inclusive fitness**. The concept reminds us that when it comes down to evolutionary success, there are many favorable ways an individual can advance its own genetic survival into future generations—by promoting one's own offspring, by funneling aid to close relatives, or inclusively by doing a combination of both.

As mentioned earlier (chapter 7), ultimately evolutionary success is measured by reproductive success. Simple survival alone is not enough, if the organism fails to reproduce and pass along its genes to the next generation. Genetically speaking, parental care is very "selfish." Parents may exhaust themselves in rearing young and expose themselves to threats when defending their young offspring. But in such a draining and risky effort, they increase the chances that their genotype will survive and persist in the next generation. Consider this simple example.

Consider This—

SEX—What Good Is It?

An amorous person might find recreational value in sex; a poet finds love; an aristocrat propagates a family name. Biologically, sexually reproducing organisms ensure continuance of their genotypes into future generations. But why share the benefit? A child produced by means of sexual reproduction receives half its genes from its father and half from its mother. Or, stated the other way around, only half of a parent's genotype is in each offspring. Or stated another way still, every time a diploid organism sexually reproduces, it cuts its genetic contribution to the next generation by half. It does not matter how many offspring are produced. The total contribution is still halved.

Never mind the recreational value; biologically, you lose by half if you reproduce sexu-

ally, and such a losing behavior certainly would be evolutionarily disadvantageous. But sexual reproduction is common—sex does not seem to be going out of style, so there must be some evolutionary compensation, something to offset the loss. Reproduction by asexual reproduction produces clones—no variety, just more of the same. Offspring born of sexual reproduction are a new genetic mix of the two parents; they bring more variety and more possibilities in a varied environment. But is the added variety enough to compensate for losing half of one's genetic contribution to the next generation? Generally, this is not known. Fundamentally, and in adaptive terms, sex is still a mystery. Perhaps the poets are right.

When diploid organisms produce offspring, each offspring receives half of its genotype from each parent (figure 12.2). The same applies equally to the father and mother, but let's simplify this example by just looking at the mother's genetic investment. If she has three offspring, and each survives in turn to produce three offspring, then half of the mother's genotype is in each offspring, one-quarter in grandchildren, one-eighth in great-grandchildren, and so on, diminishing down through the generations. The evolutionary value of her parental care is that such care aids the survival of offspring carrying her own genes.

At two extremes are two possible genetic outcomes of parental care when a predator threatens. At one extreme, the female thwarts the predator's attack saving her young but she herself is killed. In this example (figure 12.3a), the surviving amount of her genotype is **1 ½**. At the other extreme, the female bolts when a predator attacks—saving herself and gaining the chance to breed again safely, but losing all current offspring. At that time, the surviving amount of her genotype is hers alone, just **1** (figure 12.3b), less than that from a successful defense even if she perishes. From a genetic standpoint, she does better in the extreme by defending and perishing. Strictly speaking, her parental care is "selfish" because it aids survival of her *own* genes—not those of the species and not those of the group.

This simple example illustrates the significant outcome in terms of reproductive success—getting your genotype into the next generation. Certainly, many intermediate outcomes are possible between these extremes. The female defends, survives, and loses only one offspring; she deserts and two young escape on their own; she produces more than three, increasing her total investment in offspring, and thereby increases the benefits of parental care. The best outcome genetically would be for both female and all offspring to survive.

Coefficient of Relationship

Because her offspring carry forward her genotype, the female enhances her fitness by advancing the success of her offspring. The amount of effort and risk she spends on others will depend on the amount of genetic relationship she shares with them. It is highly advantageous to promote one's own genes, and disadvantageous to spend time and energy on another's with whom one has no shared, genetic investment. This idea is expressed in the term **coefficient of relationship**. Here, a coefficient is just a mathematical expression of a biological principle. It expresses the degree or fraction of shared, identical genes between two individuals—for example, between mother and offspring or between brother and sister. As a female's offspring become more distantly

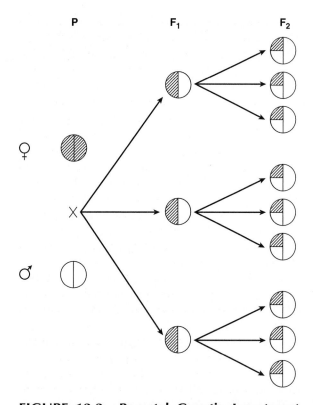

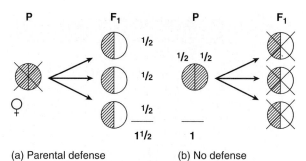

(a) Parental defense   **(b) No defense**

FIGURE 12.3   Parental Care   (a) The female parent defends and saves her three offspring, but she dies in the effort. Nevertheless, because half of her genotype is carried in the offspring (**F₁**), a total of 1½ of her genotype survives. (b) No parental defense and the female survives, but all offspring (**F₁**) perish, leaving a total of only one (1) of her genotype surviving (the same applies to males).

FIGURE 12.2   Parental Genetic Investment
When diploid organisms (**P**) produce offspring, each offspring (**F₁**) receives half of its genotype from each parent. In turn, these offspring (only the female is shown) produce progeny (**F₂**) carrying a quarter of the genotype of the original parents. The genetic contribution of the female is followed, although the same outcome applies to the male. Each individual is diploid, represented by the divided half; the amount of the original female's genetic contribution (cross-hatch) is followed here.

In animals, parental care is not a conscious decision, but, like anatomical traits, is genetically based. Kin selection is based on individual reproductive success, overall—inclusive fitness. Behavioral traits such as parental care persist, or not, based on their favorable, or unfavorable, treatment by kin selection. Suppose we removed eggs from parent birds and replaced them with eggs from an unrelated bird species. The switched eggs would likely receive parental care and the hatched young would enjoy protection from unsuspecting foster parents because the genes for such parental behaviors were still present in the host parents.

Some animals exploit this innate parental behavior of other species. When a nest with new eggs is temporarily left unattended, the female of another bird species may slip in and lay her eggs, unbeknownst to the host, leaving the host parents with the expensive task of raising the foster young. Such birds are *brood parasites*; they include cuckoos, finches, honeyguides, cowbirds, and perhaps one species of duck. The trespassing cuckoo female may actually pitch out the eggs of the host, leaving only her own eggs behind for the host parents to incubate and raise. Further, cuckoo chicks often hatch first, grow rapidly, and if any host nestlings still remain, the cuckoo chicks may muscle them out of the nest, sending them to certain death on the forest floor below (figure 12.4). This leaves the young cuckoo as the sole occupant of the nest and single recipient of the full, undivided attentions of the still apparently unsuspecting host parents.

related (grandchildren, great-grandchildren, etc.), the proportion of the original female's genotype in each becomes less and less, gradually diminishing the genetic benefit derived by assisting their survival.

The evolutionary significance of this was stated dramatically in 1932 by the population geneticist J. B. S. Haldane, who declared rhetorically that he would willingly lay down his life for *two brothers* or for *eight first cousins*. Human ethics aside, evolutionarily he was quite correct. Statistically, Haldane and each brother shared half of their alleles, so, genetically, one J. B. S. Haldane was equal to two brothers; similarly, he shared with each first cousin one-eighth of their alleles, making one Haldane equal to eight first cousins, and so on in diminishing returns of relatedness.

At the other extreme is the group and its collective characteristics. This is an old idea, with life still in it. Serious attempts have been mounted to find mechanisms by which group characteristics prevail over individual fitness—group selection. But care is needed. Groups of individuals evolve under the agency of familiar natural selection. In the group, individual fitness comes from the favorable action of natural selection upon individuals engaged in beneficial behaviors. Individuals engaged in behaviors producing successful propagation of individual genotypes into future generations have higher fitness than those not practicing such advantageous behavior. This is kin selection, a kind of individual selection, based on inclusive fitness. Here, the advantages still go to the individual, despite the fact that the individual lives in a group. The real issue is whether or not group good triumphs over individual benefits. Or, stated the other way around, do prosperous and adapted individuals relinquish their advantages for the good of the group?

The problem with claims of group selection (members of a genetically anonymous group—low coefficient of relationship) is twofold. First, there is no mechanism—no culling process—to selectively eliminate contending groups. There is, in short, no known selective agent operating on groups in nature. Second, group selection does not address how to contend with "cheaters"—individuals enjoying benefits at the expense of the group good. If members of a community share close genetic relationships, then fitness is inclusive and kin selection operates at the level of the individual or individual genotype, favoring traits and behaviors that promote success of related neighbors. But if members of a community are not closely related—a low coefficient of relationship—then kin selection does not occur and the genetically anonymous members are subject to individual selection, but no mechanism arises to promote group good over individual benefits.

Perhaps the strongest statement that can be made on behalf of group selection is that it may occur in mathematical simulations and in some limited laboratory experiments. But if it were an important biological process—one that is instrumental in determining the survival of groups based on group characteristics—then it should be widespread and evident in nature. Alas, no clear-cut examples of group selection have been found in nature. It is easy to understand why—no selective agent—there is no way to thwart cheaters that operate at the group level against benefits.

FIGURE 12.4   Brood Parasite   The large young cuckoo (right) has evicted the smaller young of the host meadow pipit (left) and begs for food.

Some parasitized species have evolved defenses. They build nests in concealed locations, and if spotted, they attack the brood parasite female before she lays her eggs. Sometimes the host parents detect the distinctive eggs of the brood parasite and evict them. In return, brood parasites evolve better deceptions. Their egg color may match that of the host; parasitic young might mimic the mouth, color, pattern, and begging calls of the host.

LEVELS OF SELECTION

Certainly, selective agents act on phenotypes—on the individual. But might there be selective forces acting at lower levels or higher levels? Some biologists think so. For example, Richard Dawkins argues that selection acts directly on DNA, the genetic material itself. The organism itself is just the tool to get the job done. Like the old refrain, a chicken is just the egg's way of making another egg; the organism is just the DNA's way of making more DNA. A humbling thought, but a little severe. The chicken may make more eggs, but first it must confront the discipline of natural selection. If the chicken fails, there are no more eggs. Similarly, if the organism fails, there is no more DNA from that organism. Organisms matter in this evolutionary contest, not just the DNA that assembles them in the first place.

Our critical, scientific alarms should go off when we hear the narrator of a nature program declare that some trait or other arose to "benefit the species" or is for the "good of the group." Such statements imply some sort of group selection, selection for a group trait that prevails over individual advantages. Usually, this is just sloppy and careless narration. Most likely, the trait is the beneficiary of kin selection, not group selection. You will have to draw your own conclusions. But despite its disappointing history, group selection still survives as an intellectual concept deployed by some, especially when looking at large-scale evolutionary events.

MICROEVOLUTION AND MACROEVOLUTION

Let's keep it simple, at least to start. The difference between microevolution and macroevolution depends upon how close to or how far away we view evolutionary events (figure 12.5). **Microevolution** is the evolutionary event seen close up, and zoomed in; we are concerned with patterns of change within a population or species. It is also a short-term perspective. Clinal variation, ring species, and changes in allele frequencies in or between populations are examples of microevolution. **Macroevolution** is an evolutionary event seen zoomed back, the overall pattern; the origin of species and higher-level taxa, such as families or classes, are the concerns of macroevolution. It is also a long-term view, attending to such phenomena as the rise, diversification, and demise of dinosaurs; the origin of birds; and the appearance of mammals.

Each perspective brings its insights. Macroevolution has been largely the province of paleontologists, who take a longer view of events unfolding through geologic time. By standing back from single species and looking at groups of species, the paleontologist focuses upon the origin of new genera, families, and even whole classes of organisms. Big changes characterize macroevolution. Microevolution focuses in on populations. Here is where the mechanism of change was sorted out. Acting upon variation within a population, natural selection culls those with unfavorable characteristics from those with favorable characteristics, leading to adaptive adjustment of the population to environmental demands.

Well back in the nineteenth century, paleontologists recognized that species appear "abruptly" in the fossil record, persist for a long time often with no apparent change, then just as "abruptly" disappear. This geologic pattern seemed to speak of discontinuity between species but was attributed to an artifact of geologic preservation—the incompleteness of the fossil record at the gaps. These gaps would disappear, it was thought, when intermediate fossils were eventually uncovered.

Quantum Evolution

Certainly, the fossil record is often frustrating in its capricious preservation of organisms from the past. But in the middle of the twentieth century, G. G. Simpson insisted that, capricious or not, this pattern was genuine, a reflection of a common feature of evolution itself. Species appeared suddenly in geological time, persisted, and then suddenly disappeared. He termed such sudden appearance, often of major new groups, **quantum evolution**. Simpson eventually attributed this pattern to the same culling mechanisms as are behind microevolution, except that they were speeded up during these short bursts of rapid change, summing up to macroevolutionary changes (figure 12.6).

Biologists generally agree that there are different patterns of evolution on different time scales—micro- and macroevolution—but not everyone is convinced the engine driving both is the same. At the microevolutionary level, natural selection acting on variation within a population may be gradual and continuous, adding one character to another, but something big seems necessary to drive big changes, and something sudden to produce sudden changes.

In Simpson's day, a menagerie of ideas already existed to describe and explain *saltation* (big jump) patterns of evolution. Early in the twentieth century, some geneticists invoked "mutationism" as an alternative to natural selection to account for the sudden and spontaneous origin of new species: A species enjoyed a relatively tranquil immutable period, but suddenly changed, producing the discontinuous appearance of a new species. Some biologists returned to Lamarck to find a quick way to add a new, major, acquired part to members of an existing species. Others found explanations within the organisms themselves, as if the direction or pattern of evolutionary change was driven by internal engines playing out a predetermined script with hops and skips in it.

Punctuated Equilibrium

In the early 1970s, two biologists, Niles Eldredge and Stephen Jay Gould, returned to the issue of saltational events and rates of change in the fossil record raised several decades earlier by Simpson (see figure 12.6). They

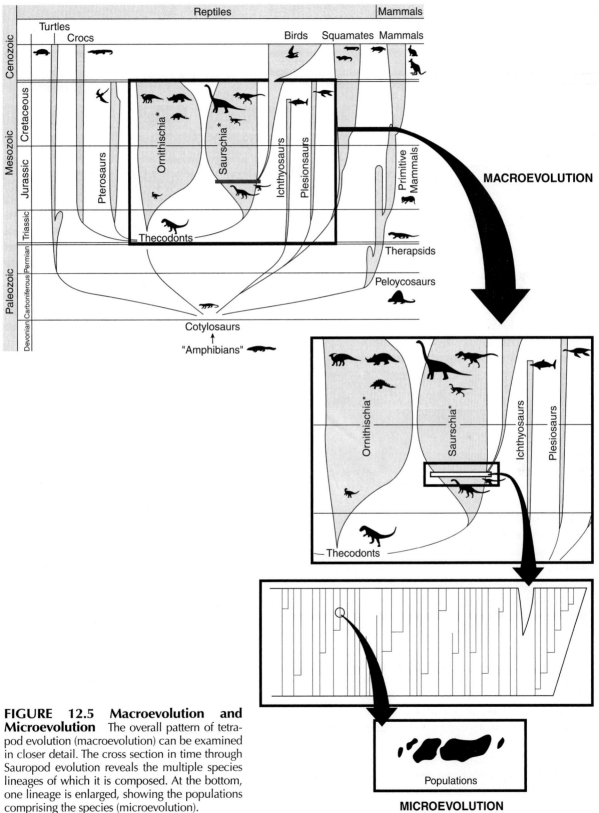

FIGURE 12.5 Macroevolution and Microevolution The overall pattern of tetrapod evolution (macroevolution) can be examined in closer detail. The cross section in time through Sauropod evolution reveals the multiple species lineages of which it is composed. At the bottom, one lineage is enlarged, showing the populations comprising the species (microevolution).

FIGURE 12.6 Quantum Evolution Within taeniodonts, a group of extinct placental mammals, two lineages evolved. One was the original group of taeniodonts, the conoryctines that survived into the late Paleocene; the other lineage was the stylinodonts, which evolved rapidly (quantum evolution) across a transition to a new adaptive zone (lifestyle). Compared to the beaver-sized conoryctines, the bear-sized stylinodonts evolved specialized dentition especially suited to rough and highly abrasive foods, well-developed claws, and strong muscles suggesting a digging foraging style. (After Simpson 1953.)

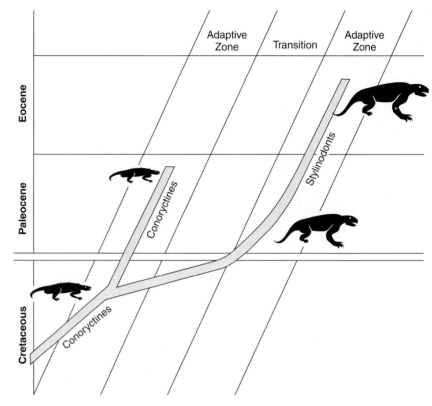

coined the term **punctuated equilibrium** to describe both a pattern and, eventually, a process. To Eldredge and Gould, the "gaps" in the fossil record did not result from discontinuities of preservation, but instead represent real biological events marked by rapid evolutionary change. A species persisted, more or less unchanged, for long periods which then were suddenly, in geological terms, punctuated by rapid change. This was punctuated equilibrium—long periods of little change (equilibrium) interrupted (punctuated) by sudden change. The punctuated moment is marked by speciation, thereby producing new lineages or, technically, clades; hence, the pattern is known as **cladogenesis.** This is in contrast to **phyletic evolution (anagenesis),** wherein a species undergoes transformation through time, eventually becoming quite distinct but without benefit of frequent speciation along the way (figure 12.7). Punctuated equilibrium, heralded as a bold new idea, would eventually have a bold, new engine to drive it—namely, species selection. This was proposed by Steven Stanley, for whom species become "individuals" and speciation and extinction equate with birth and death, respectively. The resulting process acts directly on species, leading to

their differential survival. This is **species selection.** Microevolution and macroevolution were decoupled; natural selection acts on individuals (microevolution), species selection on species (macroevolution).

Consequences of Punctuated Equilibrium

Today, punctuated equilibrium describes both a pattern (cladogenesis) and invokes a process (species selection). As to pattern, some fossil groups do exhibit cladogenesis, but others do not. The plausibility of the process is more in doubt, but bears on the issues of micro- and macroevolution.

Species selection is a close cousin to group selection, visited above and as such is vulnerable to the same criticisms. Up to a point, the contentions over species selection are a matter of personal philosophy. By one view, species don't do anything, only individuals do: only the sum total of individual outcomes makes a species; a species as an "individual" then is a mental metaphor, not a biological reality. By the opposite view, coherence among individuals comes from interbreeding, producing

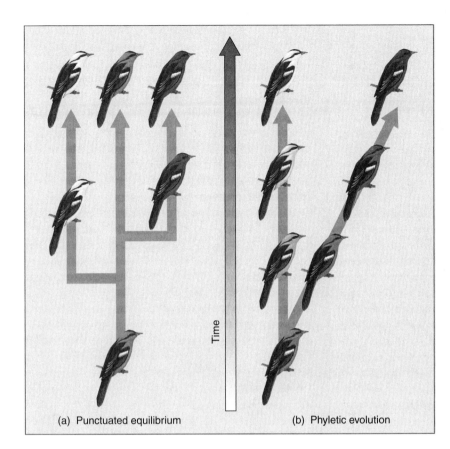

FIGURE 12.7   Patterns of Macroevolution (a) Punctuated equilibrium (cladogenesis) envisions evolution characterized by long periods of more or less unchanging species persistence, suddenly interrupted by speciation. (b) Phyletic evolution (anagenesis) envisions evolution occurring through continuous transformations through time.

(a) Punctuated equilibrium

(b) Phyletic evolution

Time

a reproductive community that has properties of an individual—birth (speciation) and death (extinction)—and so can be treated like an individual.

Even given the benefit of a doubt, species selection is more often claimed than confirmed. In fact, there are no clear-cut, unambiguous examples in nature. Since it would be a process operating over many thousands to millions of years, this is not too surprising. Still, the lack of such examples diminishes its plausibility. Further, it is hard to imagine what the selective agent might be, acting on the species as an "individual" rather than collectively through individual outcomes. Certainly, selective agents act on individual organisms (individual selection), as examples from this and other chapters confirm. But after almost three decades of looking, essentially no evidence exists for selective agents acting at higher levels of organization—groups, species, or higher taxa. It may be time to move on.

Punctuated equilibrium did emphasize that cladogenesis is one of, if not the most common pattern, of large-scale evolutionary change. However, the engine that drives it is not "species selection"; instead, it is the sum-mary results of the process of individual selection. Microevolution and macroevolution remain coupled. Macroevolution results from repeated rounds of microevolution—natural selection of individual organisms.

RAPID EVOLUTION

The debate brought about by punctuated equilibrium was over pattern (cladogenesis versus anagenesis) and process (species selection versus individual selection), but it also renewed interest in rates of change—rapid versus gradual changes. Darwin himself clung to the view that evolution is gradual, which explains why the amount of time available was for him, such an important issue (see chapter 2). Descent with modification by means of natural selection produces continuous change. Logically then, for Darwin, gaps in the fossil record are necessarily artifacts of preservation. Simpson recognized that discontinuities in the fossil record, especially between higher taxa, betray instead an unrecognized mode and tempo to evolution. For Simpson, quantum evolution was phyletic evolution speeded up, propelling a species from one adaptive zone to another, quickly crossing a transitional zone between (see figure 12.6). Punctuated equilibrium, too, celebrated rapid evolution, a result of frequent speciation, but not necessarily driven by accelerated adaptive processes (figure 12.8).

But how do big changes occur in relatively short periods of geologic time? Certainly, as we quest after big causes for big changes, no compelling evidence as yet exists for us to abandon the neo-Darwinian conditions and mechanisms we have followed to this point. Only if we meet real-world conditions where these explanations fail, only then must we abandon or amend these scientific explanations and cast about for alterna-

tives to the neo-Darwinian view. That would be exciting. At such critical moments, when a traditional view falters and a new explanation beckons, then science reveals its intellectually creative and personally satisfying rewards. But equally compelling is the discovery of greater depth in a traditional view. This brings the traditional view more richness, and brings us who contemplate it an exquisite pleasure to savor. That is the delight we shall discover next.

On the Edge

Earlier (in chapter 9) when we examined the process of speciation, we observed that when a barrier forms to di-

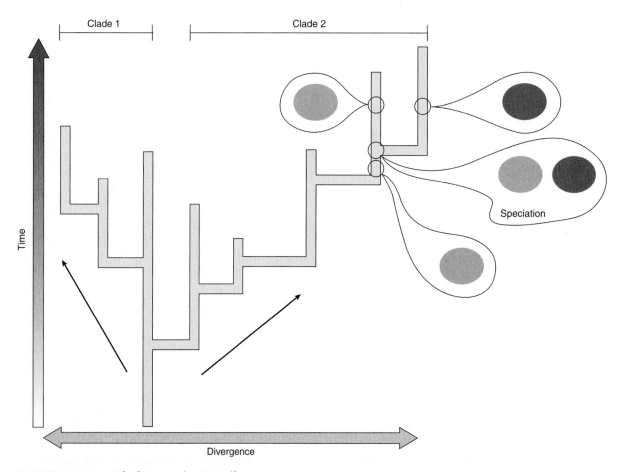

FIGURE 12.8   Cladogenesis, Details   Where sudden changes occur, they can be represented with an angular, branching phylogenetic tree (cf. phyletic evolution, figure 12.7). Each independent lineage produced is a clade, shown here as Clade 1 and Clade 2. Vertical sections represent more or less unchanging persistence of a species; branch points represent the time of speciation where populations diverge and become two distinct species. Time runs upward; species divergence is indicated along the horizontal scale. The balloons show details of the phylogeny in a species before speciation (light shading), at a branching point of speciation wherein two species form (light and dark shading), and the subsequent fate of each species thereafter.

vide a population, it often produces many fragmented and isolated groups out of the original, single population. Because isolated groups are smaller, favorable variation may collect there without being swamped by the overwhelming variety in the larger population, and new, favorable characteristics come quickly to predominate. This gives rare features a presence, denied within the larger population, and in a sense these characteristics are "seen" and "saved" locally by natural selection.

Contraction of a major population may strand a small population isolated within its borders. Or these isolated populations may be fragments, on the margins of the major species range—**peripheral isolates** (figure 12.9). Even within the large ocean of varied alleles in the major population, a new allele, bringing advantages to the individuals that carry it, will likely but slowly spread under the favorable effects of natural selection. But this process is accelerated in isolated populations because that is where new characteristics collect—they avoid being swamped, and their advantageous features more quickly accumulate.

Genetic Drift

Many peripheral isolates are populations at the extremes of a species' geographic range. Recall the example of the yarrow plant (chapter 9), whose range stretched from sea level at one end to high montane meadows at the other. Here, at the extremes of a species' range, individuals meet extreme conditions, relative to those of the major populations in the middle. At the margins, under relatively more severe environmental conditions, mortality may be high and population size fluctuates dramatically. In small populations, chance alone may determine which individuals, and the alleles they carry, prevail and survive. Alleles at low frequency in the large major population may by blind luck come to predominate in these smaller populations. Such random fluctuations of an allele resulting from

chance alone is **genetic drift**. Chance, not selection, rules. A depleted peripheral environment may be seeded repeatedly by a few dispersing colonists from the major population. Rare alleles may arrive with the few colonists and thereby become common in undiluted isolation. This is the *founder effect*—a type of genetic drift in which rare, pioneering alleles by chance become common in new populations they establish (see also chapter 6). A related type of genetic drift results from the *bottleneck effect*, wherein only a few out of the whole population breed, restricting genetic variability to just those individuals randomly enjoying reproductive success (figure 12.10; see also chapter 6).

Small, Isolated Populations

Life on the edge is risky. In small, isolated populations, chance events can wipe out a whole population. Disease, crash of the food supply, drought, or floods may be weathered by the major population but can decimate small populations without the critical numbers to recover. On the other hand, isolated habitats can be a kind of genetic sanctuary where random events and extreme conditions can shelter new but adaptive features that prosper and build to significant numbers. From here, colonists may disperse and even displace members of the major, ancestral population. Many biologists feel that this is a common feature in the origin of new species, with large consequences for macroevolutionary patterns.

Small populations, especially those living at the margins of the range, inhabit a narrow geographic range; they are largely allopatrically separated from the major "parental" population. Through genetic drift and differential selective pressures, they quickly build distinctive suites of unique characteristics. Full speciation may occur and/or adaptively superior characteristics may lead to their replacement of the ancestral population when they disperse out of isolation. When this occurs, two consequences for larger patterns of evolution

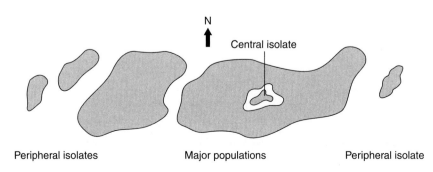

N

Central isolate

Peripheral isolates   Major populations   Peripheral isolate

FIGURE 12.9 Peripheral Isolates The geographic range of a species may include major populations fragmented into several smaller, isolated populations. These isolated populations can be peripheral to the main populations, or they may occur within a main population that has contracted around it. (After Bock 1979.)

FIGURE 12.10   Genetic Drift—Bottleneck Effect In the large "parent" population, there are about equal numbers of light and dark individuals. But by chance, the restricted few that pass the bottleneck may be mostly of one type of individual, such as dark. Consequently, the next generation produced by this limited sample of survivors will be predominantly dark. Random events produce changes in this evolving population.

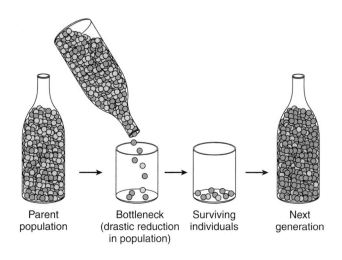

Parent
population

Bottleneck
(drastic reduction
in population)

Surviving
individuals

Next
generation

result: First, most new species likely appear on the margins of a geographic range where the small numbers of the new species are unlikely to leave significant fossil evidence of its occurrence and transitional changes. Second, the displacement of the ancestral species by the new species is rapid and, in the more protracted fossil record, appears as a "gap" between species.

A hypothetical example of the origin of new species out of small, isolated populations can be illustrated by a specific and detailed phylogeny (figure 12.11). An evolving lineage produces distinct species through time, A–H. Each species arises in isolated populations, under the chance and selective events just described, then spreads, occupying an expanded geographic range of its own, indicated by the expanded "balloons" (figure 12.11). However, if only the fossil record supplied our data of past species, then the apparent macroevolutionary pattern might look quite different. The fossil record would likely preserve a smattering of each species where it was most abundant, within its major populations. But the fossil record would be unlikely to catch the events in rapidly evolving, isolated populations. The fossil record would look as if discontinuities—gaps—occurred between species.

Many early paleontologists argued that with more diligence, and a bit of paleontological luck, the gaps

FIGURE 12.11   Fossils and Phylogeny Time passes vertically, and geographic distribution is shown horizontally. (a) The "balloons" indicate temporal and geographic distribution of species A–H. Gaps between species, as they might appear in the fossil record, are indicated by short vertical brackets at the left. *Location 1* indicates a fossil excavation site that preserves only part of the actual history of this evolving group. (b) The pattern of evolution expressed as a phylogenetic tree. Note that most new species originate in a peripheral, isolated part of the distribution (e.g., species B) or from a central, isolated population (e.g., species E).

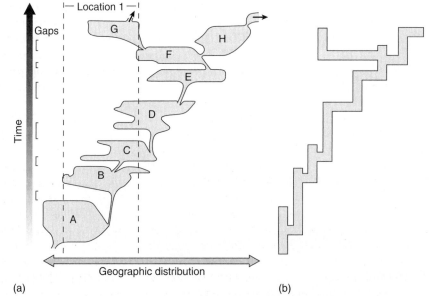

(a)

(b)

would be filled in. Quantum evolution, and later punctuated equilibrium, took a different view. The proponents of each argued instead that the "gaps" betrayed a real change in the underlying processes. Occasionally, processes were rapid, producing rapid changes speeding evolving organisms through transitional zones (quantum evolution) or through speciation events (punctuated equilibrium).

Certainly, the fossil record can be capricious. Suppose the surviving rocks represent only part of the full historical geological range of this hypothetical lineage (figure 12.11). Suppose, for example, that the only geological location available for exploration is restricted to *Location 1* (figure 12.11a), a narrow cross-section back through time. We might find remains of species A, B, C, and maybe D, but not any others until we reach G. The macroevolutionary pattern implied by such a fossil record would look quite different from the actual microevolutionary pattern behind it. Such apparent discrepancies between macro- and microevolutionary patterns have tempted some biologists to "decouple" the two, abandon natural selection, and look for alternative or additional explanations for overreaching patterns in the fossil record. The theory of species selection was one such effort. But the rapid events of natural selection and genetic drift, acting on individuals in populations, are sufficient to account for the macroevolutionary patterns. In fact, rapid changes occur in other ways as well within a microevolutionary context.

Macro Changes at Micro Levels

Evolution is remodeling—descent with modification. Consequently, each new species is built on the old, not from scratch. The refashioned ancestor with a few new, adaptive traits is the foundation of a new species.

Preadaptation

Often, old traits serve new functions. Feathers in birds, or in their immediate ancestors, evolved initially as insulation to conserve body heat. Like hair in mammals, feathers were an indispensable, energy-conserving feature. Flight came later in birds. Immediate ancestors to birds were ground- or tree-dwelling, reptilelike animals. As flight became more important, feathers already present for insulation were co-opted into aerodynamic surfaces—wings—to serve flight. Formally, we are talking about **preadaptation,** meaning that a structure or behavior possesses the necessary form and function *before* (hence, *pre-*) being remodeled into a new role it later serves. A preadapted part can do

the job before the job arrives. In birds, feathers arose from reptile scales to serve as insulation. But once present, and playing a favorable role in heat conservation, they were, in a sense, available for subsequent chores.

The idea of preadaptation does not imply anticipation. Feathers did not evolve at one time for service millions of years later in flight. They evolved initially for their advantages of the moment (insulation) not for their role in the distant future (flight). Looking back on this, we see that subsequent changes in lifestyles led to the appropriation of feathers (insulation) for aerodynamic surfaces (wings).

Vertebrate jaws evolved from the gill arches that preceded them; legs evolved from fins; penguin flippers evolved from wings of ancestors; dolphin fins evolved from legs; and so it goes. Examples of preadaptation abound, and the term captures an essential feature of evolution—namely, remodeling. If new features were built from the ground up, it would take immeasurably longer stretches of time for evolution to build new species. Each new species would await the arrival, simultaneously, of thousands to millions of new mutations producing each new body part, building the new species one piece at a time—integrating it, testing it, trying it. By comparison, descent with modification is fast. If new traits fare well in old bodies, then a new species arises from the ancestor.

Embryonic Changes

Another way to produce rapid changes is through major adjustments during embryonic development, based on genetic mutations that affect embryology. Lizards are reptiles, and some lizard species are legless. Of course, this alone does not make the legless lizards into snakes. Legless lizards still have lizard features of underlying bony anatomy, and they have eyelids and external ear openings that snakes lack. Arms and legs are so much a part of our anatomy, it is hard to imagine the benefits of living without them. But ground-dwelling lizards negotiate tight, crowded spaces, slipping between loose rock and dense bush. The absence of obstructing limbs allows the sleeked-down body to slip efficiently through such tight habitats. This major anatomical change from limbed ancestors to limbless descendants seems based on a major change in the underlying embryology.

In lizards with limbs, an early embryonic gathering of cells called a *somite* grows downward along the sides of the embryo at sites where fore- and hindlimbs are to form (figure 12.12). Here, the somite's lower growing tip meets special cells—mesenchymal cells—and together

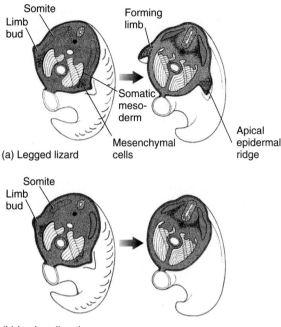

(a) Legged lizard

(b) Legless lizard

FIGURE 12.12 Lizards—From Limbed to Limbless Cross sections through the posterior end of the lizard embryo are depicted. (a) Legged lizard. The somite, an embryonic population of formative cells, grows downward to meet mesenchymal cells, which together stimulate the sprouting of the limb bud capped by an apical epidermal ridge. (b) Legless lizard. The somites fail to grow downward, thereby failing to initiate the limb bud, which regresses, producing an embryo that is limbless.

they initiate a "limb bud." Subsequently, as the lizard embryo matures inside the egg, these limb buds grow outward to sprout the limbs, which are ready by the time the young lizard hatches. In legless lizards, early embryonic events unfold similarly except that the lower tip of the somite fails to grow downward into the area of the prospective limb. This single change at this critical embryonic moment denies the limb bud the stimulation it requires to grow. These embryonic limb buds subsequently regress, and the lizard is born limbless.

Here a major adaptive change occurs from limbed to limbless, but the foundation of this change is basically a single, critical change in the underlying embryology. Within limbed lizard ancestors, a mutation occurred that interrupted the downward growth of the somites during embryology. Young were born without limbs, a new variation within the population. In the environment the

limbless young realized some competitive advantages (sleekness) over others with limbs (obstructions), and survived. Other changes were to follow, including changes in movement. And, of course, not all lizard environments would be favorable to limbless individuals, and they would perish there. However, where limbless features are advantageous, the shift from limbed to limbless occurred rapidly thanks to this fortuitous but critical mutation affecting the embryology.

Hox Genes

A more direct way to produce rapid, major changes in morphology is by gene action, especially by master control genes called *Hox* genes (see chapter 8). In animals, some genes, like military generals, control large armies of other genes. Through their action on other suites of genes, *Hox* genes regulate the appearance of major body parts, such as body regions, legs, antennae, and wings. A simple change in one of these master control *Hox* genes can produce a major change in body design. For example, limbless snakes arose from limbed lizard ancestors. Pythons are primitive snakes, legless of course, but with rudimentary hindlimbs. In snakes, the *Hox* genes that regulate forelimb development have deactivated normal forelimb development, leading to the absence of forelimbs. Specifically, it is proposed that the *Hox* genes controlling the expression of the chest region in lizard ancestors expanded their domain or sphere of influence. In a sense, the body of a snake is an expanded chest (figure 12.13). As the *Hox* gene for the chest zipped its domain through the body, limb development was simultaneously suppressed, producing the characteristic limbless condition of snakes today. As with limbless lizards, other traits were to follow, thereby consolidating the limbless condition of snakes. But the basic snake body was produced with just a few, but major, gene changes.

Evolutionary Significance

Such large-scale changes in design—from limbed lizards to limbless snakes—need not be built slowly, one small, single gene mutation at a time. Evolutionary modifications need not wait for a hundred gene mutations, each eliminating one finger, one joint, one muscle, one nerve, one part of the forelimb at a time; then another hundred to do the same for the hindlimb; and so on. Instead, big changes in morphology can be initiated by the relatively few but important master control genes, leading to rapid and big evolutionary changes.

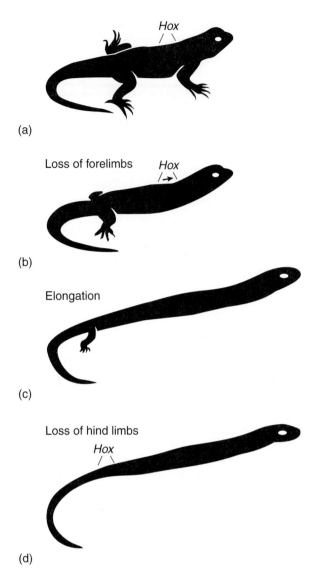

(a)

Loss of forelimbs

(b)

Elongation

(c)

Loss of hind limbs

Hox

(d)

FIGURE 12.13 ***Hox* Genes and Rapid Evolution—Lizards to Snakes** *Hox* genes associated with the chest region in lizards (a), expand their influence, leading to loss of forelimbs (b). By other changes in embryology, more vertebrae are added to the vertebral column, producing an elongated body (c). Either by a shift in influence of other *Hox* genes and/or by changes in limb bud growth (for example, see figure 12.12), hindlimbs are lost and an essentially modern snake body is produced (d). These steps may have occurred in a different order. Certainly, other changes accompanied these three basic steps to consolidate and integrate them. But apparently the major steps from lizard to snake are built upon only a few gene or embryonic modifications. Different *Hox* genes *(Hox)* are indicated at locations wherein mutations in them are hypothesized to produce a change in body design.

OVERVIEW

Life in groups has its advantages. Potential mates are close at hand; cooperative behaviors, such as by wolves in packs or lionesses in hunting prides, increase hunting success; where predators threaten, the vigilant eyes of many neighbors add spotters scanning for danger. But in groups, some individuals exhibit "altruistic" behaviors—curious behaviors because these seemed, when first described, to be actions detrimental to the individual but beneficial to the group. In some situations, selection acts on individuals (individual selection); in other situations, it seems to act at the level of the group (group selection).

However, group selection, as originally envisioned, fails on two counts. First, group selection includes no plausible culling mechanism, no selective agent that "sees" or acts directly upon group traits. Consequently, selection at the level of the group cannot act against individual "cheaters" within the group—those individuals that engage in behaviors beneficial to themselves but detrimental to the overall group. Second, most supposed cases of group selection observed in nature, in fact, when closely studied, collapse down to a special case of individual selection—kin selection. Ultimately, evolutionary success is measured by reproductive success, the passing of one's genetic endowment to future generations. And your genetic endowment is included not just in your immediate offspring, but also in your relatives—your kin. Relatives carry, in proportion to their relatedness, some of your genotype. Aiding close kin aids your own fitness.

An individual's genetic relatedness tapers off rapidly in adjacent populations, and for individuals in distant populations, it diminishes to near insignificance. Consequently, kin selection acts where the degree of relatedness (coefficient of relatedness) is high, which is usually within local populations. Certainly, a species is a reproductive community; gene alleles spread through the range of a species. But overall, there is no selective agent that acts on the species as a whole, in part because the coefficient of relatedness is so low. Instead, individual selection prevails.

Microevolution is biological evolution seen close up and in detail; macroevolution is evolution seen zoomed back, in its general overall patterns. Macroevolution, with its long-term perspective, seems to spot saltational events—jumps—producing discontinuities or perceived gaps in the recovered fossil record. Some attributed these gaps to non-Darwinian events. But most biologists, Darwin included, simply blamed the gaps on poor preservation. G. G. Simpson was one of the first to recognize that these saltational events between groups represented not an artifact of preservation, but a genuine characteristic of evolution, which he termed quantum evolution. Quantum evolution recognized both a pattern (anagenesis leading to big jumps between groups) and a process (speeding up of microevolutionary events during short bursts of rapid change); for Simpson, natural selection operates on individuals within populations. Punctuated equilibrium was also proposed as a way to account for saltational events, but it emphasized a different pattern—cladogenesis, the production of many separate lineages through the splitting of ancestral groups. Punctuated equilibrium recognized a pattern (cladogenesis producing big jumps between groups) and a process (macroevolutionary events proceeding via species selection); according to this theory, selection operates at the level of species. Alas, like the theory of group selection, the theory of species selection disappoints, failing to provide unequivocal examples from nature or a plausible selective agent acting at the species level.

Attempts to account for macroevolutionary patterns stimulated interest in rapid evolution itself. Individuals in small, isolated populations experience extreme conditions where natural selection and/or genetic drift may allow distinctive traits to collect. Favorable traits, that in the larger gene pool of the major population may be swamped, can build rapidly to significant numbers in small, isolated populations. The rapid changes that collect in isolated populations can produce new species that eventually replace ancestors. This kind of microevolutionary event accounts for macroevolutionary patterns.

Besides ecological situations favoring rapid changes, major adaptive shifts can occur within an evolving lineage. Preadaptation is the evolutionary process wherein ancestral structures come to serve in new ways—descent with modification. New species are not built from the ground up, but by remodeling of existing species. Building on ancestors is a faster way of constructing new species. Simple changes in embryonic development can produce profound and major changes in the young organism. A simple change in lizard limb embryology produces a major alteration—limblessness. Such new variation must still prove itself beneficial and survive natural selection. But if the variation arises in a favorable environmental setting, then such limbless individuals, bearing a major anatomical change, would survive and enjoy reproductive success. A major adaptive change would occur, arising suddenly as a consequence of a very simple embryonic modification.

Master control genes—*Hox* genes—control banks of genes that in turn manage the assembly of an organism. The change in just one or several of these master control genes can produce a leveraged effect, producing sudden, major anatomical changes within an organism.

Big changes can occur rapidly. Evolution is not always a slow and stately process. Sometimes it proceeds in quickened steps. The stage can be set with a fortuitous coincidence of environmental events—genetic drift, isolated populations, intense selection. In isolation, new and favored varieties build to critical numbers, then spread to outcompete ancestors. A new species replaces an old. Building the new by remodeling the old also accelerates evolutionary change. New species are not constructed from the ground up—one small part, then another, and another, and so on. Instead, the successful ancestor is remodeled, equipped with new traits that find adaptive favor. Embryonic modifications or gene mutations, if fortuitously striking critical morphological sites, produce major new anatomical varieties that depart drastically from their relatives. If these radically modified individuals are born into a hospitable environment, then survival is their good fortune and rapid evolutionary change the consequence. To the question, and sometimes layman's critique: "How could major and rapid evolutionary changes ever occur?" We can now easily answer, "In lots of ways."

Selected References

Bock, W. J. 1979. The synthetic explanation of macroevolutionary change—A reductionist approach. *Bull. Carnegie Mus. Nat. Hist.* 13:20–69.

Cohn, M. J., and C. Tickle. 1999. Developmental basis of limblessness and axial patterning in snakes. *Nature* 399:474–79.

Dawkins, R. 1976. *The selfish gene*. New York: Oxford University Press.

Eldredge, N. 1989. *Macroevolutionary dynamics*. New York: McGraw-Hill.

Eldredge, N., and S. J. Gould. 1972. Punctuated equilibria: An alternative to phyletic gradualism.

In *Models in paleobiology*, edited by T. J. M. Schopf. San Francisco: Freeman, Cooper, pp. 82–115.

Frazzetta, T. H. 1975. *Complex adaptations in evolving populations*. Sunderland Mass: Sinauer Associates.

Hamilton, W. D. 1971. The selection of selfish and altruistic behavior in some extreme models. In *Man and beast: Comparative social behavior*, edited by J. E. Eisenberg and W. S. Dillon. Washington, D.C.: Smithsonian Institution Press, pp. 57–91.

Lee, M. S. Y., and M. W. Caldwell. 1998. Anatomy and relationships of *Pachyrhachis problematicus*, a primitive snake with hindlimbs. *Phil. Trans. Royal Soc.* 353:1521–52.

Sherman, P. W. 1981. Kinship, demography, and Belding's ground squirrel nepotism. *Beh. Ecol. Sociobiol.* 8:251–59.

Simpson, G. G. 1953. *The major features of evolution*. New York: Columbia University Press.

Stanley, S. M. 1975. A theory of evolution above the species level. *Proc. Natl. Acad. Sci.* 72:646–50.

Wiens, J. J., and J. L. Slingluff. 2001. How lizards turn into snakes: A phylogenetic analysis of body-form evolution in anguid lizards. *Evolution* 55:2303–18.

Williams, G. C. 1992. *Natural selection: Domains, levels, and challenges*. Oxford: Oxford University Press.

Learning Online

Testing and Study Resources

Visit this textbook's website at **www.mhhe.com/evolution** (click on the book's title) to take advantage of practice quizzing, study/writing tips, timely news articles, and additional URLs for research on the topics in this chapter.

"More than any other time in history, mankind faces a crossroads. One path leads to despair and utter hopelessness. The other, to total extinction. Let us pray we have the wisdom to choose correctly."

Woody Allen, actor, director, comedian, philosopher

Extinctions

Mauritius

INTRODUCTION

Nothing lasts forever. Species, too, become extinct. What may be surprising is how predominant **extinction** has been in the history of life on Earth. The paleontologist G. G. Simpson estimated that of all the species of plants and animals to evolve since the Cambrian 544 million years ago, more than 99% are extinct today. *Lingula*, a strange shelled marine invertebrate, is a conspicuous exception, having survived unchanged in its anatomy since the Cambrian (figure 13.1a). The horseshoe crab is another survivor (figure 13.1b). Horseshoe "crabs" are arthropods, members of the same group as lobsters, crayfish, and crabs, but themselves are most closely related to spiders. A spiked tail projects from a solid, domed shell protectively arching over the body and jointed legs. They appear about 435 million years ago and, as a group, have endured, easily outdistancing trilobites and dinosaurs. Horseshoe crabs occupy marine shorelines where they have prevailed through ice ages, and shifting continents, and worldwide traumas that followed pummeling by meteors or asteroids. Nothing obvious in the anatomy or ecology of either *Lingula* or horseshoe crabs stands out to explain why they survived when almost all of their contemporaries perished.

Organic evolution is two-sided. Speciation produces new species; extinction eliminates them. The character and pattern of life around us today owe as much to what has been lost to extinctions as to what has been produced by speciation. Had the dinosaurs not become extinct at the end of the Mesozoic, mammals may never have experienced the opportunity to radiate as they did during the Cenozoic. And without the ascent of mammals, our own evolution as humans may have never occurred. It is intriguing to speculate what the world would be like if dinosaurs had survived. Would the bipedal, upright postures of some dinosaurs have freed their forelimbs from the mundane task of body support and opened the possibility for investigative behavior and the enlarged brain behind it? Would intelligent life then have evolved, not in a mammal's body (humans), but instead among the prosperous reptiles? Of course we do not know. But without the dinosaur extinctions, and those additional extinctions before and after, the life on Earth we meet today would look quite different.

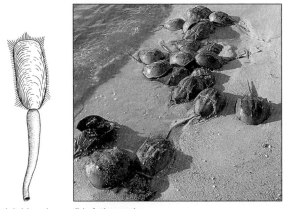

(a) *Lingula*   (b) *Arthropoda*

FIGURE 13.1   Survivors   (a) *Lingula*. A marine organism (brachiopod) that has lived in vertical burrows in sand and mud since the Silurian. (b) Horseshoe crab, an inhabitant of marine shores since the Ordovician.

On a long-term time scale, the "balance of nature" is dynamic—species come and go. In geological time, all species are but passing actors on the stage of life. Some go out with a bang, some with a whimper. Some taxonomic groups are carried out as part of large, widespread **catastrophic** or **mass extinctions.** Catastrophic extinctions include the loss of species from many different groups, take large numbers of species, and occur abruptly over a relatively short period of geological time. In **uniform** or **background extinctions,** members of taxonomic groups are lost gradually over long time periods without abrupt loss of large numbers. Most of the 99% of species now extinct were lost not suddenly, but slowly, as part of a stately demise, one by one, typical of uniform extinctions.

Why species become extinct is a difficult problem, not yielding to simple or easy answers. For example, horses arose in North America during the Eocene and spread eventually to Asia, Africa, and South America (see figure 6.5, in chapter 6). As the most recent glacial ice sheet went into retreat 15,000 years ago, horses became extinct in North and South America, but survived in Europe and Asia, and in Africa as zebras. Why did horses become extinct in the Americas? An obvious answer comes quickly to mind: As glaciers departed, the changing landscape and warming environment left behind habitats unfavorable to horses, and they perished. This is plausible. Intuitively, it makes easy sense. We have a correlation: change in habitat, loss of horses. We might live comfortably with this explanation were it not

for an inadvertent experiment performed by the Spanish. Four hundred years ago, the Spanish invaded North America, bringing with them a lust for riches, a thrust for religious conversion, and horses. This was the first time in almost 15,000 years that horses had set foot in the Americas. If landscape and climate were inhospitable, causing the earlier horse extinctions, then you would expect the same thing to happen again as European horses were reintroduced. Our prediction would be that again horses would meet the same inhospitable land that defeated them before, and with the same result—extinction in North America. But nothing of the sort happened. Instead, the reintroduced horses prospered under Spanish husbandry, and so did the horses that escaped into the wilds to take up a life on their own. Eventual dissatisfaction with Spanish rule led to uprisings of the Amerindians in the Southwest who suffered under the heavy-handed Spanish administration. Horses, previously denied to Amerindians, were now in their hands. Some horses escaped; others were traded to tribal nations to the north. Today, horses are reestablished in the Americas and even large feral herds prosper and grow to great numbers.

If glacier-altered landscapes and climates led to the demise of ancestral horses, then why would horses reintroduced to the same environment not face the same fate? The Spanish inadvertently tested our hypothesis of species extinctions—climate and habitat change. But a change in climate/habitat alone is not enough to explain horse extinctions in the Americas, even if at first it seems a compelling explanation. Simple and obvious explanations do not always hold for extinctions. Let's begin with uniform extinctions and see if we can be more careful in providing explanations for such loss of species.

UNIFORM EXTINCTIONS

Undeniably, our own human successes have resulted in loss of habitat and the accompanying loss of species that occupied those habitats. But so far as we know, there has been nothing like us before in the history of life on Earth. Plagues of bacteria and viruses probably have not caused the sort of devastation we have visited on other species. So if we are looking for natural causes of extinctions, before humans, we need to look to causes other than a single plague—human or bacterial.

Co-evolution

Some instances of species lost through uniform extinctions are easy to understand, at least in principle. Where two taxa are tightly coupled and interdependent in a co-evolutionary relationship, they are like a coupled set of evolutionary Siamese twins. The loss of one necessarily places the other at risk. Extinction of the island dodo bird in the seventeenth century endangered the *Calvaria* tree (see figure 10.14, in chapter 10). Its thick-coated seeds were adapted to withstand and safely pass through the grinding gizzard of the dodo digestive tract. Eaten at the mouth and exited at the far end, the seed's coat was thinned during transit; this enabled the embryonic plant within to break through the seed's coat and germinate in soil where it had been dropped. But without the dodo, there was no thinning of the seed's coat and young plants within could not germinate. The young tree embryos were imprisoned in their own protective coats. The adult *Calvaria* trees, already rooted, grew old but did not effectively replace themselves, and the species was in danger of becoming extinct like the dodo birds with which it was evolutionarily coupled.

The koala, an Australian marsupial, feeds exclusively upon the leaves of eucalyptus trees, whose toxic oils are harmful to other herbivores. The koala's digestive system neutralizes these toxins. But the koala is so specialized that it cannot feed and sustain itself on any other plant. It must eat eucalyptus. Should the various species of eucalyptus become extinct, the koala would certainly follow.

Where plants and animals are specialized and ecologically locked together, it is easy to understand how the demise of one species would be followed by the loss of its co-evolutionary partner. But instances of uniform extinction cannot always be reduced to such specialized conditions of interdependency.

Islands

Uniform extinctions occur worldwide. But the world is too big a landscape with which to start. Let's begin more modestly with a small piece of real estate—an island. Here, on a small scale, we can observe directly the actual factors that contribute to local extinction of species. Islands afford that opportunity. As we shall see, islands are not just a special situation. Events affecting species on islands also affect species over large continents. However, we shall begin first on islands to understand the processes affecting extinctions and then scale up to larger problems of extinctions covering continents.

Species–Area Relationships

The larger an island, the more species of plants and animals the island supports. Conversely, the smaller the

area, the fewer the species. For example, if we count the number of species of reptiles and amphibians on various West Indian islands and then plot these numbers versus the area of each island, we discover a direct relationship between the number of species and the area of the island (a **species–area relationship**) (figure 13.2). Why this should be so is relatively easy to understand.

In general, large islands have more substrate, more living space, more resources, and the opportunity for a more varied environment. Small islands have less of all this. Our plot in figure 13.2 shows the results. If we wished, we could visit these warm tropical islands and study these factors directly. A tempting suggestion indeed. However, let's look at the consequences of this relationship from the ease of our armchair. Let's make a simple graphic model, a representation of the complex factors affecting island species trimmed down to essentials.

What particular factors contribute to the number of species we might find on an island? If an island sits off the coast of a nearby mainland, then one factor contributing to new species' appearance on this island is the *immigration rate* of new species arriving on the island from the mainland. As this island fills up with species, the number of new species arriving should drop (figure 13.3a). Think of a specific example—perhaps ten species of birds reside on the mainland. As a result of migrations and storms, birds might arrive infrequently to an offshore island. If no birds exist on the island, then the chance that an arriving bird represents a new species is quite high and we would say that the immigration rate is high. Conversely, if all ten

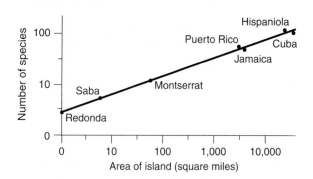

FIGURE 13.2   Species–Area Curve If the size of an island is plotted against the number of species present on it, a direct relationship usually holds—the more area, the more species. Reptile and amphibian species in the West Indies are plotted here.   (Data from MacArthur and Wilson 1967.)

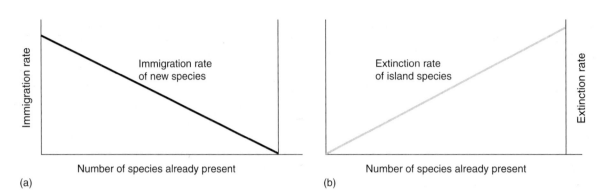

(a)

(b)

FIGURE 13.3   Immigration and Extinction on Islands   (a) Immigration curve. As colonists fill the island, the rate of arrival of new species drops. (b) Extinction curve. As colonists fill up the island, the rate at which species disappear increases.   (After MacArthur and Wilson 1967.)

mainland species already inhabit the island, then any arriving bird finds others of its species already present and the immigration rate is low—zero, in fact. The graph representing the changing immigration rate is the immigration curve (figure 13.3a).

A second factor, acting in opposition to immigration rate, is the *extinction rate*. This is the rate at which species already on the island go extinct. As the island begins to fill up with species, competition increases, resources become more scarce, and the rate of extinction therefore increases (figure 13.3b).

We are still in our armchair, contemplating island species. But if we actually measured these two curves—immigration and extinction—in nature, they might not be straight lines or they might slant at different slopes than those drawn in figure 13.3. For convenience, we shall keep the curves as straight lines. But whatever their exact shapes and slants, basically the immigration rate would decrease (slope down) and the extinction rate would increase (slope up) as the island fills with species.

If we visited this island, we would likely find fewer total number of species than the number present on the mainland. The missing species would be extinct from the island. Why this is so can be understood by putting the two curves together in the same graph (figure 13.4). This gives us a basic model of island equilibrium. Immigration and extinction act in opposite ways, tending to add or eliminate species, respectively. The trade-off between them establishes the equilibrium on the island. In our example, the rate of immigration and the rate of extinction balance out between the extremes, 0 and 10 (figure 13.4). Where these curves cross, an equilibrium is reached between opposing factors that affect the

number of species on the island. Think again of our 10 species of mainland birds. The number of island species settles around the equilibrium number, 6 species in our example. Now, if the island number drops to 3, then the immigration rate increases but the extinction rate declines, more species persist, and the equilibrium moves back to 6. Conversely, if the island number of species increases to 9, then the immigration rate decreases

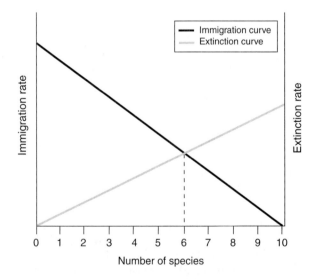

FIGURE 13.4   Species Equilibrium   Where immigration and extinction curves cross, an equilibrium number of species is reached. In this example, there are 10 species on the nearby mainland, making it possible for up to 10 species to be on the island. However, in this example the equilibrium sustainable by the island is 6.

(fewer new species arrive) and the extinction rate increases above equilibrium (more species already present are lost), and these factors contribute to a return to the equilibrium number of 6.

In actual practice, the equilibrium number will depend on the slope of these two curves, which in turn depends upon the actual ecological hospitality and resource richness of the particular island. But the general principle will be the same. Islands will reach an equilibrium number of species that is below that of the larger mainland nearby. In our bird example, the equilibrium is 6 species on the island compared to the 10 on the mainland. We have an explanation for the loss (extinction) of 4 species from this island. True, we may not know or be able to predict from this basic graph which particular species will be present or which absent. However, we can understand how the trade-off between the two opposing forces, immigration and extinction, will produce an equilibrium. More to the subject of this chapter, this graphic representation of factors acting on the island helps explain the cause of species extinctions should the number rise above equilibrium.

What happens if we change the conditions of our armchair model? Suppose we change the location of our island relative to the mainland. What happens to the equilibrium if our island is located farther from the mainland? And why?

Because it is the same island, but farther from the mainland, nothing happens to the extinction rate. The same factors that led to increasing extinctions with increasing numbers of species remain unchanged, so the extinction curve remains unchanged. However, the immigration curve adjusts. Located farther from the mainland, the chance that new species will reach the distant island decreases overall. Therefore, the immigration curve is lower overall (figure 13.5a). As a consequence, the intercept with the extinction curve shifts left and the equilibrium number declines. The farther an island is from the mainland, the fewer the number of species present because of its more distant position relative to a source of colonists. We can see exactly how this distance effect contributes to loss of possible species from our island. Distance affects the immigration rate.

Now suppose that we keep the island in the same location relative to the mainland but change the area of the island. What happens now? And why? If we make the island larger, the immigration rate increases because the island is a larger target, more easily located by

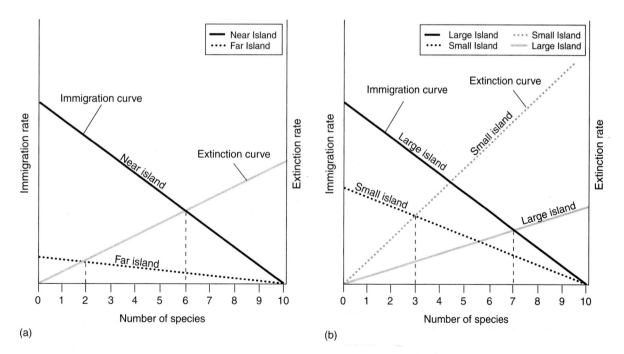

FIGURE 13.5   Distance and Area Effects on Species Equilibrium   (a) Distance effect. If our island (from figure 13.4) were moved farther from the mainland, the equilibrium would shift to the left, settling at 2, and some species would become extinct on the island. (b) Area effect. A large island reaches a higher equilibrium than a small island.

moving individuals, so the overall curve is steeper (figure 13.5b). The chance that a bird blown off course will spot the larger island increases because the larger island is a larger target. The extinction rate, however, declines overall because a larger island has more space and resources. Life is less tenuous, so the island can accommodate more species comfortably. Now the point of intercept between the steeper immigration curve and the shallower extinction curve is farther to the right. The average number of species at equilibrium increases. The reverse occurs if we make our island smaller, and extinctions would follow.

Island Insights

Early biologists studying island faunas and floras were struck by how few species were present on islands compared to the nearby mainland. Something about islands made them seem "impoverished." Of course, nothing of the sort is going on. Island populations are a matter of balance. Our basic model of species on islands helps us understand why islands usually hold fewer species than the nearby mainland. Compared to the mainland, islands offer a tenuous existence (expressed in relatively high extinction curves) and present problems with colonization (expressed in low immigration curves). Not one or the other alone, but the balance between immigration and extinction, accounts for the relatively low equilibrium number of species on islands.

Species can be marooned on more than just islands and be subjected to the same opposing forces of immigration and extinction. From the standpoint of a species, alpine mountaintops, separated bodies of water, or an isolated tree can be "islands." Foresters following an outbreak of spruce budworm may notice that a large, forested stand of trees ("mainland") is thickly infested, but many isolated trees ("islands") in a field escape colonization by the insect pest. What we have modeled on proper islands is also applicable to a variety of other situations in nature, where small parcels of habitat stand apart from large enclaves of resources elsewhere.

Most significantly, we notice the importance of area in our basic model. Reducing area reduces the number of species supported, and extinctions follow. Scaling up from islands to continents, this becomes especially important. Continents are large land masses, so not unexpectedly they support lots of species. But here is the difference. Compared to islands, immigration plays a small role in the equilibrium number of species supported on continents. Instead, immigration rate is replaced by the more important *speciation rate* that contributes to increased numbers of species. But if the area of a continent declines, then continental area is reduced and the number of species can be expected to decline. We have therefore identified one cause of extinctions—loss of living space. In turn, the loss of habitable areas can be caused by a variety of factors. Tectonic events in the Earth's crust or inundation by rising sea levels can reduce continental area. More recently, accelerated human activity has reduced habitat area with the inevitable result—accelerated loss of species.

Red Queen

In the early 1970s, Leigh Van Valen discovered a feature of extinctions that surprised even him. He expected extinctions to be abrupt, the result of some sudden, unfavorable change in the environment. What he found instead was that most groups of species became extinct gradually, at a constant rate characteristic of their taxonomic group. He discovered this by examining the fossil record of a group, beginning well back in the group's history, then following it forward through subsequent evolutionary time (figure 13.6). Tracked through time, the initial species within the group he began with were lost one by one, their numbers declining as the survey moved forward in time. Van Valen plotted the data (on semilog paper), producing a graphic description of the demise of this group. What he discovered was that rather than meeting some sudden and abrupt end, most groups declined steadily, losing species at a constant rate, until they disappeared (figure 13.7). Why this should be the pattern is not clear.

Certainly, Van Valen's numbers are skewed, as he recognized. Note that in figure 13.6, a species we begin with, C or J, does not really become extinct, but instead speciates into another species and therefore persists. But still the numbers and patterns are surprising. Van Valen speculated that, biologically, we are taking a page from Lewis Carroll's *Through the Looking Glass* and the adventures of Alice:

> Alice never could quite make out, in thinking it over afterwards, how it was that they began: all she remembers is, that they were running hand in hand, and the [Red] Queen went so fast that it was all she could do to keep up with her: and still the [Red] Queen kept crying "Faster! Faster" but Alice felt she *could not* go faster, though she had no breath left to say so.
> … Alice looked round her in great surprise. "Why I do believe we've been under this tree the whole time! Everything's just as it was!"

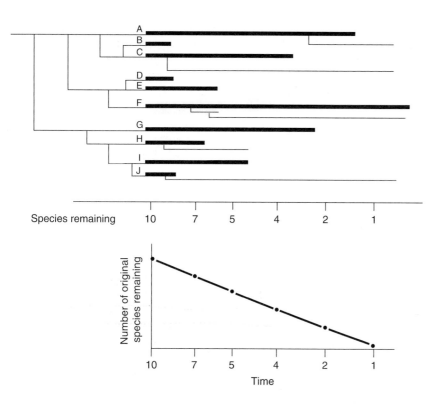

FIGURE 13.6   Phylogeny of a Group   Given is a family of fossil organisms. To follow declines in numbers over time, we pick a starting point and then follow these species forward in time (thick lines). Starting with 10 species (A–J) in this family, some become extinct, leaving 7, 5, 4, 2, and 1 remaining species on successive time intervals. Below the phylogeny, these numbers are plotted (semilog plot) showing the constant decline through time.

> "Of course it is," said the [Red] Queen. "What would you have it?"
>
> "Well, in *our* country," said Alice, still panting a little, "you'd generally get to somewhere else—if you ran very fast for a long time as we've been doing."
>
> "A slow sort of country!" said the [Red] Queen. "Now, *here*, you see, it takes all the running *you* can do, to keep in the same place. If you want to get somewhere else, you must run at least twice as fast as that."

Perhaps evolution is much the same, metaphorically, as Alice's adventure. It takes all the "running" a species can muster just to hold its own. Species continue to evolve to stay in the same place and survive. The evolutionary advance of one species is experienced as environmental deterioration by a competing species. Some species decline and others get better. As numbers decline in the face of an improved competitor, the challenged species must match the advancement or decline even more. Failing advancement, the number of members in a species falls; the chance of extinction rises with declining numbers as the dwindling species becomes more susceptible to chance events. One by one, members of a species fail to keep up with competitors,

lag behind, and finally flicker out. Eventually this leads to the observed constant rate of extinction. Like the inhabitants of the Red Queen's realm, a species must evolve faster just to stay in the same place competitively. If the species falls behind, it becomes more vulnerable to extinction.

Assessment of Uniform Extinctions

Uniform extinctions are common, perhaps the rule in living systems. Such extinctions are characterized by a constant loss of species from a group, at a rate dependent upon the group or species examined. Local "islands" help us understand the dynamics that could lead to loss of species on a regular basis. Among species of plants, mammals, insects, and marine invertebrates, the typical survival time of a species varies—marine invertebrates are the most resilient, with an average species survival time of around 30 million years, and mammal species generally are the least durable, averaging about 2 or 3 million years per species. Overall, this produces a characteristic "background" of extinctions—a low-level but constant and ongoing loss of species. At the other extreme are mass extinctions, appearing as spikes in extinctions above the background.

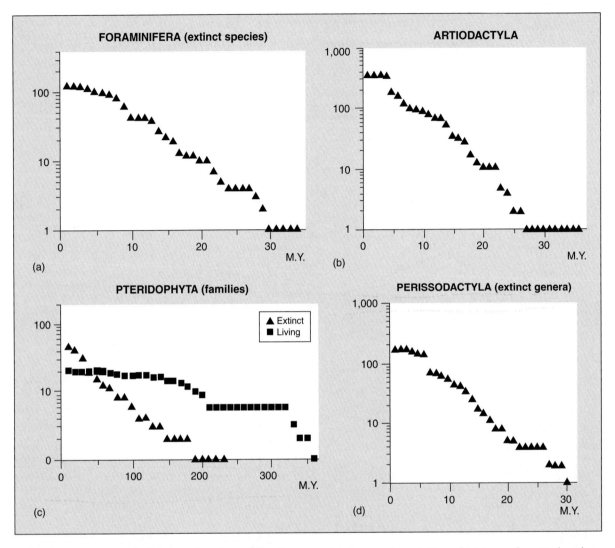

FIGURE 13.7   Evolutionary Survivorship Curves   (a) Protists: foraminifera. (b) Mammals: artiodactyla—deer, elk, and related species. (c) Pteridophyta: ferns and related species. (d) Mammals: perissodactyla—horses and related species. M.Y.: millions of years passed.   (After Van Valen 1973.)

MASS EXTINCTIONS

Perhaps when we think of extinctions, we think of mass extinctions. Everyone has their favorite. For most, it is probably the dinosaurs! After millions of years of pre-eminence and success, a group suddenly and dramatically becomes extinct. How could this happen? Such a special evolutionary drama seems to require special evolutionary explanations. However, the first task is to decide when such mass extinctions occur—not an easy job.

One attempt to mark mass extinction events starts reasonably. Select, at random, ten samples of genera from a large data set of fossils, then plot the maximum and minimum numbers becoming extinct over available intervals of time. The result is a graph of peak extinctions over the Phanerozoic, almost 600 million years (figure 13.8). Five common extinction peaks are noted—late in or at the end of the Ordovician, Devonian, Permian, Triassic, and Cretaceous. But notice that there are other peaks that might also lay claim to being mass extinction events. Characterized this way, we might see up to a dozen mass extinctions. Plotted another way (figure 13.9), the diversity of families of tetrapods can be graphed from their debut in the Devonian to the present. Within this curve are now six dips, geological moments of high extinction among tetrapods. A more comprehen-

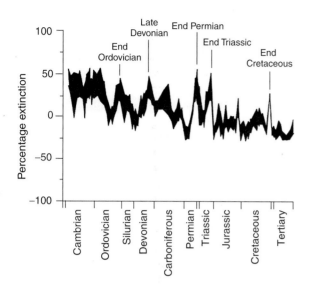

FIGURE 13.8   Extinction Episodes—Random Sample   Of more than 19,000 genera, 10 samples of 1,000 genera were chosen at random and the percentage becoming extinct was plotted. Major peaks of mass extinction are indicated.

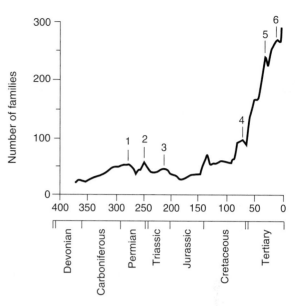

FIGURE 13.9   Extinction Episodes—Tetrapods The number of tetrapod families, the standing diversity, is plotted beginning with the first tetrapods in the Devonian. Six mass extinctions are indicated where diversity abruptly drops. Notice that the curve starts sharply upward during the Tertiary—this is a reflection of some increase in diversity, especially among birds and mammals, but also an artifact of the better preservation of fossils as we approach the present. (After Benton 1989.)

sive extinction survey, reaching back to the beginning of the Phanerozoic, is based largely upon families of marine vertebrates and invertebrates (figure 13.10). Five major, abrupt drops in the number of families are marked along the curve, and the relative magnitudes of these drops are given in parentheses.

We can search for episodes of mass extinctions in yet another way. The total extinction rate can be determined from the fossil record calculated as the number of families becoming extinct per million-year interval. When these rates are plotted statistically against geologic time, the mass extinction events stand out above "background" extinctions (figure 13.11). Five major extinctions are clearly evident. But such an approach is not without problems. Smaller, abrupt extinctions do not appear as noticeable changes in either figure 13.10 or figure 13.11. By other methods, up to five "minor" mass extinctions have been claimed for the Cambrian, two for the Jurassic, and one for the Cenozoic (end of Eocene). Should these be counted? It depends upon both preference and method. The number of mass extinctions depends on the group examined and upon how their numbers are analyzed. No matter the method or motive, mass extinctions have marked the history of life on Earth. Let's try an unmistakable and unchallenged example—dinosaur extinctions.

Dinosaur Extinctions—The Heated Debate

Dinosaurs (along with large marine reptiles and some groups of marine invertebrates) experienced mass extinctions at the end of the Cretaceous, just over 65 million years ago. But who were the dinosaurs?

In the late Paleozoic, the transition from fish to tetrapod occurred, yielding the first land vertebrates, the amphibians (see also chapter 5). From these first amphibians would arise the *cotylosaurs*, and in turn from them arose all later groups of tetrapods. The Mesozoic would become populated by these expanding tetrapod groups, especially within the reptiles (figure 13.12). Among the many groups of Mesozoic reptiles, two in particular would be especially abundant: the *ornithischia* and the *saurischia*. Together, these two reptile groups are the dinosaurs. Notice that dinosaurs are not the only group of Mesozoic reptiles, just the most abundant and diverse (figure 13.12). And the dinosaurs are the center of our story.

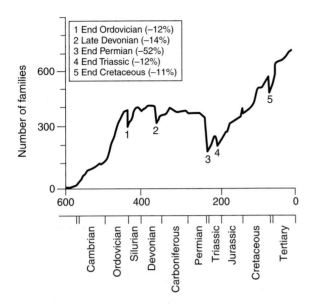

FIGURE 13.10   Extinction Episodes—Marine Animals   Plotted a slightly different way from figure 13.8, the standing diversity of marine invertebrate and vertebrate animals is shown. Five abrupt drops in diversity are numbered (1–5). The relative magnitudes of these drops were determined by measuring from the stage before to the stage after the extinction event. These relative magnitudes are given in parenthesis.   (After Raup and Sepkoski 1982.)

Many hypotheses about what caused their extinction are based on the kind of animals they were. Some suggest that dinosaurs were inferior competitors to the birds and mammals who replaced them. But such claims have been challenged on several levels. Most notably, birds and mammals were contemporaries of the dinosaurs, but never replaced them as dominant land vertebrates until after the dinosaurs became extinct. A popular view suggests that dinosaurs were warm-blooded, like birds and mammals, not cold and sluggish like lizards and snakes, and therefore physiologically equal competitors to birds and mammals.

To be exact, the issue is not really whether the blood of dinosaurs ran hot or cold. After all, on a hot day with the sun beating down, even a "cold-blooded" lizard can bask, heat its body, and, strictly speaking, have warm blood coursing through its arteries and veins. The issue is not blood temperature, hot or cold, but the source of the heat—internal or external. To clarify this debate, two useful terms need to be defined—*ectotherm* and *endotherm*. Animals that depend largely upon sunlight or radiation from the surrounding environment to heat their bodies are cold-blooded or, more accurately, **ectotherms** (heat from outside). Turtles, crocodiles, lizards, and snakes are examples. "Warm-blooded" animals produce heat inside their bodies by

FIGURE 13.11   Extinction Episodes—Families of Marine Animals   The number of marine animals going extinct per million years is plotted. The solid line (regression line) expresses the average—fewer than 8 species extinctions per million years. The parallel dashed lines are the statistical confidence limits that bound the "background" extinction rate. Elevated levels of extinction above background are "mass extinctions," indicated with the five points (1–5) corresponding to geological time intervals. (After Raup and Sepkoski 1982.)

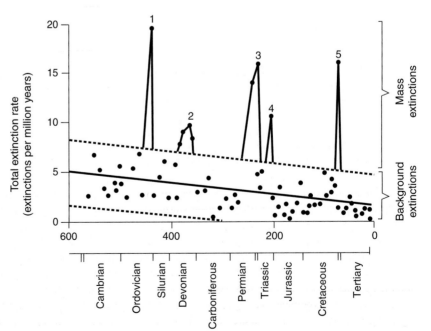

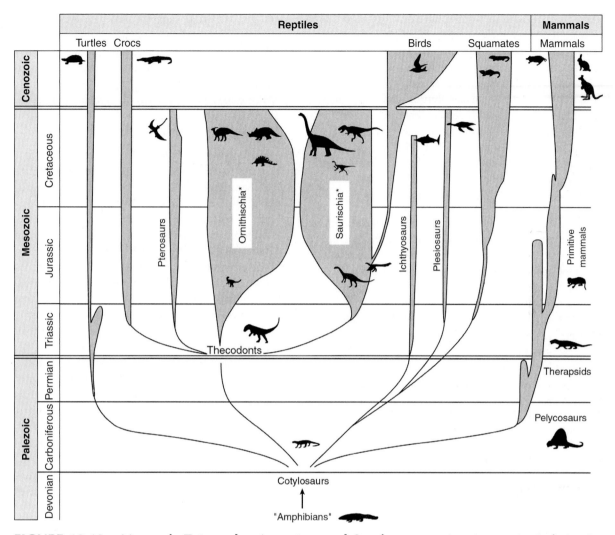

FIGURE 13.12   Mesozoic Tetrapods—Ancestors and Survivors   Together, the Ornithischia* plus the Saurischia* constitute the "dinosaurs." Notice that birds and mammals are early contemporaries of the dinosaurs.

metabolizing proteins, fats, and carbohydrates. To be more accurate, warm-blooded animals are **endotherms** (heat from within). Birds and mammals are obvious examples. The debate over "warm-blooded" dinosaurs is a debate over the source of their body heat—external or internal—not their blood temperature.

Heat for *ecto*therms is cheaply won. They need only bask in the sun. The trouble with such a lifestyle is that the sun is not available at night, nor is it always available in cold, temperate climates. By contrast, heat for *endo*therms is expensive. A digested meal, often caught at great risk and with great effort, produces fats,

proteins, and carbohydrates necessarily spent, in part, to generate heat to keep the endotherm body warm. Where endotherms have an advantage is that their activity need not be tied to heat available from the environment. These different physiologies are accompanied by different lifestyles. Ectotherms bask; on cold nights they become sluggish; and in freezing winters they hibernate. Endotherms remain metabolically active throughout each day and each season, despite the cold or inclement weather. Certainly, there are exceptions—bears and some small mammals hibernate—but endothermy usually requires continuous activity in a quest for fuel (food). Thus, the issue of warm-blooded-ness in dinosaurs is not just an issue of physiology but one also of the types of lifestyles they practiced.

Because dinosaurs are reptiles, they were for many years envisioned to be ectotherms just like their living counterparts—lizards, snakes, turtles, and crocodiles. There were occasional dissenters from this view, but the first person to assemble an embracing case for endothermic dinosaurs was Robert Bakker. He and several other paleontologists eventually marshaled four principal lines of evidence on behalf of dinosaur endothermy. Let's look at the arguments.

1. *Insulation*. First, some mid- to late-Mesozoic reptiles had surface insulation, or at least they seemed to. For ectotherms, a surface insulation would only block absorption of the sun's warm rays and interfere with efficient basking. But for endotherms, a surface layer holding in their internally generated heat might be an expected adaptation. Unfortunately, soft insulation is rarely preserved, but in a few fossils of the Mesozoic, impressions in the surrounding rock were suggestive of an insulating layer that resembled hair in some (pterodactyls, therapsids) and feathers in others *(Archaeopteryx)*. In fact, feathers likely first arose as thermal insulation and only later evolved into aerodynamic surfaces. Apparently, then, some Mesozoic reptiles appeared to have had surface insulation like that of endotherms rather than bare skin like that of ectotherms. (We will see in the next section that in fact this was a mistaken reading of the fossil impressions.)

2. *Large and temperate*. Second, large Mesozoic reptiles are found in temperate regions. Today, large reptiles such as great land tortoises and crocodiles do not occur in temperate regions—they live in warm tropical or subtropical climates. The only modern reptilian inhabitants of temperate regions are small or slender turtles, lizards, and snakes. The reason is easy to understand. When winter arrives in temperate regions and freezing cold settles in, these small, ectothermic reptiles squeeze themselves into deep crevices or small burrows where they safely hibernate until spring and escape the freezing temperatures of winter. On the other hand, for a large and bulky animal, there are no large and commodious cracks or crannies into which they can retreat to avoid the bite of winter cold. Large animals must be endothermic to survive in temperate climates. Thus, the presence of large reptiles in temperate climates

of the Mesozoic suggests that they were warm-blooded. This argument went something like this: Wolves, coyotes, elk, deer, moose, bison, and other large temperate-climate mammals today must be endothermic to survive in cold regions; thus, the large Mesozoic reptiles, living in temperate regions, would also have to be endothermic to see them through freezing winter temperatures.

3. *Predator-to-prey ratios*. The ratio of predator species to prey species argues for endothermic dinosaurs. Endotherms, in a sense, have their metabolic furnaces turned up all the time, day in and day out, to hold a high body temperature. Therefore, compared to an equal-size ectothermic predator, an endothermic predator requires more "fuel" in the form of more prey, to keep its metabolic furnaces stoked. Bakker reasoned that there should be few predators but lots of prey (lots of fuel to feed the few predators) in ecosystems dominated by endothermic reptiles. But if ectothermic reptiles dominated, then less heating fuel (prey) would be needed and, so, proportionately more predators should be present. By selecting strata that stepped through the rise of dinosaurs, Bakker compiled the ratios. From their ancestors into dinosaurs, the ratio of predator to prey should drop if dinosaurs were evolving an endothermic physiology. That happens. As this ratio was followed from early reptiles, to pre-dinosaurs, and to dinosaurs, it dropped. As the time horizon reaches dinosaurs, the strata hold proportionately fewer predator species and more prey species.

4. *Bone structure*. The microarchitecture of dinosaur bone is similar to that of endothermic mammals, not to that of ectothermic reptiles. Bones of ectothermic reptiles show growth rings like those of trees and for much the same reason: they grow in seasonal spurts. Endothermic mammals, with constant body temperature year round, lack such growth rings in their bones. When various groups of dinosaurs were first examined, the microarchitecture of their bones seemed to tell a clear story—no growth rings.

Critics of this four-part case for warm-blooded dinosaurs challenge the various lines of evidence. Insulating hair or feathers were present in a very few Mesozoic reptiles, and even in a few likely dinosaurs, but other than a few exceptions, the fossil record preserves no evidence of an insulating coat of feathers in most dinosaurs. Fossils from so-called temperate regions may actually come from subtropical deposits or represent dinosaurs that migrated to mild climates during cold winter months. Calculations of predator-to-prey ratios are not so straightforward and may reflect capricious fossil collection. A nearly constant body temperature may result from large size alone; therefore, evidence from the microarchitecture of bone need not reflect an endothermic physiology. The debate continues.

The important point to keep before us is that dinosaurs were in their own right an extraordinary group. These active animals occupied almost every conceivable terrestrial habitat. Their social systems were complex, and the adults of some species were enormous. If dinosaurs were endotherms, their complete demise at the end of the Mesozoic can only be more mysterious and the loss of the awesome splendor of this group all the more intriguing. This view of "hot-blooded" dinosaurs has reached the movie theatres, with menacing, active, hot-breath dinosaurs dashing across the screen and gobbling up simple scientists on remote islands or panicked pedestrians in city streets. But does this initial, four-part case of promising evidence withstand the challenge of careful critique?

Dinosaurs: The Sequel—After *Jurassic Park*

Seldom does the first announcement of a new theory meet with instant scientific acceptance. We should be professional skeptics until the evidence is evaluated, independently examined, and checked again. When scientists do this, the result is often the appearance of a new perspective, different from any of the theories that guided us in the first place. Dinosaurs, warm- or cold-blooded, may be an example.

Insulation

Contrary to first reports, Mesozoic reptiles lacked hair. Recently described pterosaur (flying reptile) fossils, extraordinarily well preserved, show that the flight membranes stretched across their forearms were internally supported by an exquisite network of ordered connective tissue. Superficially, this produced a fine-lined pattern on the skin, which initially was mistaken for hair. In fact, no hair was present. Instead, the fuzzy pattern was an artifact of preservation produced by the frayed connective tissue upon disintegration. This internal wing webbing, reacting to air pressure, permitted the wing membrane to

shape itself into an aerodynamic surface meeting demands while in flight. But it was not insulation.

Bone Structure

Although some dinosaurs seem to lack growth rings typical of ectotherms, and therefore meet one prediction of endothermy, some early birds do not. Bones from primitive Cretaceous birds, enanthiornithines, show evidence of annual growth rings like those of ectothermic animals. If these birds were ectotherms, then likely their immediate known relative, *Archaeopteryx* was ectothermic as well, as were the primitive saurischian dinosaurs from which *Archaeopteryx* presumably evolved. Finally, recent examination of bone from an early dinosaur sauropod, *Massospondylus*, revealed faint growth rings, at least in this particular dinosaur. Growth rings in bones, present or absent, may eventually speak to dinosaur physiology (cold- or warm-blooded), but currently it is not clear what, if anything, growth rings may imply about an animal's physiology. Further, it is not clear how widespread growth rings were in dinosaurs and their ancestors. On the other side of this debate, evidence is building directly against the theory of dinosaur endothermy.

Noses

Turbinates are thin folds of bone in the nose, covered in life by moist membranes, across which air is directed entering and departing from the lungs (figure 13.13). These moist membranes, supported by the turbinates, warm and humidify entering air and dehumidify departing air, thereby recovering water otherwise lost when air is exhaled. Where the breathing rate is high to support endothermy, turbinates are present in the nasal passage. Mammals have them; birds have them, or at least birds have analogous bones. If dinosaurs were endothermic like mammals and birds, then their noses should be similarly packed with turbinates. But dinosaur noses apparently lacked turbinates. CAT (computer-aided tomography) scans of dinosaur fossils show no evidence of these respiratory turbinates.

Today, the evidence for dinosaurs being warm-blooded is much less convincing than it was at first. Nevertheless, dinosaurs were remarkable. Their growth rates from young to adult were apparently high, much like the growth rate of birds and mammals. And they seemed built for active lives. The debate has not yet cooled; dinosaur heresies may heat up again. For our purposes in this chapter on extinctions, dinosaur physiology, hot or

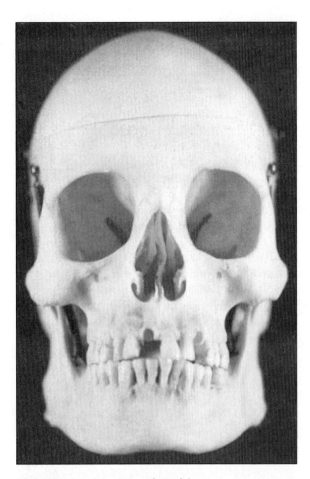

FIGURE 13.13   Nasal Turbinates Frontal view of a human skull. The paired orbits bound the single teardrop-shaped nasal opening between. Seen deep in the nasal opening are the vertical, thin turbinate bones.

cold, will not enlighten us on their demise. Their physiology is still unknown, and we shall have to look elsewhere to explain their extinctions.

Causes of Dinosaur Extinctions

It has been a great pastime in popular culture and cinema to guess at the causes of dinosaur extinctions—from a great plague to overhunting by aliens in flying saucers. Scientific attempts include many hypotheses. One hypothesis suggests that the rise of angiosperms (flowering plants) indirectly led to dinosaurs' demise. Flowering plants arose at about the same time as small mammals diversified, about 130 million years ago. Flowers supported insects (pollinators), and insects were the major food of small early mammals. This gave mammals the edge, or so goes the argument. But this correlation between the rise of flowering plants and demise of the dinosaurs is more likely a coincidence, and a logical stretch to go from flowering plants to insects to small mammals to toppling large dinosaurs.

Another proposal, in various forms, suggests that birds, or mammals, or both—being warm-blooded—enjoyed competitive advantages over the sluggish dinosaurs and outcompeted them. Even if dinosaurs prove to be ectotherms, many were large and active, and they likely held their body heat from day to day simply because of their massive size. Further, to see ectotherms as inferior to endotherms is a human conceit. Most vertebrates of the world today are in fact ectotherms, surviving quite nicely alongside endothermic competitors. Perhaps the most compelling evidence against this hypothesis is that mammals, and at least modern families of birds, never became predominant until *after* the demise of the dinosaurs. Recall that both birds and mammals were early contemporaries of the dinosaurs, and remained so for millions of years (see figure 13.12). But not until after the mass extinctions of the dinosaurs did the great diversifying occur in these groups. The pattern of bird and mammal evolution is to fill in behind the declining dinosaurs, not push them into evolutionary oblivion.

Mass extinctions are difficult to examine, and certainly more than dinosaurs were affected. Let's look to more general causes than just what might have led to dinosaur extinctions.

MASS EXTINCTIONS—CASE STUDIES

Let's look at some test cases of extinctions. We have at least five major mass extinctions during the Phanerozoic from which to choose (see figures 13.8 to 13.11). The first occurred at the end of Ordovician, with 85% of marine species becoming extinct within about 10 million years. Second was the late-Devonian extinction, which lasted less than 3 million years and took an estimated 83% of marine species. The third extinction occurred at the end of the Permian, and saw a loss of 53% of marine invertebrate families. This late-Permian extinction emptied marine habitats around the world and was considerably more devastating than even the dinosaur extinctions. Because most species were concentrated in a few expiring families, it is estimated that these vanishing families contained 96% of all species that perished. But it is unlikely you have heard much of it because the Permian extinction occurred before many large, charismatic land vertebrates had evolved, so little notice has been taken of it by the popular press.

Consider This—

The North Pole Is Headed South

Sea turtles, and birds, and compasses know north from south because of Earth's magnetic poles. The molten core, composed mostly of iron, flowing within the Earth produces the magnetic field around the Earth, a dipole—north and south. This field deflects charged cosmic and solar particles, which often follow the lines of the magnetic fields to the poles, where they taper and slant into the earth. When these charged particles collide with the upper atmosphere, it becomes luminescent, producing the glowing northern and southern lights.

Less well known is that this protective magnetic field has been weakening for the last 150 years. Some scientists think this decline anticipates a pole reversal, where the magnetic north and south poles switch locations. This would leave compass needles pointing toward Antarctica. What it would do to animals that navigate by the magnetic fields, we do not know. Such reversals have happened before; the last time was about 740,000 years ago. Hundreds of these reversals have occurred since the Paleozoic, but there is no pattern to their onset. The interval between reversals is quite varied, from 5,000 to 50 million years. But the reversals are over quickly—a mere thousand years or so.

During the time when the polarity flip-flops, the intensity of radiation reaching the Earth's surface intensifies, and animals navigating by magnetic fields lose their compass. Oddly perhaps, no major extinctions correlate with past polarity reversals. We can at least rule out magnetic pole reversals, by themselves, as agents in mass extinctions.

The fourth mass extinction occurred at the end of the Triassic, lasted 4 million years, and took many land vertebrates, but up to 80% of all marine species. The fifth extinction was the dinosaur extinctions, occurring at the end of the Cretaceous, taking all the dinosaurs, many large marine reptiles, small foraminifera, and many other marine species, over a period of about 1 million years.

Let's pick two extinctions, the Cretaceous and Permian, because they are two of the most dramatic, and see if they share a common cause. If so, we can then apply this to other mass extinctions to see if the hypothesized cause is applicable to mass extinction events generally. Several categories of extinction causes have been proposed: plate tectonics (mingling of species, trophic stability, changes in sea level), ice ages, and cosmic collisions.

Plate Tectonics

The solid crustal surface of the earth encases a molten core (see also chapter 6, figure 6.13). The crustal surface is fractured into large plates. The upwelling of molten rock along suture zones between these plates pushes the plates slowly apart, building them along these edges as the lava hardens into new rocky crust. At their other edges, where they are pushed into adjacent plates, one plate dips down to be remelted and the other plate overrides it. Mountain ranges often build along such "collision" zones. Continents are the high parts of the plates that project above the surrounding seas. As the plates move, they carry their continental projections into new locations, sometimes gathering them all together in one supercontinent, other times carrying them into isolation. This slow movement of the continents affects favorably, or sometimes unfavorably, the conditions under which plant and animal life exists on and around the edges of the continents.

Mingling of Species

The continents that would become North and South America separated at the end of the Cretaceous and remained separated for most of the subsequent Cenozoic, until about 2 or 3 million years ago. By then, plate tectonics had moved them again closer together, and produced an uplifted land bridge between them—the isthmus of Panama—the first land connection between the two continents in more than 60 million years (figure 13.14). During their separation, a distinct fauna and flora evolved on each. Among the mammals, placental mammals prospered on the North American continent, and a mix of different placental mammals plus marsupials prospered in South America. When the Panama

FIGURE 13.14 The Great Exchange — North and South America Separate faunas and floras evolved on these continents when they were separate during the Cenozoic. About 2 to 3 million years ago, the Isthmus of Panama formed, providing a land bridge between the continents that became a route of migration and exchange between the continents. Among the placental mammals, many arising in North America dispersed south, and many originating in South America dispersed north.

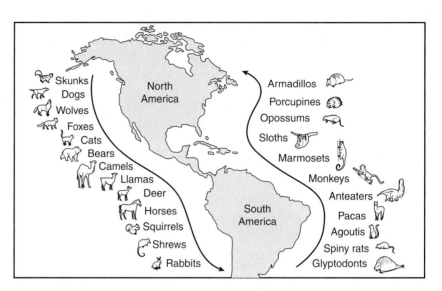

land isthmus formed, the respective assortments of animals tended now and for the first time to extend their ranges. Today, almost half of the medium to large mammalian species in South America originated in North America; and about one-quarter of those in North America originated in South America. Cougars and camels (llamas) moved north to south; porcupines and opossums moved south to north. But overall, there were large-scale extinctions on both continents. It is not hard to imagine why.

In the community of competitors that evolved over the preceding period of 60 million years, the sifting and sorting guided by natural selection had occurred. On each of the separated continents, organisms were accommodated to each other. But with the sudden, on a geologic time scale, appearance of new competitors arriving across the land bridge, the co-evolutionary dynamics were suddeningly changed. There were new competitors, and little time to adapt. Extinctions followed.

Prediction:

Could such an abrupt appearance of new species disembarking from another joined continent explain the general episodes of mass extinctions? If long-separated continents came together, they would suddenly introduce species heretofore unknown to each other on the settled faunas of each continent. We would predict that at moments in Earth history when isolated continents made first contact, mass extinctions should occur. Let's put this to the test, using our two test cases.

Test Cases

1. At the end of the Permian, plate tectonics had brought all the major continents of the world together into a supercontinent called Pangaea ("all land") (figure 13.15a). The continents, previously separated, now were assembled into one large continent. By our hypothesis, mass extinctions should occur. That happened. The late Permian extinctions correlate with the joining of continents.

2. Unfortunately, in the other test case—the dinosaur extinctions at the end of the Cretaceous—the continents were actually moving apart, separating them (figure 13.15b). The hypothesis that mass extinctions should be caused by the joining of continents does not correlate with dinosaur extinctions; the continents were separating, not joining, at the end of the Cretaceous. Let's try another hypothesis.

Trophic Stability

Trophic means food or nourishment. This hypothesis suggests that the *stability* of trophic resources affects species survival. The more stable the trophic resources, the more species; conversely, the more unstable the trophic resources, the fewer the species. We noted earlier in chapter 9 that species diversity correlates directly with latitudinal stability of trophic resources. In temperate regions, there is a high degree of seasonal fluctuations in solar energy from summer to winter; hence, there is a similar variation in photosynthesis, resulting in trophic instability and correspondingly low numbers

(a) Permo-Triassic

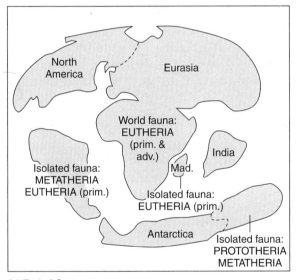

(b) End of Cretaceous

FIGURE 13.15   Continental Drift   (a) Pangaea formed at the end of the Permian as the result of previously separate continental landmasses fusing together into this single supercontinent. (b) Continents broke up again at the end of the Cretaceous.

of species. In tropical regions, by comparison, there is little seasonal fluctuation, relatively constant trophic stability, and high species numbers. The difference, temperate to tropics, is not temperature per se, but fluctuations in productivity.

Continental movement should also affect trophic stability. When continents come together, uniting in a single, large supercontinent, trophic resources are unstable, leading to low numbers of species. When the same landmass is broken up into separate continents, trophic resources are more stable and more species, overall, are supported. The reason is that when a large continent breaks up into separate, smaller continents, the surrounding ocean has more influence in moderating the continental climate. The interiors of smaller continents are closer to the ocean, and so the extreme fluctuations in climate through the seasons are buffered, bringing trophic stability even to the core of a small continent. Conversely, when separate continents join together into a large supercontinent, marine mitigation of climate lessens and there is greater trophic instability within the continent.

In addition to trophic stability, breaking up a large continent into smaller continents also adds more intertidal area around the perimeter of the continents. This is a simple matter of geometry. Taking the same block of land and breaking it into two parts adds extra perimeter area where they pull apart (figure 13.16). For marine organisms living in the intertidal region, this change in geometry increases the area available for occupation. Earlier in this chapter we observed the consequences of the species–area relationship: the more area, the more species. Conversely, as continents come together, area is lost, and fewer species should be supported. (Please NOTE: This hypothesis does *not* suggest that when continents come together, organisms in between get squashed. Even the slowest snail can out pace the "collision" of continents measured in inches over a century.)

Predictions:

When plate tectonics bring continents together into one supercontinent, mass extinctions should occur (increased instability of trophic resources, loss of some perimeter area). When a supercontinent fragments into separate continents, mass extinctions should be absent (increased trophic stability, additional perimeter area). Does the hypothesis of trophic stability correlate with extinctions in our two test cases?

Test Cases

1. The end of the Permian was a time when continents, previously separate, came together forming Pangaea, the supercontinent (figure 13.15a). This correlates with our prediction: the joining of land masses should lead to catastrophic extinctions.

2. However, the end of the Cretaceous was to see the supercontinent of the Mesozoic, Pangaea, fragmenting into the smaller and more recognizable

FIGURE 13.16 Perimeter Area When a landmass (a) is broken in two (b), this adds area along the perimeter where they split. This adds to the intertidal area. When two landmasses (b) are brought together (a), this results in loss of available intertidal area.

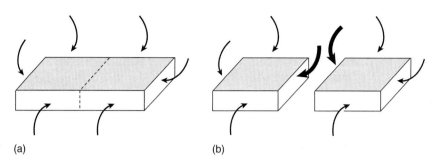

(a)   (b)

continents we know today (figure 13.15b). Just the opposite of our prediction.

The trophic stability hypothesis does not help explain the demise of the dinosaurs. Don't despair. Let's try another hypothesis.

Changes in Sea Level

Sea level marks the line where the ocean surface washes up against the edge of a continent. It forms the shallow intertidal region, which may move up or down on the continent, thereby changing sea level. Sea-level changes may result from the influence of plate tectonic activity. The shape and depth of the ocean basins of the world change as ridge-building waxes and wanes along the deep abyssal trenches. If the ocean basins deepen, sea level along the continental edges drops; if the ocean basins become more shallow, sea level rises. Sea level also is affected by glacial activity, which we will discuss next. For now, let's just examine the hypothesized consequences of sea-level changes, whatever the cause.

A drop in sea level has three consequences, all of them leading to loss of species. First, shallow marine habitat is lost. Most continents are characterized by continental shelves, gently sloping platforms leading out to deeper water where they then drop off into the vast abyssal depths. When sea level is high, these shelves are covered and provide large areas for marine organisms to occupy. Because the shelves slope gradually, a small drop in sea level exposes a large area that previously supported intertidal communities (figure 13.17). Here again, the species–area relationship informs us about the consequences—loss of area, loss of species. (Please NOTE: This hypothesis does *not* suggest that intertidal marine organisms are left high and dry. They do not awake one morning to find that the nearest seashore has moved several miles away. Sea level drops over a period of thousands of years, and even the slowest snail can march along with the retreating shoreline.) Second, a drop in sea level indirectly leads to more severe terrestrial climates of the interior continent. Effectively, the sea retreats out on the continental shelf, moving it farther away from the interior of the continent and thereby reducing the moderating effect of the departing ocean on interior climate. Third, the groundwater table drops, indirectly leading to dry, even drought, conditions within the interior of the continent. A rise in sea level reverses these consequences with the opposite effects, favoring increased diversity of organisms.

FIGURE 13.17 Continental Shelves Because of the sloped geometry of the continents, a drop in sea level along a continental shelf eliminates large areas of intertidal habitat (dark shaded) compared to a similar drop in sea level against a steeply sloped drop-off area (light shaded).

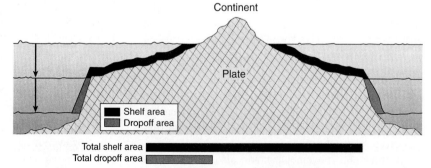

Continent

Plate

■ Shelf area
■ Dropoff area

Total shelf area ▬▬▬▬▬▬▬▬▬▬▬▬▬
Total dropoff area ▬▬▬▬

Predictions:

Mass extinctions should correlate with a drop in sea level. Conversely, a rise in sea level should not coincide with mass extinctions.

Test Cases

1. At the end of the Permian, the sea level dropped significantly and did not recover until well into the next geologic period (figure 13.18).

2. Sea level also dropped at the end of the Cretaceous, by some estimates by as much as 150 to 200 meters (about 500 to 650 feet). Surviving marine deposits of the Cretaceous are not easily interpreted for sea levels, but what can be gathered speaks of significant drops in sea levels correlating with dinosaur extinctions.

Not bad. In both of our test cases, mass extinctions correlate with a drop in sea levels. Let's try another hypothesis.

Ice Ages

We are currently, right now, in an **ice age**. That's right. We are now living during an ice age event—one of at least three, and perhaps more, major ice ages in Earth's climate during the Phanerozoic. An ice age is made up of cycles wherein there is a **glacial phase** (climate cools, ice sheets form and spread) and an **interglacial phase** (climate temporarily warms, ice sheets retreat).

The question of what causes an ice age is not settled. Some geologists point to a "wobbling" of the Earth on its axis: the wobbling temporarily moves the Earth into a less favorable orientation to receive heat from the sun, and general cooling sets in. Others point to plate tectonics that might move continents into positions where they intercept and disrupt the flow of warm tropical waters to polar regions. Blocked from warming ocean waters, the poles cool further, and their icy chill spreads toward the equator. Winter snows last later into the spring, and snows return earlier in the fall. As more snow collects, it reflects more sunlight back into space, and the climate cooling deepens. The weight of the collected snow compacts into ice—glaciers. Glaciers form first in high montane regions and near the poles, and these move to lower elevations and toward the equator as they grow under continued climate deterioration. Tropical species crowd around the narrowing and chilled, but still relatively warm, equatorial belt; temperate species follow the forests and climate in advance of the growing ice sheets.

Deep ice cores taken from old glaciers in Greenland and Antarctica give an almost annual record of climate conditions over the last 250,000 years. Pollen and oxygen isotopes captured in each annual cycle of the core bear witness to the annual flora and temperatures, respectively. Pollen is hardy and its microscopic appearance is unique to the plant or group of plants from which

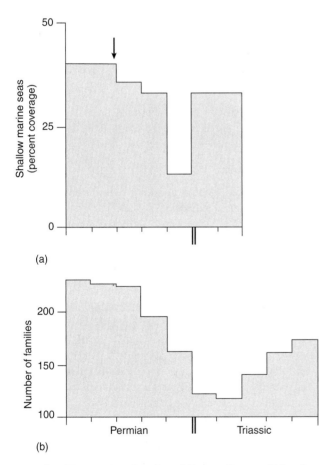

(a)

(b)

FIGURE 13.18   Sea Level Drop, Permo-Triassic
Note that as the sea level dropped (a) at the end of the Permian, there was a corresponding loss of marine animals (b). Sea level is expressed as a percentage of possible coverage, which began falling about 30 to 35 million years ago (solid arrow) before the end of the Permian. Estimates of absolute drop in level range from 300 to 600 feet (100–200 m). Tick marks along horizontal axes represent geologic stages of varying duration.   (After Schopf, 1974.)

it came—a kind of botanical fingerprint. Wind-blown pollen settling on the glaciers captures a sample of the world's flora of a particular year, at a particular time. If most of the pollen is from temperate plants or from tropical plants, then indirectly the pollen tells us that the world's climate of the day was cold or warm, respectively. Oxygen comes in several varieties, called isotopes: oxygen-18 (O^{18}) and oxygen-16 (O^{16}). Their ratio to one another in the atmosphere changes with several factors, including temperature. Like a thermometer, the ratio goes up or down with the temperature. These isotopes are captured in each year's snow accumulation in glaciers—locked into the ice. An ice core cut back through the accumulated layers is a sample back in time of these isotope ratios. Each layer's isotope ratios can be used to calculate the successive climate temperatures, often over thousands of years. Perhaps most surprising, and unsettling, is the evidence in these ice-core samples that the change from a glacial to interglacial phase, or back again, might have occurred within the space of a century, or even within a few decades. If this is true, then we are apparently still within an ice age, and glaciers might start to return within the space of a generation.

The current ice age began about 30 to 40 million years ago, as the culmination of a general cooling trend throughout the Cenozoic, marked by the return of large polar ice caps. Glacial and interglacial phases of the present ice age have been of varying intensity and duration. We are currently enjoying an interglacial phase (a general warming) that began only about 12,000 years ago, with the end of the previous glacial phase, which had itself lasted about 100,000 years and was preceded by an interglacial period lasting 20,000 years, and so on back through vacillating cycles over the 30 to 40 million years of the current ice age.

Even during the current, relatively warm interglacial phase, there have been short-term intervals of cooling. Before Christopher Columbus "discovered" North America in 1492 and made it known to western Europe, the Vikings striking out of northern Europe reached and colonized Greenland. From there, Viking colonists reached North America, probably first settling in the northeastern shores of Nova Scotia in the year 1000. Unfortunately for these colonists, the world soon entered a "little ice age," a period of short-term cooling that lasted from about 1200 to 1850. Stressed by such deterioration of climate and cut off from trade with Europe by winter ice, the colonists eventually succumbed to poor harvests and competing Eskimo peoples, and disappeared from Greenland and the

"New World" well before Columbus ever set sail. A historical record of this little ice age survives. In Europe at this time, the paintings of artists, such as the Flemish painter Brueghel, chronicle the record of cold and frozen-over winter rivers.

Prediction:

An ice age is characterized by a major change in climate, with glacial phases accompanied by a drop in average temperatures, and deterioration of climate, and attended by buildup of great sheets of glacial ice. During interglacial phases, climate recovers and warms, and glaciers retreat. Such dramatic environmental changes must stress existing faunas and floras, perhaps correlating with times of mass extinctions. Therefore, we predict that ice ages should correlate with mass extinctions.

Test Cases

Our two test cases—end of Permian and Cretaceous (plus the other three major extinction episodes)—are compared to times of Phanerozoic ice ages (figure 13.19). Of the five episodes of mass extinctions, only the end of the Ordovician is roughly correlated with an ice age. Ice ages, therefore, do not seem to be a common cause of mass extinctions.

Cosmic Collisions

Asteroids, comets, meteors, death stars—all hold favorite places in the imaginations of many scientists and moguls of popular culture when casting about for causes of mass extinctions. Until recently, convincing evidence supported the proposal that earthly extinctions occurred on a regular schedule, about every 26 million years. If true, then we should really know of this and its cause, especially if we are scheduled for another visitation of this extinction event in our lifetimes.

Indirect evidence suggests that cosmic events do occasionally afflict Earth and its sheltered assortment of organisms. The idea that cosmic events occasionally decimate earthly mortals is an old one, biblical in scope. The first scientific evidence for this came in the early 1980s, when iridium spikes, big jumps in iridium concentration, were documented in rocky formations at the very end of the Cretaceous. Iridium, a metallic element, is rare in the Earth's crust but up to 10 thousand times more common in asteroids and meteors. The presence of such elevated levels of iridium in Earth rocks suggests a

PHANEROZOIC										Eons
Paleozoic						Mesozoic			Cenozoic	Eras
Cambrian	Ordovician	Silurian	Devonian	Carboniferous	Permian	Triassic	Jurassic	Cretaceous	Tertiary	Periods
										Ice Age
										Mass Extinctions

FIGURE 13.19   Phanerozoic Extinctions and Ice Ages   Ice ages are indicated along this geologic time line for comparison to five mass extinction episodes. Note that there is no tight correlation between ice ages and mass extinctions.

pulverizing impact of an asteroid or meteorite. Along with iridium, there is also evidence of shocked quartz, a mineral formed under high heat and pressure (from an impact?), at the end of Cretaceous time horizon. An impact crater at this same time horizon and about 100 miles (150 km) across has been found, partially submerged by current seas, along the Yucatan Peninsula of Mexico. The object that produced it is estimated to have been almost 6 miles (10 km) wide.

The collision scenario goes as follows: The asteroid (or comet) certainly smashed plants and animals un-

lucky enough to be under its direct impact, but its most devastating effects were broadcast and felt worldwide. Pulverized rubble from the disintegrated asteroid directly generated massive, thick dust clouds sent skyward; the fiery impact further ignited continent-wide forest fires, adding smoke to the darkening atmosphere. These immense, thick, suffocating clouds of debris encircled the globe within days or perhaps a few weeks (figure 13.20). This devastation was accentuated by tidal waves crashing onto innocent shoreline communities of unsuspecting plants and animals. Beneath this

FIGURE 13.20   Asteroid Collision with the Earth   This figure depicts events on the young planet Earth when, 4 billion years ago, still hot from its cosmic birth, it endured a pummeling by rocky debris. This bombardment waned but did not cease entirely. A similar strike by an asteroid is thought by some scientists to have caused the extinction of dinosaurs 65 million years ago.

choking cloud, sunlight was reduced or sealed off. Seeds or spores might have survived, but plants were snuffed out, taking down the herbivorous dinosaurs with them, with carnivorous dinosaurs soon to follow.

Predictions:

If a major cosmic missile struck the Earth—asteroid, comet, meteor—then the devastation would extend beyond the impact site as smoke and dust encircled the globe to choke off sunlight and would generate tsunamis that crashed on distant unsuspecting shorelines. Worldwide plant and animal communities would be eliminated and mass extinctions across all exposed groups should result.

Test Cases

Sounds plausible doesn't it? So what is the problem? Perhaps nothing. This may, in fact, be the cause of dinosaur extinctions. So far, no such similar telltale iridium spike has been discovered at the end of the Permian, but it might just await being found in a well-preserved rock formation. And there is some evidence of an iridium spike at the end of the Devonian, correlating with mass extinctions then. Some have argued, further, that such collisions with the Earth have a regularity, a periodicity, related to celestial causes.

The asteroid belt between Mars and Jupiter contains an arsenal of rocky missiles, thousands of catastrophic sizes, circling in irregular paths. A little tug from Jupiter's strong gravitational field could send any one of these into an encounter with Earth. As for comets, these travel in great attenuated orbits, carrying them well outside the last planet of our solar system before returning in what has been proposed as "comet storms" raining down on the hapless Earth crossing their path. Estimates of intervals between such events range from 26 million years to 100 million years. So giddy had become the argument, and so seemingly overwhelming the evidence, by the late 1980s, that in the prestigious journal *Science* an author declared that "the end seems to be in sight [to finally prove the impact theory]."

Critique of Collision Theory

This is all great fun, and eventually much of it may prove to be true. But let's turn a skeptic's eye on the theory of cosmic collisions. As we saw at the beginning of this chapter, it is difficult, and somewhat arbitrary, to clearly identify episodes of mass extinction. This ambiguity leads to difficulties when we test hypotheses about extinction cycles. But even if periodic cosmic events rain down on defenseless plants and animals, this theory is not particularly persuasive for dinosaur extinctions in particular.

First of all, it is too fast. The scenario of extinctions—from collision to choking dust cloud to deaths—is completed within years, at most a century. In the popular mind, it is as if the dinosaurs started into Friday on a binge of extinctions and never made it to Monday morning. So far as we can tell at the moment, dinosaur extinctions occurred over a span of about 1 million years—rapid in geologic time, but not a long weekend. Perhaps new finds of better geologic preservations will document an extinction episode for the dinosaurs measured in weeks and not millennia. But at the moment, the interval of decline is 1 million years. Second, dinosaur numbers were already dwindling several million years before the asteroid impact occurred. This suggests that something other than a cosmic collision was at work earlier to drop their numbers. Third, the catastrophe was too selective. Such a catastrophic event, as envisioned in the cosmic collision theory, would certainly affect all exposed groups of organisms. Apparently this did not happen. Dinosaurs, the preeminent land vertebrate of the late Mesozoic, took it in the chin, as did large marine reptiles such as mosasaurs (distant relatives of lizards) and some marine organisms such as ammonoids (soft-bodied animals living in coiled shells). But oddly, other dinosaur contemporaries did not. Amphibians showed no special mass extinctions, nor did snakes or lizards or turtles or crocodiles or mammals, all of which were out and about and just as susceptible to the proposed environmental devastation following an asteroid shock. If the worldwide shocks from an asteroid brought down the dinosaurs, then other groups should have fallen as well. This did not happen.

Others have suggested that prolonged volcanism might be the cause. Not one or two big eruptions, but perhaps an episode of prolonged volcanic activity lasting several millions of years at the end of the Cretaceous. This could have produced a persistent, but not choking, cloud of dust, placing dinosaurs under extended assault from which they might have dwindled and then eventually succumbed.

OVERVIEW

We have covered much in this big chapter because there is much to tell. After all, most of the other chapters of this book are devoted to the origin of species and their evolution. By comparison, this one chapter dedicated to extinctions, the other side of the evolutionary coin, seems downright skimpy. In addition to reviewing extinctions, this chapter should also cause you to pause and reflect on the all-too-common use of handy fashion in science, which then seeps into popular talk-show culture, magazines, movies, and other media where plausibility is measured more by how frequently a theory is repeated than by how solid are the facts that support it. Let's start with the science.

At least two types of extinction occur—uniform and catastrophic extinctions. With uniform extinctions, species within a group are lost at a constant rate. These extinctions are small-scale, geologically speaking. A few species flicker out, but these losses are not part of a broad-scale collapse of faunas or floras. Viewed on a small scale, events on small pieces of real estate, such as islands, engage our intuition to understand why species are added or lost. Immigration rates oppose extinction rates, shifting the equilibrium one way or another. Over larger areas, such as continental landmasses, rates of origination (speciation) and extinction balance to produce an equilibrium. Imposed on this balance is the advance of one competitor compared to another. To stay competitive, it takes all the evolving a species can do just to stay even with its competitors—the Red Queen effect. Fall behind, and survival is in doubt. Fall even farther behind, and numbers dwindle until chance events intervene and species are lost. As a consequence of uniform extinctions, there is a general background extinction that occurs, against which sudden and large mass extinctions occur.

Catastrophic (mass) extinctions are characterized by the loss of large numbers of species, across various groups, during a relatively short period of geologic time. The proposed causes are many—a drop in sea level, continental linkages, climate changes, cosmic events, and volcanism all seem to be contenders as most likely causes; ice ages do not. The five episodes of mass extinctions need not be the result of one and the same cause. Among the candidates, a drop in sea level seems the most common culprit across the various episodes of mass extinctions, but it certainly is not alone. Multiple events may converge to produce and deepen the insults to plant and animal communities, accelerating their rapid demise.

If mass extinctions arrive on a schedule, then pinpointing the causes is of more interest than simple scientific curiosity. If the Earth periodically passes through the path of a long-term streaking comet, or if Jupiter's conjunction with the asteroid belt hurls one of the large rocks off course, then these events can be observed, future dangers predicted, and perhaps the threats to our survival can be intercepted or averted.

Whatever their causes may prove to be, mass extinctions are events beyond the reach of natural selection. Mass extinctions are infrequent. Millions of years pass before survivors could put the survival traits that spared them extinction to use again. Further, little can be done to armor an animal living at ground zero to protect it from an asteroid's direct impact. The eruption of Mount St. Helens in 1980 provides an example of recovery following a local extinction of surrounding plant and animal communities. Recovery was relatively fast, but selective. Plants already adapted to occupation of disturbed areas moved in quickly to recolonize the desolate, ash-covered landscape. Some salamanders living in the cooling and protective waters of ponds on the margins of the blast area survived to reestablish populations. The unusual features of these plants and animals that served them in recovery from a volcanic eruption were happy and fortuitous accidents of evolutionary history. These features were fashioned by natural selection for less-radical circumstances, but they served when volcanic devastation struck. Ahead of time, there was no selection for volcano-resistant traits.

Similarly, the causes of worldwide extinctions strike with no warning and no preparation. What can be done? Not much. So, here we meet infrequent, but natural events that drastically affect life on Earth, eliminating much of it and leaving the future to the lucky survivors. Extinction events clear the field, eliminate the predominant species that receive the assaults, and the survivors carry on. Without such chance events and random elimination of species, the world would be a different place. Extinction of the dinosaurs cleared the way for mammals. Had dinosaurs persisted, mammals, and ourselves in particular, might never have found our evolutionary moment. An intelligent, bipedal, and conscious vertebrate may have evolved not in a mammalian body, but on a reptilian mold; or it might not have evolved at all.

Careful interpretation, qualified conclusions, and evidence are usually the first causalities when a scientific idea enters popular culture. Perhaps asteroids killed the dinosaurs, then again, perhaps not. The same goes for dinosaur physiology. Perhaps dinosaurs were "hot-blooded," then again, perhaps not. These questions are open, but you would not guess this from the way the issues circulate in popular culture through magazines, movies, television, and the Internet.

Further, science is skeptical, and adamantly so. Why? Because many plausible ideas melt before the critics' heat. Good science works that way when it is at its best. An idea is proposed. Then it is examined, more evidence is gathered, assumptions are challenged, alternative interpretations are tested. Look at the issue of hot-blooded dinosaurs, an issue that has unfolded during much of your own lifetime. At first the evidence seemed compelling, even overwhelming: Dinosaurs were warm-blooded. So when Hollywood made movies about them, animated dinosaurs expelled hot breaths, warmed in transit through their lungs inside warm bodies. Then the evidence was challenged. Some evidence simply collapsed (neither dinosaurs nor their reptilian contemporaries were hair-covered). Contrary evidence suggests that they were not warm-blooded (absence of turbinates). Our interpretation has swung in the opposite direction. They were ectotherms, or close to that.

The larger point is this: Fashions come and go in science as well as in pop culture. Sinister motives and quiet conspiracies need not be invoked. This is just the natural process of challenge and counterchallenge, the centerpiece of active science. Those who first marshaled the case for endothermy in dinosaurs certainly had a vested interest in the hypothesis. But what persuaded many scientists was what, at first, seemed to be the compelling evidence: predator-to-prey ratios, insulation, latitudinal gradients, bone structure. Science works in both directions. Those that propose serious hypotheses; those that are professional skeptics.

Scientific hypotheses are perpetual works in progress, not edicts written on eternal stone. Often popular culture does not appreciate the tentativeness of first hypotheses. Notice the slide of dinosaurs (hot-blooded) into movies *(Jurassic Park)*. The careful attention dinosaurs received during the debate did have the side effect of clarifying their lifestyles (more active) and evolutionary relationships (dinosaurs and birds). Notice the various hypotheses of mass extinctions—this debate is still in progress before us. But the all-too-easy view, that an asteroid impact killed the dinosaurs, needs some professional skepticism. At the very least, the scenario of events needs some qualified modification.

Certainly, science has settled some issues about dinosaurs and about extinctions. But the point to make to end this chapter is this: Despite the confident words used to express hypotheses, the final arbitrator is not how loud one yells or how many Hollywood movies adopt the hypothesis. In the end, the final arbitrator is the evidence, carefully considered. With all the remarkable findings in genetics and medicine paraded before us in the media, it is surprising to me that fundamental and basic questions in biology still remain unanswered. Were the dinosaurs endotherms or ectotherms? We don't know. What caused the five episodes of mass extinctions? We don't know. The answers may be no farther away than a new discovery or bright idea.

Selected References

Bakker, R. T. 1986. *The dinosaur heresies*. New York: William Morrow.

Benton, M. J. 1989. Mass extinctions among tetrapods and the quality of the fossil record. *Phil. Trans. Royal. Soc. London* 325:369–86.

Gibbs, W. W. 2001. On the termination of species. *Sci. Amer.* 285(May):40–49.

Kerr, R. A. 1995. A volcanic crisis for ancient life? *Science* 270:27–28.

Kring, D. A., and D. D. Durda. 2003. The day the world burned. *Sci. Amer.* 289(December): 98–105.

Lemon, R. R. 1993. *Vanished worlds: An introduction to historical geology*. Dubuque, IA: Wm. C. Brown.

MacArthur, R. H., and E. O. Wilson. 1967. *The theory of island biogeography*. Princeton: Princeton University Press.

McGhee, George R., Jr. 1996. *The late Devonian mass extinctions*. New York: Columbia University Press.

Raup, D. M. 1991. *Extinction: Bad genes or bad luck?* New York: W. W. Norton.

Raup, D. M., and J. J. Sepkoski, Jr. 1982. Mass extinctions in the marine fossil record. *Science* 215:1501–3.

Ruben, J. A., W. J. Hillenius, N. R. Geist, A. Leitch, T. D. Jones, P. J. Currie, J. R. Horner, and G. Espe. 1996. The metabolic status of some late Cretaceous dinosaurs. *Science* 273:1204–7.

Schopf, T. 1974. Permo-Triassic extinctions: Relation to sea-floor spreading. *J Geol.* 82:129–43.

Stanley, S. M. 1987. *Extinction*. New York: W. H. Freeman.

Van Valen, L. 1973. A new evolutionary law. *Evol. Theory* 1:1–30.

Learning Online

Testing and Study Resources

Visit this textbook's website at **www.mhhe.com/evolution** (click on the book's title) to take advantage of practice quizzing, study/writing tips, timely news articles, and additional URLs for research on the topics in this chapter.

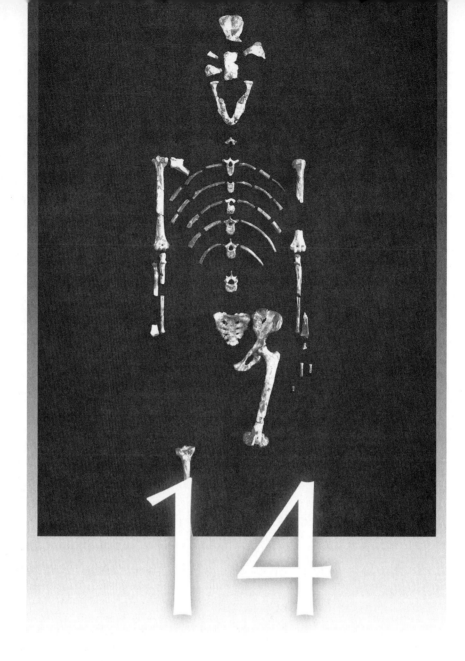

"Molecular evidence suggests that our common ancestor with chimpanzees lived, in Africa, between five and seven million years ago, say half a million generations ago. This is not long by evolutionary standards. . . . in your left hand you hold the right hand of your mother. In turn she holds the hand of her mother, your grandmother. Your grandmother holds her mother's hand, and so on. . . . How far do we have to go until we reach our common ancestor with the chimpanzees? It is a surprisingly short way. Allowing one yard per person, we arrive at the ancestor we share with chimpanzees in under 300 miles."

Richard Dawkins, biologist

Human Evolution:
The Early Years

INTRODUCTION

Our own evolution is daunting—not because it is different or intrinsically more complicated than what we have seen already, but because it is us. The subject of human evolution has always been, and remains, contentious. Darwin faced the bitterness of public belligerence because he eventually and directly included humans in his discussion of descent with modification. In *The Origin of Species* (1859), he promised that "light would be thrown on the origin of man and his history." In 1871 he delivered, with the publication of *The Descent of Man and Selection in Relation to Sex*.

Certainly, there is the "hard" evidence serving the study of human evolution—the bones, teeth, and tools—the physical evidence that fossilizes. But often what intrigues us most about our evolution is the "soft" stuff—language, culture, and religion—the parts that make us human, our human nature, which does not fossilize for direct inspection. Guesses and claims about human nature often are unreferenced to hard, physical evidence. To some, humans are an inevitability, the final pinnacle of life; to others, nature preaches moral messages—subjugation of the "unfit," survival of the fittest—justifying cruel human actions that imitate nature. Some people believe that culture nurtures our adult behaviors; others believe that genes dictate our basic lusts. Wild speculation occasionally rules, especially in the popular culture delivered in magazines at grocery store checkout lanes. Vigilance is required. Claims about evolution and human nature vary in credibility. Even careful scientists enjoy an occasional freewheeling speculation, and these fanciful conjectures get strewn and scattered about as established "fact." When recognized as whimsical flights of fancy, such speculations can loosen up old, smug dogma and encourage new, fresh ideas. But ideas come and go, especially within the emotion-charged arena of human evolution. A skeptical and critical eye will serve us well. If we keep our wits about us, we will not be seduced by flawed but fashionable ideas eloquently paraded before us; instead we will enjoy the wonderful evolutionary insights into our own human character.

Already we are well equipped to sensibly consider our own evolution. Some uncomfortable issues await, but not staggering new evolutionary processes to discover. Natural selection did not stop with the twentieth or twenty-first century. Humans are not exempt from evolution's reach. The same processes that regulate biological change in other organisms apply to us. Or at least they did until we began to intervene medically and genetically in our own destiny—but this is getting ahead of the story.

Let's take comfort in the view that the basic evolutionary principles already developed apply to our own human history. Evolution proceeds with the preservation of favorable characteristics by means of natural selection. Meeting directly some confusing issues now will help us anticipate and then avoid unnecessary and empty controversies later.

"NEW" ANCESTORS

Monthly, it seems, an anthropologist finds a "new" human fossil that provokes instant optimism, a claim for central location in hominid history, and controversy. In the relatively brief time during which I wrote, reviewed, and revised this chapter, two new and earlier hominid fossils from Africa were reported (we will meet them later in this chapter); and almost as quickly, interpretation of both has been adamantly challenged. Certainly, some hyped new hominid fossils will eventually earn a coveted place in hominid history. Until then, the fossils will be controversial. Such controversy is not a challenge to Darwin, but a bid for preeminence among fossil finds. Anthropologists engaged in such a debate do not challenge the fact or the mechanism of evolution. They are engaged, usually, in debate over the *course* of evolution (see chapter 1): Did or did not the fossil they discovered lie on the main line of hominid evolution?

Recall that a species, without reproductive data, becomes a morphospecies or a paleospecies. Whether a fresh fossil find represents merely a variety of an already described species or is an entirely new species can become an arbitrary decision based on the discretion of the anthropologist involved. Much premium and status attend the claim that a discovered fossil is, in fact, a new species, not a variety. I do not mean this as a raw cynical comment. Anthropologists are susceptible to normal flights of overoptimism and the quite human hope that upon the new fossil a professional reputation will grow. When first discovered, fossil finds of "Java Man" were christened *Pithecanthropus erectus* and "Peking Man" *Sinanthropus pekinensis*.

Following more sober investigation and less giddy claims, both now are recognized as varieties of the same species, *Homo erectus*. Giving each new fossil a new scientific name plays well as a media sound bite and soothes anxious funding agencies that support the work of anthropologists in hot, dusty regions of the world. We should be a cautious audience when word arrives of an anthropologist with a new species to peddle to the public. A spirited and, alas, sometimes unseemly dispute among anthropologists over a contentious fossil is not evidence of deep dissatisfaction with Darwinian evolution in general. Rather, it is a scrappy debate, within a Darwinian context, over the course of that evolution.

PITFALLS

We should not be alarmed when we come upon the complexity of the human species. We have already seen complexity, and we have seen how it evolves—in manageable steps (e.g., mimicry). But other misunderstandings stand in our path. One is the entrenched idea of our own inevitability.

Human Inevitability

The view persists that humans are an evolutionary inevitability. Before me as I write is a copy of a poster, not too old, from a well-known natural history museum in the United States. It depicts a pyramid of species stacked one on the back of the preceding, leading from squashy sponges to fishes and up a ladder of life to "Man" at the top. The intended point is that humans are the pinnacle of evolution—the top of the heap. Yet such a view is certainly mistaken. The millions of other species with which today we share this world are themselves equal survivors. Like ourselves, the species that share this world are simply the most recent productions of a long evolutionary history.

A deeper claim, sometimes made unintentionally, is that humans are at the top because we are an evolutionary certainty—the endpoint of an ancient inevitability. But humans were not preordained to evolve at the top, or anywhere for that matter. As mammals, we come out of the synapsid radiation, an independent lineage from reptiles. However, if relative numbers are indicative of future promise, our mammalian ancestors during the Mesozoic dwindled to such alarmingly low numbers of species that, in retrospect, it might look as if they were on the brink of extinction (figure 14.1). If that had happened, where would we be today? Nowhere. We would never have evolved, and the biological world of today would be a quite different place.

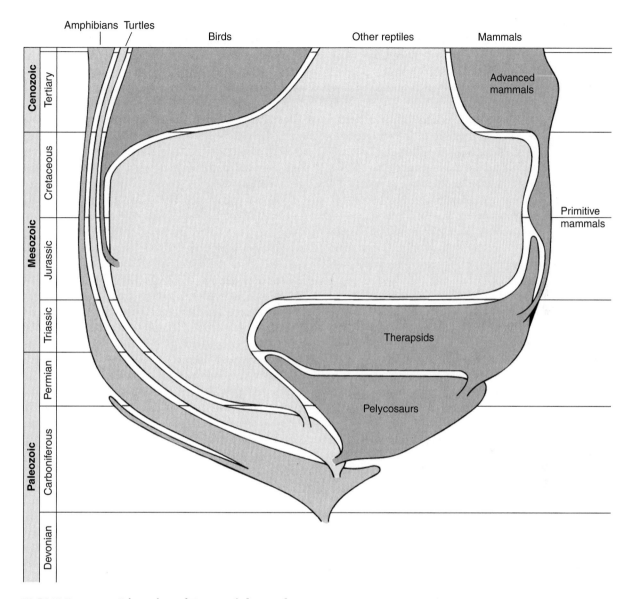

FIGURE 14.1 Diversity of Terrestrial Vertebrates The width of the "balloon" expresses the relative diversity of each group. Notice that during the Mesozoic, the numbers of primitive mammals shrunk significantly compared to the predominant reptile groups of the time, which included the dinosaurs and allied groups (see also figure 13.12).

Now you know this already—if, that is, you have kept up with old Hollywood movies. It is becoming increasingly difficult to avoid the Christmas movie *It's a Wonderful Life*. The main character, played by the actor Jimmy Stewart, despairs at his financial and personal ineffectiveness. In the middle of a snowstorm, he stumbles to a bridge over a freezing river and apparently contemplates jumping in. He is saved by an angel who turns him away from hopelessness and returns him successfully to his satisfying life. The angel does this by showing Jimmy Stewart's character the alternative consequences that would have unfolded had he never been part of the town's community.

Similarly, if mammals had flickered out during the Mesozoic, then none would have survived to produce the diversity of mammals living today, and none would have persevered to produce primates, ourselves in particular. Had that happened, we, *Homo sapiens*, would never have appeared on Earth—a somber thought. What then might have been the foundation upon which a conscious, sentient species evolved? Perhaps an intelligent species would have emerged instead among the reptiles. Upright and two-footed (bipedal), hominids colonized the open savannahs and later spread to most regions of the globe. In our absence, would a lineage of bipedal reptiles have followed our evolutionary course from savannahs to cities? If bipedal, the reptiles would benefit from having forearms that were free for inquisitive exploration and to manipulate objects of fascination within the environment, and we can suppose that relatively big brains would eventually follow. But lucky for us, mammals did persist, and primates evolved, ourselves amongst them. The point is that humans are not an evolutionary inevitability. We were not, at the beginning of life, an evolutionary guarantee later is life's history. Chance and selection played dice, and we won. Our existence is privilege, not certain.

Nature versus Nurture

We should settle, or at least reconcile, an old dichotomy—nature versus nurture. The dichotomy rises from this debate: Do behaviors arise from genes (nature) or from culture (nurture)? During the mid twentieth century, behavioral science tended to be divided between those taking one side or the other. Taking the "nature" side were those biologists who believed that adaptive behavior is shaped by natural selection and inherited to match the environmental circumstances in which an organism lived. Taking the "nurture" side were those biologists, especially many psychologists, who believed that humans in particular are born as blank slates with no predetermined behaviors in place: after birth, the cultural experiences the young meet determines their behaviors as an adult—behaviors were learned, not innate.

Today, a welcome synthesis of these two extreme, polar views prevails: Both environment (nurture) and inheritance (nature) shape final behavior—human behavior, like all behavior, is built from both. Neither nature nor nurture alone provides a complete explanation of a human behavior; the two are partners.

Nature and nurture contribute in different ways to a finished behavior. For example, consider an analogy: the building of a house. The detailed blueprints (nature/genes) direct the assemblage of available building materials (nurture/environment) into a completed home (completed behavior, phenotype). If the building materials—wood, concrete, metals, nails, workers—are unavailable, the blueprints remain no more than sheets of

blue paper. If the blueprints are unavailable, the building materials remain only a disorganized pile of lumber and bricks and idle workers. Both are necessary to build a house. Further, it would seem strange to ask how much of each contributes to the finished product? Are blueprints 60%, materials 40%? Or is the ratio 10% to 90%? The blueprints determine what materials are used, when they are used, and how much of each is used. The question is strange because both blueprints and materials contribute, but in such different ways that it is pointless to try to determine a proportion of each.

So it is too with building behaviors. Both innate and learned components go into the finished behavior. Both genes and environment contribute to the finished product, but in different ways. The environment, consisting both of the natural world and of culture, holds information that may be incorporated into the final behavior. In adaptive terms, learning depends upon environmental predictability and upon the chance of mistakes. But aspects of learning are also coded into the genes. The genes determine what is learned, when it is learned, and how much is learned.

The nature/nurture issue is not likely to divide into helpful insights about our evolution. Both genes and environment contribute to the finished product. The survival value of traits concerns us most, not their sources—genes or culture. A characteristic arising in genes or in culture must survive the test of natural selection. This matchup against natural selection, not a feature's source, determines a feature's evolutionary success and settles its biological destiny. The fate of an adaptation rides on advantages bestowed within a challenging and threatening environment, not by its source—nature or nurture.

Innate or Learned Behavior?

If we avoid harsh dichotomies, it can be useful to discuss behavior as being innate or learned, provided we keep in mind the fact that eventually both genes and environment interact in the development of every behavior. By **innate behaviors** we mean species-specific behaviors that are expressed in a highly stereotyped and complete manner the first time they are deployed. By **learned behaviors** we mean behaviors built up during the lifetime of successful and unsuccessful experiences of an organism. Within limits, innate behaviors can be modified by learning. For example, when an adult seagull returns with food, a newly hatched chick responds with an innate feeding behavior—it flails its small bill in the direction of the adult's large bill. When the chick's errant flailing strikes the adult's bill, this stimulates the release of the food to the chick. After a few days' experience, the chick's innate behavior of bill flailing for food is perfected into effective and directed begging. The chick no longer thrashes out randomly, but specifically times its head turning and exacting grip of the parent's beak to elicit and receive the food. The chick has learned; the innate behavior is modified.

Innate Behavior and Biased Learning

Human behaviors are assembled from inherited instructions in the genes, modified by historical practices of society, and completed by resources from the environment. Therefore, a behavior is built from many sources.

Imprinting   Innate behaviors are those ready to serve when needed; they are in place, "hardwired" in, and more or less complete when first performed. **Biased learning** is restricted learning—the ability to learn and modify behavior from a limited array of environmental information. The research on imprinting by Konrad Lorenz and others shows that imprinting has both innate and biased components (figure 14.2).

When a mother greylag goose gets up and walks off, her new young trail close behind. This "following behavior" includes an innate and learned component. The *following* is innate; the attachment to a *particular* adult is learned. The young goslings come "preprogrammed" to imprint on their parent. To avoid mistakes, this behavior must be right the first time. If the young gosling had to learn what was a bush or tree, from what was their supportive parent, much time would be wasted. There is certainty in the greylag breeding system—young arrive into a predictable world. They can expect to hatch into a world where parents are present from the start. However, the particular parents the young first greet cannot be anticipated. Attachment to particular parents cannot be preprogrammed fully into the genes because particular parents cannot be predicted ahead of time. The particular mother to whom the goslings form an attachment must be learned, after hatching, and the uncertainty resolved.

Speaking Up   Human language is another example of a behavior developed from both innate and learned components. If human language were encoded into genes and spoken fully and completely at birth, this would certainly accelerate child development. But that is risky business. Suppose for a moment that genes coded for English and the newly born child spoke im-

peccable English immediately. But the society into which this English-speaking child was birthed cannot be anticipated ahead of time. If born into a society speaking Bantu, the child's language skills in English would have no value and most likely would be quite disadvantageous. Today, there are more than 3,000 human languages. The particular language spoken in the particular society into which the child will be born cannot be predicted ahead of time. Full language cannot be carried in the genes with the expectation of matching exactly with one of these thousands of languages. Instead, children learn a language after entering the society of their birth. The particular language is an open component learned from culture.

On the other hand, it is certain that a human child will learn *some* language. The child will be born into a society with some spoken language, but which one is not known ahead of time. Consequently, the ability to learn a spoken language is an innate ability. Young humans are born ready to go, to learn the spoken language

FIGURE 14.2   Goslings' Following Behavior
Konrad Lorenz himself raised these greylag goslings from first hatching, so it was to him that they imprinted, expressing their normal behavior of following their "parent."

of their society. The general language ability is built in; the particular language is learned (figure 14.3).

Humans are not the only organisms to receive adaptive traits, at least in part, through culture. How this occurs helps us appreciate the importance of culture in transmitting beneficial behaviors. Among wolves, elements of hunting skills are taught and improved by experience. Among primates, young learn safe foods from adult example. Among small songbirds, vocal renditions of surprising richness and complexity are modified through learned experience in some species, but appear innately in others. The meadowlark, a denizen of open country in North America, produces several songs during the spring breeding season. One of its songs begins with a single note, called an accent note, followed by a series of notes (figure 14.4a). If a male meadowlark is hand-reared from hatching to adult in complete sound isolation—never exposed to vocal culture—as an adult he still develops songs identical in every note to songs of compatriots never heard (figure 14.4a). Thus, here the trait is largely self-contained, innately passed to each generation. Such behavior is relatively unaffected by the environment and ready to go when needed. However, in other species of songbirds, such as the chaffinch, a sound-isolated young male develops into an adult with an incomplete vocal repertoire. Notes are sung, but the tune is different from the songs of fellow chaffinches. In the culturally deprived chaffinch the trait develops incompletely (figure 14.4b on p. 236). If exposed at this formative time to the song of another species, the chaffinch picks up

elements of this alien song culture. Here, within the chaffinch, the behavior incorporates elements supplied by culture. Such a behavior is learned; experience provided by immediate culture contributes to the character of the adult song. Remove culture, the trait is passed imperfectly, and the adult song is deficient.

Therefore, in approaching the evolution of humans and their behavior, we will not be too concerned with whether a behavior is learned or genetically defined. Instead, we will focus upon the adaptive advantage of the behavior, regardless of whether it is passed by genes or culture to the next generation.

PRIMATES

Humans are primates. Primates, in turn, are mammals whose closest living relatives include bats (Chiroptera), a gliding Asian mammal (Dermoptera), and a small Asian group, the tree-shrews (Scandentia) (see chapter 5, figure 5.15). Many subtle and esoteric features of the skull and skeleton contribute to the suite of characters that, together, define the primates. A few familiar features can be noted.

Primate Features

Primates have grasping hands, adaptations for gripping boughs and branches during movement through the tree tops by arm swings, a method of navigation referred to as **brachiation**. Other mammals are also arboreal—tree dwelling—but navigate by different methods. Squirrels dig sharp claws into rough bark. Essentially, cats do the same. The tree sloth hangs from its claws. Primates brachiate, use alternating grasps of branches to swing from branch to branch. Primates have *nails*, not claws or hooves. Nails cover the tips of fingers and toes, protecting them from blows and firming up the friction grip of the roughened, calloused hand. Rabbits and other prey animals have eyes more to the side of the head, each eye scanning a larger sweep of the brush and sky for incoming threats. Primates have *stereoscopic vision*, wherein the two eyes scan an overlapping part of the visual field before them. This improves the precision of depth perception. As a good primate, you can confirm this yourself. Cover one eye. This eliminates one visual field and its overlap with the open eye. You lose easy depth perception, and the world before you tends to flatten. You can navigate about the room with just a single eye to guide you, but it is more difficult. Working together, two eyes gather slightly different but

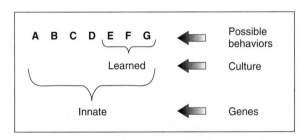

FIGURE 14.3   Nature and Nurture—Speech
Human language comes in a variety of dialects, here represented as *A–G*. In humans, genes (nature) provide a person with the innate ability to speak a language, but the culture (nurture) into which the person is born provides the particular language learned. The final behavior is built with guidance from genes and culture.

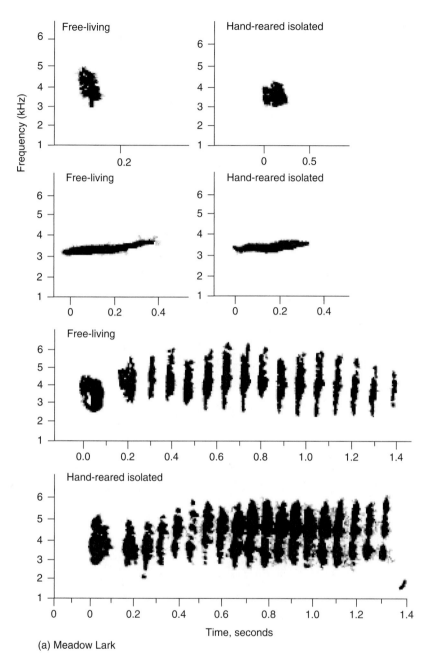

(a) Meadow Lark

FIGURE 14.4   Nature and Nurture—Bird Song (a) Three calls of the male meadowlark are shown, one set produced by a free-living individual and one produced by a hand-reared male kept isolated from ever hearing the songs of another male. The meadowlark's genetic program is sufficient to produce normal songs. (b) The free-living male chaffinch produces a complex song, but if raised in song isolation, its song is much different. If the chaffinch is exposed to the song of a tree pipit, then the chaffinch song picks up some of this vocal culture. Not only is the chaffinch genetic program insufficient to produce the normal song, but song culture (exposure to the tree pipit) can modify it. These recorded songs were played back through a sonograph to produce these visual displays of time versus frequency. ([a] is based on Lanyon 1960; [b] is based on Thorpe 1961.)

(continued on next page)

overlapping scans of the world before you. The distance between the two eyes produces enough difference in their shared images that the nervous system can turn this difference into depth perception. Stereoscopic vision, bringing acute depth perception, is especially important for active life high in the treetops. When a monkey moves from branch to branch, it may launch itself into space to span a break in the canopy, anticipating the branch ahead on the other side. Reliable depth perception is crucial, to judge the distance and safely reach the next branch.

Further, primates have *relatively* large brains. Placed next to the isolated brain of an elephant, the brain of a monkey or even a human would be smaller. But the primate brain is relatively bigger, for its body size, compared to the brains of other animals.

FIGURE 14.4
(continued)

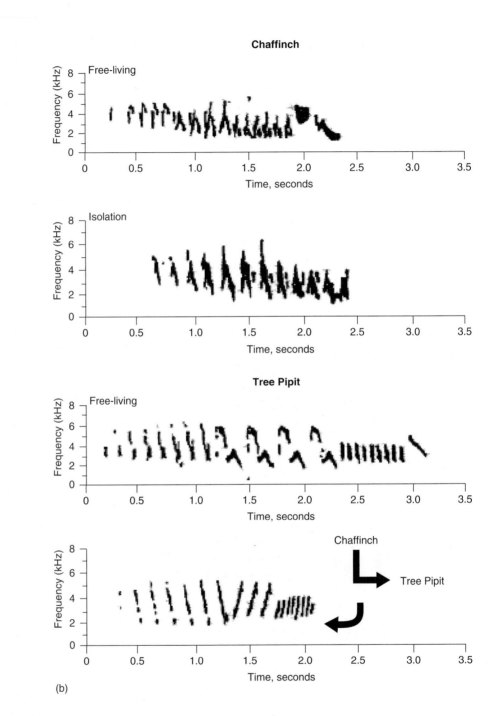

(b)

Primate Evolution

Primates evolved about 65 million years ago, just as the last of the dinosaurs were flickering out (figure 14.1; see also figure 13.12). About 40 million years ago, these early primates split into two surviving groups—the prosimians and anthropoids (table 14.1). The *prosimians* (basal primates) today include tarsiers and lemurs, which feed on fruits, leaves, and flowers. Their large eyes serve them well in their mostly nocturnal lifestyles. The *anthropoids* (derived primates) are an assortment that includes monkeys, apes, and

Table 14.1   Classification of Primates

Prosimians (basal primates)—Tarsiers, lemurs, lorises

Anthropoids (derived primates)
> **Platyrrhines**—New World monkeys, marmosets
> **Catarrhines**
>> **Old World monkeys**
>> **Hominoids**
>>> **Gibbons**
>>> **Apes**
>>>> **Pongids**—Orangutans, gorillas, chimpanzees
>>>> **Hominids**—Humans and immediate ancestors

humans. One group within them is the *platyrrhines*—marmosets and New World monkeys. About 30 million years ago, some early anthropoids made it to South America (how is still not known). There, they evolved in isolation into the arboreal New World monkeys, recognizable by their flat, spread nostrils and, in many, a grasping, prehensile tail. Another group of anthropoids, the *catarrhines*, is African in origin. Nostrils are close-set, and the tail when present, is not prehensile. The catarrhines are composed of two lineages—the Old World

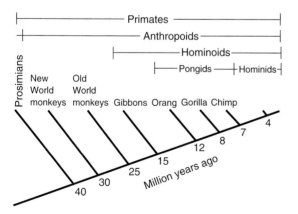

FIGURE 14.5   Primate Relationships Primates divide traditionally into two major groups. The prosimians tend to be small and nocturnal, including bush babies of Africa, the lemurs of Madagascar, and lorises and tarsiers of Southeast Asia. The anthropoids are more derived, including monkeys plus gibbons, apes, and hominids.

monkeys and hominoids. The Old World monkeys occupy tropical and subtropical areas across Africa, into Asia, and out onto some of the islands in the Pacific. The hominoids include the gibbons and the apes, including ourselves (table 14.1). Notice that there is nothing unusual in our phylogenetic position amongst the hominids (figure 14.5). We are just one of the most recent derivatives of the primate radiation. We share our contemporary place with primate survivors who also evolved within this primate radiation—other apes and monkeys. We are at once ordinary and yet remarkable.

THE COURSE OF HOMINID EVOLUTION

Consider our hominid ancestry (figure 14.6). During the twentieth century, the growing collection of hominid fossils, together with extrapolations back from rates of change in molecular structures, pointed to a hominid divergence from chimpanzees at around 7 million years ago. In the early years of the twenty-first century, anthropologists, working in Africa, uncovered fossils that promised to mark this point of divergence. These oldest hominids, so far, date to between 6 and 7 million years ago. One, *Sahelanthropus tchadensis*, is known from a nearly complete skull and a number of fragmentary pieces of the lower jaw and teeth. The other fossil, *Orrorin tugenensis*, consists of pieces of arm, thigh, and jawbones plus several teeth. So far, fossil fragments of both are scant, preventing any conclusion about their basic lifestyles (savannah or tree climbing) or body posture (four-footed or standing bipedal). *S. tchadensis*, in particular, fits these expectations of lying at the divergence of hominids from chimpanzees, being in possession of traits common to both. It has a relatively small brain (chimp-like) but brow ridges and small canine teeth (hominid-like). This mixture of features, along with its fossil age (6 to 7 million years), fits expectations of being close to the common ancestor of chimpanzees and hominids. *O. tugenensis*, occurring at a similar geologic age, carries similar expectations, but its fossil remains are even more fragmentary. Advocates promote one or both of these fossils as common ancestors to chimps and hominids; informed skeptics see other possibilities, including the possibility that one or both of these early fossil species may be an independent offshoot of early hominid or even chimpanzee history. I place them (figure 14.6) in chronological order, but

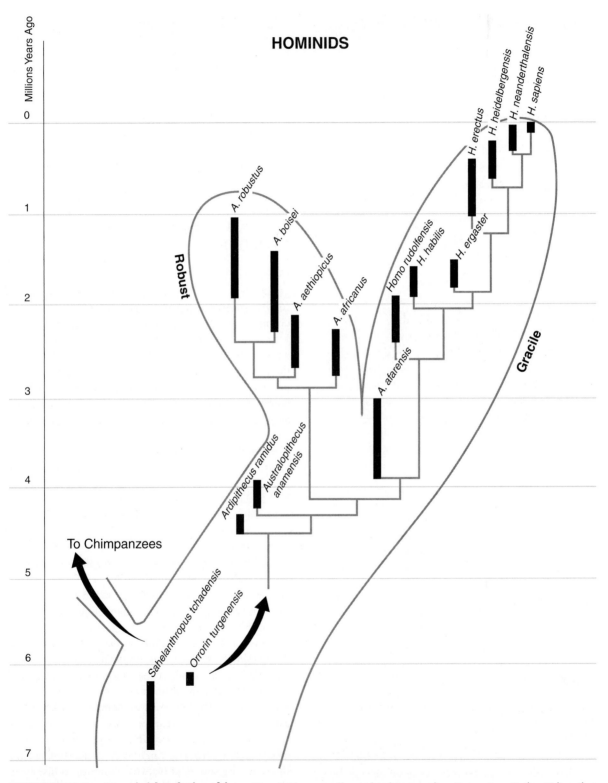

FIGURE 14.6   Hominid Relationships   Hominids generally evolved in two directions—in a "robust" line that became extinct about 1 million years ago, and in the "gracile" line that continues down to modern *Homo sapiens*.

Consider This—

Hominid or Hominin?

The rising fashion in primate taxonomy is to name groups on the basis of genetic similarities. Orangutans diverged some 11 to 13 million years ago, and in recognition of this are placed in the subfamily Ponginae. The African apes—gorillas and chimpanzees—and humans are genetically more closely related to each other and so are placed together in the other subfamily, the Homininae. Within the

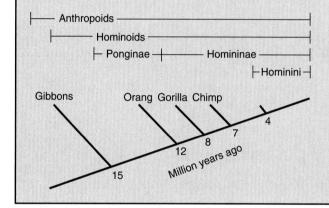

Homininae, humans and our fossil ancestors are in a subdivision referred to taxonomically as a "tribe," namely the tribe Hominini (*hominin* in English conversation). Unfortunately, such a system based primarily upon genetic relationships ignores the phenotype and its evolutionary significance. For example, chimps and humans share about 98% of the same genes, but are considerably different in morphology and in social systems. In particular, chimps and humans differ substantially in their phenotypes. As emphasized in chapter 8, a relatively small change in the genotype can produce a profound change in the phenotype and, hence, in the phenotype's adaptive scale, in its departure from lifestyles of ancestors, and in its general evolutionary success. A genetically based taxonomy may be in our future. But for the present, the traditional taxonomy does a better job of recognizing major phenotypic and, hence, adaptive changes. The useful term *hominid* survives to serve and inform us a while longer.

reserve judgment about their connection to later hominids until the two species are better known.

Discovered in the early 1990s, *Ardipithecus ramidus* lived almost 4.5 million years ago, also in Africa. Its scientific name means, approximately, "ground ape" or "basal ape," representing the optimism of the anthropologists at the time that here was the base of the hominid tree. *Australopithecus anamensis* soon followed, 4.2 to 3.9 million years ago. Thereafter, the subsequent course of hominid evolution produced two major branches: the *robust* hominid line, which became extinct about 1 million years ago, and the *gracile* hominid line, which persists up to the present in ourselves.

Hominid Features

Consider our features. We share more with our primate cousins than we possess uniquely (figure 14.7). We have the features of brachiators even though our ancestors stopped regularly swinging through the trees several million years ago.

Long Arms

If you are sitting, pause a moment and let both arms drop to your sides. Notice that your hands hang at or below your hips. If you placed your dog's or cat's forearms back along its sides (assuming they would permit it), its front paws would not reach so far. Ours is a carry-forward of long arms from brachiating ancestors (figure 14.8).

Collarbone

Your collarbone is the stout bone at the top of your chest. Unbutton a collar button on a shirt or blouse and you can reach this bone, running from sternum to shoulder joint at the arm. In dogs and cats, the collarbone is vestigial, allowing the forelimb large excursions during long, striding gaits. In primates, the collarbone (clavicle), together with the flat shoulder blade (scapula) on the back, is large to stabilize the shoulder joint and prevent dislocation of the joint during arm swings when the weight of the body hangs from the arm (figure 14.8).

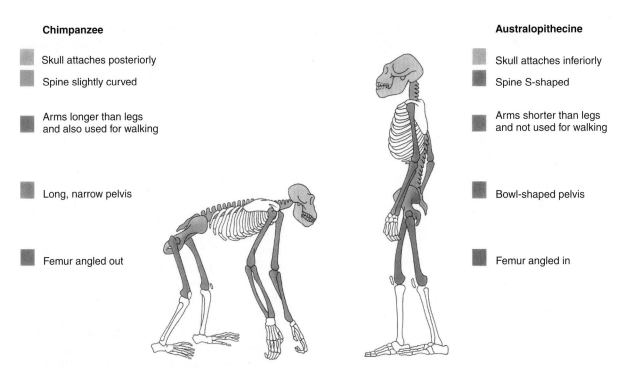

Chimpanzee

Skull attaches posteriorly

Spine slightly curved

Arms longer than legs
and also used for walking

Long, narrow pelvis

Femur angled out

Australopithecine

Skull attaches inferiorly

Spine S-shaped

Arms shorter than legs
and not used for walking

Bowl-shaped pelvis

Femur angled in

FIGURE 14.7   Ape and Hominid Skeletons   In apes versus hominids, respectively, the backbone joins the back of the skull vs. the bottom; the backbone is arched vs. S-shaped; arms are long in both, but shorter than hindlimbs in the ape compared to the hominid; hips are long vs. bowl-shaped; femur (upper leg bone) is angled out vs. in.

Grasping Hands

Fingers hook a branch; the thumb opposes for a secure grip. You used this grip as a kid when climbing about on a playground jungle-gym. Today, you may use it at the end of the school term when you grab the handle of your fully packed suitcase to head home.

Locomotion

Unlike our primate cousins, we walk upright, comfortably on two limbs—a posture and mode of locomotion termed **bipedal**. You may have seen nature programs showing a chimp or gorilla hustling across a jungle opening, running on two hind legs. Other primates can go bipedal for short distances, but they are awkward and quickly return to quadrupedal or to knuckle-walking. Hominids are comfortable on two legs and designed for bipedal locomotion.

What is the advantage of bipedalism? Certainly not speed. Bipedal locomotion is not a faster mode of locomotion than four-footed, **quadrupedal**, running. Imagine trying to outrun an angered dog or catch a fleeing cat

over a short distance. Not likely. And a bipedal posture is intrinsically unstable, as the weight of the upper body is balanced above the hips. That is why a lineman in American football gets down on all fours in preparation for action. In that position he is stable and much harder to overturn. But bipedal posture has its advantages. Lifting the forearms from the ground and rising up on the hindlimbs also lifts the head, with its important sensory receptors—nose (smell), ears (hearing), eyes (vision). This provides for better *surveillance* of opportunities or dangers otherwise obscured by low brush in the immediate vicinity. Prairie dogs temporarily sit up on their haunches to survey dangers around them. Meerkats, carnivora of Africa, stand tall on their hindlegs during sentry duty, gaining height to scan the incoming landscape for danger. A bipedal posture assumed among early primates accomplished the same beneficial function. It also freed the forearms from the quadrupedal demand of supporting the front of the body. The job shifts to the two hindlimbs exclusively, and the forearms are available for other roles, such as carrying food or infants, or investigating objects of curiosity in the environment.

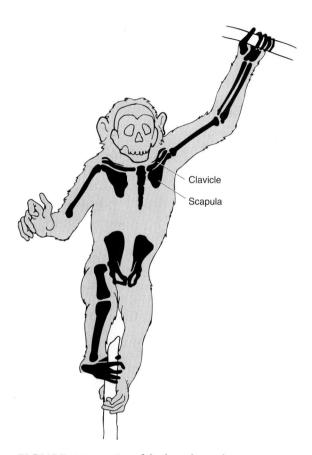

Clavicle

Scapula

FIGURE 14.8   Brachiation in Primates Loco-motion by swinging through the trees suspended from the forelimbs places special weight on the shoulders. To ac-commodate this, the shoulder joint is braced in the back by the broad scapula (shoulder blade) and in the front by the clavicle (collar bone). Hands are modified for grasping.

Other Features

We humans have *relatively* large brains, even com-pared to other primates. A hair coat over our bodies is reduced from the thick pelage of other primates to just a sparse smattering of hair. We use speech to make complex sentences, more than just a few vocal calls or grunts. How did we get this way? What was the course of our evolution?

HOMINID EVOLUTION—ON BECOMING HUMAN

One of the most important hominid fossil finds ever was in Africa, by Raymond Dart, in 1924. It turned out

not to be the oldest or most complete hominid fossil, but it was a turning point in the study of human origins, unlike any other before or since.

In the century before, human fossils had been found in Germany (1856) in the Neander Valley, from which they took their common name, Neandertals. Such finds reinforced the European expectation that still-earlier ho-minid fossils awaited discovery in deeper European soil. Certainly, the prosperous cultures and economies then in Europe testified to the promising role of the continent as the center of human origins. Further, scientists antic-ipated that these early hominid fossils—expected but then still undiscovered—would show a course of evolu-tion that began with quadrupedal ancestors with large brains. From such a beginning, some scientists of the day predicted that upright and bipedal posture would follow. The course of evolution seemed decided—first, large brains on quadrupedal bodies, then bipedal pos-tures. All was in place intellectually. It was just a matter of going and finding these fossils wherever they might be buried in Europe. Such were the scientific anticipa-tions as the twentieth century dawned. Initially, these predictions seemed confirmed by a find on English soil at a site near the hamlet of Piltdown.

Hoax

Discovered in 1911, the "Piltdown" skull held in life a large brain, large canine teeth (baboonlike), an ape-like lower jaw, and other features that fit early predictions. Forty years later, in the 1950s, radiocarbon dating would reveal that this skull was an elaborate, well-crafted fake. It was not ancient at all, but a doctored, modern human skull fitted with a filed orangutan jaw. This surreptitiously crafted hoax had been secretly slipped into the dig site outside Piltdown, by a perpe-trator unknown to this day. There, it waited for a gullible, anxious scientist to find it. But when the Pilt-down skull was "discovered" in 1911, it was not known to be a hoax. It then only confirmed and hardened the expectations already established from the earlier ho-minid finds on European soil.

Taung Skull—A Child's Story

In the 1920s, paleontologist Raymond Dart was working not in Europe, but at the southern tip of Africa, a continent away, in the Anatomy Department at the University of Witwatersrand. Nearby workmen excavating for lime blasted open a small cave at a location termed *Taung*.

Honoring Dart's requests, they packed up any fossils that might prove interesting and sent them off to him. Several days later, two cases of brecca (cemented lime, rocks, sand) holding fossils arrived at Dart's home, just as Dart was being assaulted by appeals to dash off to a wedding where he was to be the best man. Yielding to curiosity, he peered into the boxes. Within was a brain endocast, larger than that of any chimp, and the partial skull of a young hominid. Putting this together in his mind, he realized this was a very ancient hominid, older than anything found to that date. That was 1924.

Now very late for the wedding, Dart was wrenched away. Finally, with the wedding over and the reception complete, he returned home, borrowed common knitting needles to use as fine picks, and began to clear the brecca from the skulls. As the skull slowly emerged from the rocky matrix, he realized that his first impression was correct. The teeth were milk teeth—those of a *child*—teeth not yet replaced by the permanent teeth of an adult. Affectionately, the skull became known as the "Taung child," after the site. The child had died at about 3 or 4 years of age. The soft tissue of the brain quickly rotted away upon death, but minerals eventually infiltrated, filled the skull, and hardened in the empty brain cavity, forming the endocast, a facsimile of the brain. This testified to a relatively large brain, larger than that of any closely related pongid. This was not an ape; this was an early *hominid*. The spinal cord passed through a large hole in the skull, the foramen magnum. In quadrupeds, this hole is angled, placed at the back of the skull to accommodate the angled, four-footed posture. In the Taung child, the foramen magnum was on the bottom of the skull, a position testifying to a *bipedal posture* in this early hominid.

Formally, the fossil was named *Australopithecus africanus*, roughly translated as "southern ape of Africa." The 32-year-old Dart had found, and soon described scientifically, a remarkable hominid but one much at odds with earlier expectations of European anthropologists. Here was a juvenile early hominid with a relatively large brain compared to that of pongids, but not so large as expected for early hominids. And it walked upright. Bipedal posture was an early hominid invention, not one to appear later. Perhaps most inconvenient of all, this fossil placed very early origins of hominids, not in Europe, but in Africa. Dates for this and subsequent *A. africanus* fossils fall to just under 3 million years, 2.8 to 2.4 million years ago. Note where these fall within hominid evolution (figure 14.6).

Still-earlier hominid species would be unearthed by others, but not for about 50 years. The Taung child turned the search for early hominids away from Europe and to Africa. It revised our view of what distant human ancestors looked like. And it changed expectations about our geographic origins—human ancestors had an African genesis.

Lucy—Farther Back in Time

The paperwork never ends. It gets deeper. It suffocates. It stalks you. It seems to become one's purpose in life, replacing the joyous career in biology you thought you were pursuing. Forms upon forms upon forms to complete. It goes home with you in the evening. It awaits you when you arrive in the morning. It even tracks you down. Or at least, so it seemed to anthropologist Donald Johanson, associated eventually with the Cleveland Museum of Natural History and currently affiliated with Arizona State University. But in 1973, Johanson was on expedition in eastern Africa in search of early hominid fossils. Facing either a day of paperwork or a ride with a visiting graduate student to a promising fossil site, Johanson chose the freedom ride. Four miles of jarring road, even in a Land Rover, took half an hour, but Johanson was feeling good. Even the 110 °F noon temperature did not tempt him to return to the looming paperwork back at camp. Life was glorious, and he felt lucky. Soon after arriving at the fossil dig, he noticed a bit of fossil on the ground—a hominid arm bone. Then more, and more, and more skeleton parts. Eventually, over the next three excited weeks of careful excavation, Johanson and his team recovered about 40% of a single hominid skeleton. And back at camp at the end of the day, the Beatles' song "Lucy in the Sky with Diamonds" filled the evenings and inspired the common name for this fossil, "Lucy." (Good thing "Nowhere Man" wasn't popular around the campfire.)

Scientifically, Lucy was eventually christened *Australopithecus afarensis*. The skeleton was incomplete, but when examined with an informed and careful eye, it yielded remarkable details (figure 14.9). She was older than and ancestral to Dart's Taung child, dating to just under 4 million years. She was an adult, about 3^{1}/$_{2}$ to 4 feet tall, ate both plants and animals (an omnivore), walked bipedally, and was, in fact, female. Let's look at the evidence for each of these traits:

Adult. Her third molar teeth had erupted and her milk teeth were gone—she was an adult.

Height. Assembled bones of the spinal column upon the hips, with remaining leg bones below, indicated she measured just under 4 feet in height.

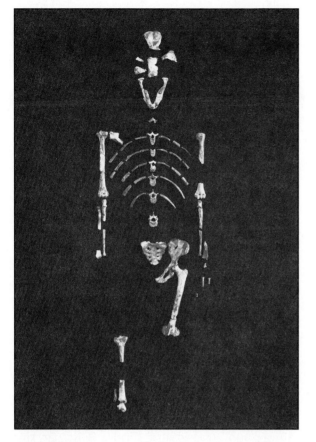

FIGURE 14.9   Lucy Skeleton, *Australopithecus afarensis*   The lower jaw and parts of the skull survive, along with some of the arm bones, ribs, backbone, left hip and upper leg bone, and some right lower leg bones.

Omnivore. Her teeth were neither those of a strict carnivore such as a cat, nor the tall, deep teeth of a dedicated herbivore such as a wildebeest. They fell in between, like those of a pig, confirming that she subsisted on both plants and meat.

Bipedal. The upper leg bone, the femur, articulated with the hip at a likely angle that suggested a bipedal posture. Several years after the find of Lucy, Mary Leakey and her team discovered preserved hominid footprints at a similar time horizon. Step by step. Walking on hindfeet only—upright. Not likely stamped out by Lucy herself and friends, but certainly by members of the same species.

Female. How did Johanson know that the skeleton he discovered was female? Lucy's bones included the left hip bone. A mirror image of this bone completes the hips, and together the hip bones define

the birth canal between them. The birth canal—the round hole in the pelvis, not the wide pelvis itself—is the opening the infant must clear at birth. In particular, the relatively large head of a hominid baby must pass through this opening. Consequently, in hominids the birth canal is large, and especially so in females, who must do the job of birthing the infant (figure 14.10). Lucy's birth canal was especially large, testifying that the skeleton Johanson discovered was that of a female.

Birthing is serious business. This is when the next generation makes its debut. Mammals generally, primates notably, and hominids in particular have relatively large brains. Consequently, the birth canals must be commodious. In chimpanzee females, the oval infant head passes through the oval birth canal. In modern humans, even infants have relatively larger brains, and larger skulls to hold them. For a human infant, passage through the birth canal is torturous, requiring decisive but gentle turns and twists to navigate the passage. Compromises in hip design have evolved to accommodate a big-headed infant through a narrow birth canal.

To understand the compromises, consider the mechanical problems. During a normal striding gait (walking), one of the two legs is off the ground, swinging forward and extended during the recovery cycle. While one leg is off the ground, the weight of the body is balanced solely on the other leg that remains in contact with the ground. Because the weight of the upper body is offset from the limb, the hips tend to drop (figure 14.11). This tendency to drop is counteracted by the gluteus maximus muscles, whose name says it all. *Gluteus* means "buttocks"; in humans, they are large—*maximus*. They constitute the "sexy buns" displayed to advantage in bikinis and brief shorts of frolicking young women and men on the beach.

You can confirm their functional role yourself. When walking across the room, grab hold of one of your gluteus maximus muscles. Notice that when one leg lifts to swing forward, the opposite gluteus maximus muscle contracts. It compensates for the tendency of the hips to drop, keeping them level. Think of this as a lever system, something like a playground seesaw. The weight of the body rotates the hips one way, the gluteus maximus contracts to tilt it in the opposite way and so compensates. Both act about the fulcrum, or balance point—the head of the upper leg bone (femur). As you know from your playground days, the longer the board between you and the fulcrum, the more your weight can lift on the opposite side. This distance to the

FIGURE 14.10 Birth Canal
Hips of the human male (a) and the human female (b). In humans, the birth of a baby with a relatively large head requires a relatively large birth canal. Note that the inner rims of the hips of a female define a larger birth canal than do those of a male.

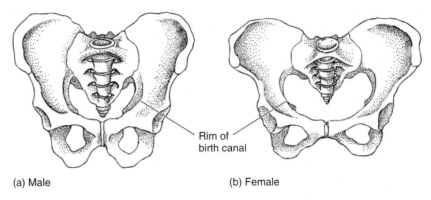

Rim of birth canal

(a) Male   (b) Female

fulcrum is the lever arm, and it expresses or is proportional to the effectiveness you have in acting on the other side of the board.

In Lucy's hip, the lever arm for the gluteus maximus muscle was almost twice the length of the lever arm in modern humans (figure 14.11). Compared to modern humans, Lucy's gluteus maximus muscle was roughly twice as effective in balancing the weight of the upper body and keeping the hips level during normal walking strides. What has happened in modern human females is that the birth canal has widened to accommodate the still-larger head of human infants. This has crowded the hip joint, forcing it outward, and shortening the lever arm of the gluteus maximus muscle (fig-

ure 14.11). In compensation, the human gluteus maximus muscle is larger, producing more force, to make up for the shorter lever arm. The result? Humans have big "buns," large gluteal muscles, at least compared to those of other primates.

Vegetarians—A Dead End

In 1938, Robert Broom purchased from a collector (who got it from a school boy) an African fossil that became named *Paranthropus robustus*, "stout near-human." In 1959, Mary Leakey discovered in Africa, and Louis Leakey described in print, *Zinjanthropus boisei*, "East African man" (named after Charles Boise, the kind

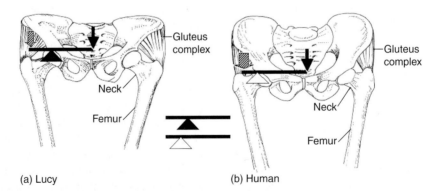

(a) Lucy   (b) Human

FIGURE 14.11   Hip Mechanics and Birth Canal   The larger birth canal of human females crowds the hip joint and reduces the lever advantage upon which the gluteus complex of muscles act. Consequently, these muscles are large in humans to compensate. (a) Hips of Lucy, *Australopithecus afarensis*. (b) Hips of human, *Homo sapiens*. When walking, the weight of the upper body (solid, vertical arrow) balances on the head of the femur (fulcrum, represented by a triangle) against the force produced by the gluteus complex (cross-hatched vertical arrow). Like a teeter-totter, the balancing of these is represented between the drawings. Note that the position of the fulcrum in the human (open triangle) is offset, making the action of the gluteus complex different than in Lucy (solid triangle). In humans, the enlarged birth canal and shortened "neck" of the upper leg bone, the femur, produces these mechanical differences in humans.

benefactor who financed the expedition). Since then, new, related fossils have been found. These early enthusiasms for giving separate names to the genera have given way to a more informative naming. Today both fossil finds belong to the same genus, *Australopithecus*. Broom's is *A. robustus* and Leakey's is *A. boisei*. Both are a stocky build of *Australopithecus*. Males, in particular, of each species sported bony crests running along the top of their skulls. These crests offered sites of attachment for very strong muscles acting on the lower jaw. The jaws were heavy and angular; the teeth broad and thick with tough enamel. These were teeth and jaws suited for grinding the coarse fiber of plant tissue. These hominids were largely, if not exclusively, herbivores, representing the robust line of hominid evolution.

Such a diet of fruits, tubers, and some leafy plants most likely kept these robust hominids in or near forests, living in small social groups. When danger threatened, they could scoot up a tree, and likely slept there at night for safety. This was a lifestyle similar to that of gorillas and some other pongids today. The males reached a height of a little over 5 feet (160 cm); females were half a foot shorter. This robust branch of hominid evolution became extinct about a million years ago, leaving no descendants (figure 14.6).

AT THE ROOT OF IT ALL—THE OLDEST HOMINIDS

Our basic human history is now generally clear. Hominids divided—one line leading to a robust group, the other to a gracile group. The immediate common ancestor to both was *A. afarensis*—Lucy—or at least that is the closest fossil so far discovered. Lucy, like many primates of today, lived on the edges of open savannahs of Africa. She and members of her group spent much of the day foraging over 3 to 4 miles of terrain. As evening approached, they retreated to the safety of trees for the night. Now, thanks to more recent finds, we know of some ancestors that precede even Lucy, deeper into our past. One, *Australopithecus anamensis*, predates Lucy, at about 4.1 million years ago. It is known so far from a few jaw fragments and bits of tibia, a forearm bone, with clear indication of bipedal posture in its legs. Earlier still is *Ardipithecus ramidus*, from about 4.4 million years ago. *Ardi* means, in the local African language, "ground-dwelling"; *ramid* means "root or basal"; and *pithecus* is Greek for "ape." Reaching almost to the

common ancestor between pongids and hominids are *Orrorin tugenensis* and *Sahelanthropus tchadensis*.

OVERVIEW

How lucky we are to see these wonderful connections between the pongid divergence and the line leading to hominids (figure 14.6). Each new find, in turn, gives focus to subsequent archaeological research. The gaps continue to fill between our group, the hominids, and pongids. Our old history yields up new discoveries each year, revealing a rich and diversified tree of early hominid ancestors. Yet humans are not one of life's imperatives, an evolutionary inevitability. Had dinosaurs persisted or mammals perished, then primate history might have been quite different or happened not at all. But primates did evolve, and within that primate diversification arose humans.

An old debate that grew up around the source of our behaviors is referred to as the nature–nurture dichotomy. Are behaviors innate (nature) or learned (nurture)? Today we see the mistake of thinking in terms of such a strong dichotomy, as human behavior is built from both. Genes (innate) and experience (learning) both contribute to the finished behavior.

We share much with our primate cousins—evidence of brachiation, fingers and toes tipped with nails, grasping hands, stereoscopic vision, and relatively large brains. In a few ways, hominids stand apart from other primates. In hominids, social systems are usually complex, posture is bipedal, and brains are especially large relative to body size. Hominids departed from our closest primate relatives, the chimpanzees, about 6 to 7 million years ago. Subsequently, the course of hominid evolution produced two major branches, one the gracile line in which modern humans arose and the other the robust line.

Raymond Dart's discovery in 1924 of the "Taung child," *Australopithecus africanus*, proved a turning point in the study of human origins. Although older hominid finds would follow, at the time it was the oldest known hominid and implied an African genesis of hominids. Attention turned to Africa. In 1973, Donald Johanson discovered "Lucy," *Australopithecus afarensis*, in eastern Africa. Lucy was an adult and lived almost 4 million years ago. In the last decade, still earlier hominid species have been recovered, all from Africa, pushing the hominid fossil record back to its divergence from chimpanzees.

The robust line of hominid evolution, also in Africa, was recognized by fossil finds by Robert Broom in 1938 and later by additional finds by Mary and Louis Leakey. The various species of robust hominids had deep jaws and grinding teeth; they were principally vegetarians. As a group, they became extinct about a million years ago.

The gracile line of hominid evolution would produce some surprises. Most were omnivores, mixing meat into their diets. In fact, the task of procuring meat would help define the social systems of some of the later gracile hominids. In chapter 15 we turn to this gracile line of evolution.

Selected References

Conroy, G. C. 1990. *Primate evolution*. New York: W. W. Norton.

Fleagle, J. G. 1988. *Primate adaptation and evolution*. New York: Academic Press.

Fleagle, J. G., and R. F. Kay. 1997. Platyrrhines, catarrhines, and the fossil record. In *New World primates: Ecology, evolution, and behavior*, edited by W. G. Kinzey. New York: Aldine de Gruyter.

Fossey, D. 1988. *Gorillas in the mist*. New York: Houghton Mifflin.

Goodall, J. 2000. *In the shadow of man*. New York: Houghton Mifflin.

Johanson, D. C., and M. Edey. 1981. *Lucy: The beginnings of humankind*. New York: Simon & Schuster.

Johanson, D. C., and B. Edgar. 1996. *From Lucy to language*. New York: Simon & Schuster.

Park, M. A. 2002. *Biological anthropology*. 3d ed. New York: McGraw-Hill.

Stein, P. L., and B. M. Rowe. 2003. *Physical anthropology*. 8th ed. New York: McGraw-Hill.

Willis, D. 1992. *The Leakey family: Leaders in search of human origins*. Facts on File.

Learning Online

Testing and Study Resources

Visit the textbook's website at **www.mhhe.com/ evolution** (click on the book's title) to take advantage of practice quizzing, study/writing tips, timely news articles, and additional URLs for research on the topics in this chapter.

15

"We are the product of 4.5 [4.6] billion years of fortuitous, slow, biological evolution. There is no reason to think that the evolutionary process has stopped. Man is a transitional animal. He is not the climax of creation."

Carl Sagan, astronomer

Human Evolution:
Building Modern Humans

INTRODUCTION

Try this thought experiment. You are out camping and trying to drive a tent peg into the ground, but no hammer is at hand. What tool do you use? A big rock, a log, a convenient stick within reach? So it was with our ancestors. A small, struggling animal within their grasp needed to be dispatched quickly; a dead animal needed skinning. What tool did they use? Any heavy or sharp rock lying conveniently close at hand, within easy reach. Any natural object in the environment was a potential tool. Sensible enough, but this makes it difficult to identify the first "tools" used by hominids. Any natural stick or stone might serve. Such tools would betray no evidence of human crafting, reveal no evidence of their use in meeting human needs, and thus bear no witness to human tool use. Even chimpanzees will reach for a stick or rock to use in gathering foods. Within hominids, fashioned tools—those crafted deliberately to serve hominid needs—are first recognized at about 2.5 million years ago.

In the 1960s, Mary Leakey identified and characterized tools from about 1.8 million years ago that bore clear evidence of having been deliberately manufactured. These were from sites she was working in the Olduvai Gorge in east Africa, and therefore they were named Oldowan tools (*Oldowan*, adjective, from *Olduvai*). These tools are usually attributed to *Homo habilis*, 1.9 to 1.6 million years ago. Still earlier tools of the Oldowan type, dating to almost 2.5 million years ago, have been recovered elsewhere in Africa. Who made them is still debated. Let's start with *Homo habilis* (figure 15.1).

Finds of *Homo habilis* were indirect at first. Louis Leakey claimed to have discovered "tools," simply formed and shaped, but implements fashioned for practical chores. These consisted of hand-held choppers and stone flakes. In the early 1960s, bits of fossil jawbones were found, followed by skull parts. The scientific name *Homo habilis* means "handy" and "Man," reference to the use of tools. Brains of *H. habilis* were about 30% larger than those of *Australopithecus*, there was pronounced sexual dimorphism, but the skeletons preserved close affinities to australopithecines before them. In fact, such general skeletal similarities led many at first to

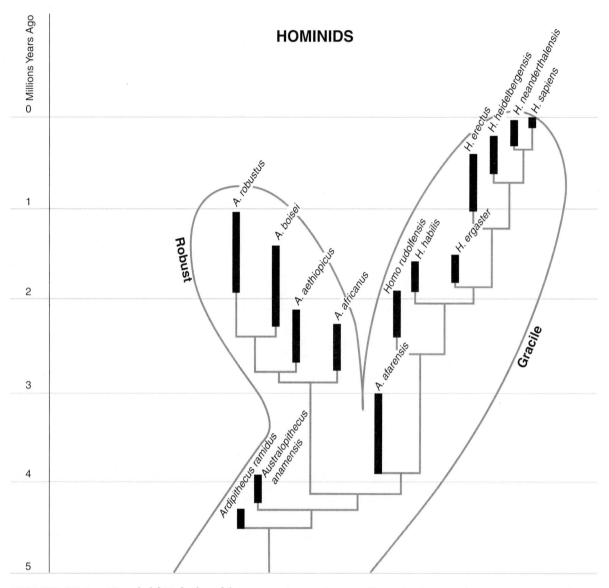

FIGURE 15.1   Hominid Relationships   Later hominids generally evolved in two directions—one a "robust" line that became extinct about 1 million years ago, and the other the "gracile" line continuing down to modern *Homo sapiens* (see also figure 14.6).

question whether this fossil was *Homo*, along the line to modern humans.

More recent finds at about this same time horizon identify a slightly later species, *Homo ergaster* (1.9–1.5 mya), and slightly earlier species, *Homo rudolfensis* (2.4–1.9 mya). Both are similar in anatomy, leading some to argue that they are just varieties of *Homo erectus* and *Homo habilis*, respectively. Others, as we have

done here, treat them as separate species, part of an early radiation of the gracile line of hominid evolution.

Intentionally fashioned tools, at 2.5 million years ago, coincide with various fossils of australopithecines who could be their makers. But more likely, these tools were made by one of the *Homo* species, most likely *H. rudolfensis*. The use of manufactured tools marks an important moment in hominid evolution because it implies a new level of mental ability—animation of current efforts to meet a future use. Rocks such as lavas and quartzites, which tended to fracture and flake along sharp edges, were selected, transported distances to the expected kill or slaughter site, and there fashioned purposefully for the task. Archaeologists date the beginning of the **Paleolithic** (Old Stone Age) from this time. From this time onward, tools produced by more sophisticated manufacture characterize hominid evolution (table 15.1). Within the Paleolithic, three major cultural stages are recognized: Lower (first stone tools), Middle (stone tools of Neandertals and their contemporaries), and Upper (stone tools of *Homo sapiens*). The **Neolithic** (New Stone Age) is marked culturally by the advent of farming in human activity.

ON TO MODERN HOMINIDS

Let's return to the gracile line of evolution where we left it in chapter 14, and the making of modern hominids. *H. habilis*, like its ancestral species, was also bipedal, a means of locomotion invented much earlier, dating at least back to the earliest australopithecines several million years before. Like its ancestors, *habilis*

was a denizen of Africa. To these features, *H. habilis* evolved a larger brain and used fashioned tools it manufactured. Each development—an enlarging brain and the manufacture of tools—drove the other, resulting in a different kind of hominid with a more complex social and hunting system. This is the *hunting hypothesis*—the view that procuring meat from small animals stalked or scavenged led to increased mental abilities, coordination of effort, and aggressive behavior. Stone tools could be quickly made at the site of the kill, the animal butchered, and edible parts carried back to other members of their local social unit.

H. habilis, together with related species of *rudolfensis* and *ergaster*, account for about a million years of hominid evolution, and were the earliest members of *Homo*. Their arms were long, and their skeletons suggest that they retained close affinities to australopithecines. That was not true for the hominid species that followed. In fact, the *Homo* species to follow included lots of surprises (Table 15.2).

Out of Africa

Homo erectus was one of the next *Homo* species. Not the least of its surprises was its location—China, India,

Table 15.1   Recent Hominid Cultures

Use of stone tools generally defines the Paleolithic, in which cultural use of tools specifically defines the Lower (*H. habilis/erectus* culture), Middle (Neandertal culture), and Upper (modern human culture). The advent of farming marks the Neolithic.

	Defining economic practice	Hominid species present
Neolithic (New Stone Age)	**Farming**	
(present – 10,000 years ago)		
(Mesolithic)	**Transitional**	
Paleolithic (Old Stone Age)		
Upper	**Hunter-gatherer**	*Homo sapiens*
10,000 – 40,000 years ago)		(modern)
Middle	**Hunter-gatherer**	*H. sapiens*
(40,000 – 160,000 years ago)		(Archaic)
		H. neanderthalensis
Lower	**Hunter-gatherer**	*H. erectus*
(160,000 – 2.6 mya)		*H. habilis*
		H. rudolfensis

Approximate years are indicated for each stage. Stages are defined culturally, but the hominid species occurring at these times are indicated to the right.

Table 15.2 General Terminology for Hominids

Basal Hominids

? Orrorin turgenensis

? Sahelanthropus tchadensis

Ardipithecus ramidus

Australopithecines

Australopithecus anamensis

Australopithecus afarensis

Robust-type Hominids

Australopithecus aethiopicus

Australopithecus boisei

Australopithecus robustus

Gracile-type Hominids

Homo rudolfensis

Homo habilis

Homo ergaster

Homo erectus

Homo heidelbergensis

Homo neanderthalensis

Homo sapiens

Archaic *Homo sapiens*

Modern *Homo sapiens*

? = Currently uncertain placement in hominids. May be closer to chimpanzees.

Southeast Asia. *Erectus*, or its immediate ancestor *H. ergaster*, was the first of the hominids to migrate out of the African center of hominid origin. Unlike the earlier *Homo habilis*, *H. erectus* from the neck down was similar in stature and anatomy to modern humans, and its brain was far larger, almost 50% larger, than the brain of *habilis* before it. Large quantities of animal bones and numerous stone tools are found at *erectus* occupation sites; many of the animals were large, not easily taken down by a solo hunter. Vegetable foods were certainly gathered, but hunting game was a central feature of its social system. Some sites are associated with charcoal, the result of a natural brush fire sweeping through a campsite or a fire deliberately struck by *erectus* for warmth or to cook its food.

The first *erectus* fossils were discovered as the result of a mistake, a stubborn Dutchman, and extraordinary good luck. All this happened in the late nineteenth century. The mistake grew from the then current view

that a single, direct ancestor connected apes with modern humans. This connecting species was predicted to be a blend of modern ape and human characters, an intermediate. However, to find the ancestor common to humans and apes (pongids), we would have to travel back through a variety of fossils, to almost 7 million years ago (figure 15.2). There is no direct, straight link between the two modern groups, apes and humans. But in the late nineteenth century this mistaken view held great sway in scientific circles. This was the legendary "missing link" view of human origins. Although it was a serious misreading of hominid history, it energized many an anthropologist of the time.

One individual thus energized was the Dutchman Eugène Dubois, a physician by trade. As a young boy, he had heard of the Neandertal finds in Germany, but he reasoned that older hominids awaited discovery outside

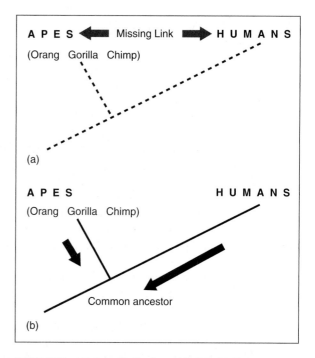

FIGURE 15.2 "Missing Link" Fallacy (a) A misreading of primate evolution led to the view that midway between modern apes (such as orangutans, gorillas, chimpanzees) and modern humans there existed a direct, intermediate ape that was a blend of ape and human traits. (b) As current phylogeny shows, no such direct intermediate occurs, or is expected. Instead, apes and humans trace their ancestry back along independent lineages to a common ancestor about 6 to 7 million years ago (see also figure 14.6).

of Europe, in warmer places untouched by past glaciation that would have destroyed their fossil traces.

By good luck, Dubois picked Sumatra, in Indonesia, to look for this manlike ape, the "missing link." In this he was encouraged by Alfred Wallace, who had traveled there, and by the presence of the living orangutan, which seemed to fit some of the intermediate anatomical criteria of a missing link.

Unable to raise expedition funds, Dubois joined the Dutch Army as a military physician and was posted to Sumatra. Two years of off-duty searching turned up no manlike apes and gave him a bad case of malaria. To recover, Dubois was invalided to Java and placed on inactive duty. There he explored further along the Solo River and turned up some promising fossils. The Dutch government was now interested in the work and loaned him convict labor to help him dig. Unfortunately, this enterprising workforce stole fossils and sold them on the side to Chinese traders, who in turn ground these "dragon bones" into a powder prized in China as an ingredient in folk medicines and aphrodisiacs.

After about a year on the river, Dubois enjoyed extraordinary luck. In 1891 his team unearthed a skullcap. A year later, upstream, the team unearthed a leg bone. Together, the bones suggested a fossil with a brain larger than that of an orangutan, an upright posture, and an age far older than any hominid then known. Dubois had found his manlike ape, dubbed "Java Man," but a doubting scientific community did not agree. Off to Europe with his finds, expecting to be received in triumph, he was met instead with deep skepticism—he had misidentified the bones; they were from different species; they were too ancient. Dubois even tangled with friends. Those sympathetic saw these bones as evidence of a hominid ancestor leading back in time, but not a direct missing link across to modern apes. Dubois himself was adamantly inflexible. His mind was closed. He clung to the missing-link theory. All sides were against him. In stubborn disgust, he hid the bones in the floorboards of his dining room and for the remaining 30 years of his life refused to let anyone see them.

Meanwhile, other *erectus* finds were accumulating. In the 1920s, a series of hominid fossils from the same time horizon were uncovered outside Beijing, China (then called Peking), along with crude tools. The fossils—skullcaps, mandibles, facial fragments, teeth—were from approximately two dozen different individuals in all (figure 15.3). Together the bones gave a relatively complete reconstruction of *erectus*, termed at the time "Peking Man." With the outbreak of World War II, all

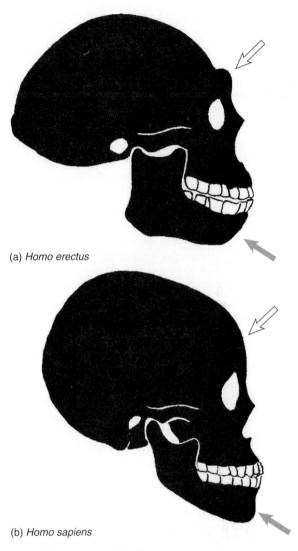

(a) *Homo erectus*

(b) *Homo sapiens*

FIGURE 15.3   Skulls (a) *Homo erectus*; (b) a modern human, *Homo sapiens*. In *H. erectus*, note the prominent brow ridges (open arrow) and their absence in *H. sapiens*. Chin (solid arrow) is recessed in *H. erectus*, and protruding in *H. sapiens*.

bones were packed up in December 1941 for shipment to a safe haven in the United States. Somewhere along the way, these valuable fossils vanished; perhaps they were pulverized into powder and sold as ingredients for folk medicine.

Still earlier related fossils occur in Africa, dating from about 1.9 million years ago. These African fossils had been placed in their own species, *H. ergaster*, leaving *H. erectus* as the occupant of Asia. However, *ergaster* fossils recently found in Africa look a lot like the

Asian *erectus*, thereby linking the two. If this recent find is *erectus*, then the fossils from Africa to Asia were in fact one species (*H. erectus*), not two (*H. erectus* in Asia and *H. ergaster* in Africa). Thus, *H. erectus* (plus *H. ergaster*) likely arose in Africa, with large brains and tall stature, some reaching 6 feet in height. Scattered and abundant remains of animals among the *erectus* finds testify that hunting was a central part of *erectus* culture and way of life, although we should not see this as a sophisticated and highly coordinated social activity. Likely, small groups spread out casually, surrounding or cornering prey, and others moved in for the kill. Tools and perhaps fire were used within daily activities. At this stage in hominid evolution, body hair was likely reduced and sweat glands increased, allowing improved heat loss and evaporative cooling for an active hunter. So equipped, *erectus* moved out of Africa, spreading to China and Southeast Asia, and possibly into Europe, although fossil evidence there is so far unknown. Sexual dimorphism between the genders decreased. *Erectus* emerged 1.2 million years ago (*ergaster*, 1.9 mya) and survived up until 400,000 years ago, although we might expect subsequent finds to turn up varieties or even different contemporary species related to *erectus*.

Today *erectus* still beguiles. Spread to strategic geographic locations throughout Europe and Asia, *erectus* may have evolved there in place into modern humans. Or modern humans may have evolved elsewhere and replaced *erectus* as one of the most successful of all hominid species (figure 15.4). Before treating the demise of *erectus*, let's finish our survey.

Out of Africa—Again

From the beginning, no one knew quite what to do with Neandertals. This reflects their wonderful intermediate position between *H. erectus* before and modern *H. sapiens* later. When their fossils were discovered in 1856 in the Neander Valley in Germany, the first thought of some German scientists was that these were bones of modern humans, perhaps the remains of a wounded Cossack horseman who fell during the Napoleonic wars that raged through Europe earlier in that same century. As the twentieth century began, Neandertal reputations took a turn for the worse; they were viewed as brutish, coarse cavemen, and subhuman. By the 1960s, attitudes changed again and Neandertal reputations were looking up. Clean-shaven and suited up, it was claimed, they could have walked down the streets of New York City unnoticed, without drawing a crowd. Perhaps, in New York City.

Consider This—

Evolving Language — -*thal* or -*tal*?
The first partial Neandertal skeleton turned up in Germany's Neander Valley in 1856 and was called and spelled Neander*thal*—*thal* meaning "valley." The scientific name today is *Homo neanderthalensis*, or, if one views it as a human subspecies, *Homo sapiens neanderthalensis*—both with the *th*. But in about 1900, German's silent *h* was dropped in the spelling of some words. Following this change, the common name can be spelled acceptably *Neanderthal* or *Neandertal*, with or without the *h*.

Their close affinity to modern humans is retained in the scientific name *Homo sapiens neanderthalensis*. This makes them a human subspecies. But they were different. Brow ridges were prominent; the projecting chin of modern humans was absent; a low skullcap keel ran along the top of the adult male skull; teeth and details of the skeleton set them apart. Here we treat them as their own species, *Homo neanderthalensis*.

Neandertals appear to have branched off 500,000 years ago from the ancestral line leading to modern humans, specifically from the earlier *H. heidelbergensis*. The oldest fossil of *heidelbergensis* is from Ethiopia around 600,000 years ago and the most recent is from Europe 200,000 years ago, suggesting that they too originated in Africa, contemporaries of late *erectus*, and spread northward, surviving up to and briefly overlapping with the early Neandertals. Like Neandertals, *heidelbergensis* had brow ridges and a recessed chin. However, *heidelbergensis*'s forehead and nasal bones anticipate those of *H. sapiens*, and it seems less specialized than Neandertals, leading some to see *heidelbergensis* as the more likely ancestor to both Neandertals and modern humans (see figure 15.4).

Neandertals were hominids that preceded and then for a time coexisted with modern humans, *Homo sapiens* (figure 15.4). Neandertals appear as fossils dating to more than 300,000 years ago and survived down to 30,000 years ago, flourishing from western to eastern Europe. They left useful artifacts, such as practical tools fashioned to kill, to butcher, to dig, and to prepare the foods they ate and the weapons they used to defend themselves. Simpler but clearly crafted tools predate

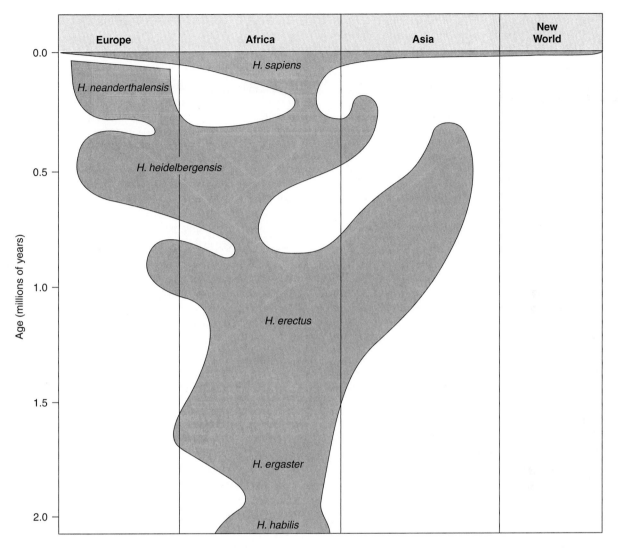

FIGURE 15.4   Out of Africa Various species of gracile hominids originated in Africa and then spread to other continents. This figure summarizes the approximate times and geographic locations of hominid species.

Neandertals by several million years through intermediate ancestors reaching back to *Homo habilis* in Africa 1.9 million years ago. However, the fashioned tools left by Neandertals, and all before them, were utilitarian implements—practical tools to perform a job in food getting or defense.

Among living primitive peoples, artifacts are both utilitarian and symbolic. Burial of deceased humans is encouraged by religious doctrines, in particular by a belief in a life after death. The discovery, then, that at least some Neandertals intentionally buried their dead gains significance. At Le Moustier in southern France, a Ne-

andertal youth was uncovered who had been laid in a sleeping position on his right side, head reposed on his arm, and surrounded by animal bones and stone tools. At another site in southern France, a Neandertal cemetery, presumably serving one family or local tribe, contained two adults laid head-to-head, three children, and a very young or stillborn baby. Child burials are known from other sites as well. On the one hand, these Neandertal burials were very simple. Art and adornments were absent. The grave goods, such as they were—animal remains, practical tools—could have been present already by accident or as debris from an animal slaughter for food. On the other hand, if these were deliberate burials, if the bodies were thoughtfully arranged, and if the grave goods were intentionally placed, then Neandertals had religion, or a rudimentary religion, 70,000 and perhaps 300,000 years ago.

In stature, Neandertals were short and stocky, up to 200 pounds (90 kg) of robust strength. Skeletons included short limbs, barrel chests, and skulls that bore brow ridges. Most Neandertals lived a periglacial life on the edges of and southward from the ice sheets that then filled northern Europe and Asia. Large bones of wooly rhinoceros, mammoth, and wild cattle are found at Neandertal sites, testifying to their bold, predatory lifestyle and skillful hunting abilities. They were super carnivores, with tools to match, but faced competition from a species of large wolf, from lions, and from the huge cave bear (larger than even the current-day Kodiak grizzly).

Balancing this harsh life was a supportive social group that greeted returning hunters or warriors, and lent aid and opportunity to heal from wounds suffered and traumas inflicted. Some Neandertal bones carry evidence of severe, even debilitating damage—scars, gouges, amputations. But these bones also show evidence of healing, which would be possible only if the injured individuals could convalesce and mend in the safety of a hospitable social group that nurtured them, giving them time to heal.

Neandertal culture was more elaborate than that of earlier hominids, but its simple, standard toolkit changed very little during the almost 300,000-year span of the species. Neandertals disappeared in Europe shortly after the first appearance there of modern humans. The timing is suspicious. Perhaps Neandertals were simply outcompeted, or exterminated, or just absorbed into the gene pool of modern humans. Later, we shall turn to modern genetics and paleontology to consider these questions. First, let's finish this hominid history. Now we reach us, at least in a more primitive or archaic form.

Homo sapiens—Out of Africa a Third Time

Modern humans originated in Africa, or so the current evidence testifies, and later dispersed from there to other continents. Possible *H. sapiens* fossils are scattered across Africa at various sites, with reported ages of 260,000 years ago or older, but problems of preservation, incompleteness, and chronology make such ages uncertain. This changed in 1997, when the sharp eyes and the good luck of anthropologist Tim White and colleagues spotted a fossil hippopotamus skull jutting out of ancient sand. Unusually heavy El Niño rains had pounded Ethiopia, in East Africa. For White, the luck of these torrential rains was to wash away the last embracing layers of earth holding the hippo skull, and temporarily drive away the flooded local people and their livestock, saving the newly exposed, fragile fossils from trampling. The hippo skull bore cuts and gashes inflicted with a stone tool, either from killing or scavenging, clear evidence of the presence of hominids. Follow-up excavation soon was rewarded with two partial adult skulls, probably males, and one skull of a child, 6 to 7 years old, smashed into more than 200 tiny pieces. The cleaning, reassembly, and study were painstaking, but worthwhile, because these proved to be anatomically modern humans, *Homo sapiens*. Radioisotope dating places their age at about 160,000 years.

Further work unearthed skull pieces from seven other *H. sapiens*, stone artifacts, and a general picture of the habitat in which these humans lived. They frequented the shores of a shallow lake inhabited by catfish and crocodiles as well as the hippos. The surrounding grasslands supported antelopes and zebras as well as these humans. White's fossil skulls show evidence of ritualistic treatment—the overlying skin, muscles, and tissue were removed with a scraping tool. The significance is not known. As with other ancient human fossil sites, this one in East Africa yielded no evidence of intentional burials, although that may be a matter of chance. These first humans were anatomically similar to us, but their accompanying simple cultures were little different from those of the hominid species before them. Hallmarks of modern humans—spoken language, art, civilization—were perhaps a hundred thousand years in the future. Culture lagged anatomy.

By about 100,000 years ago, these archaic *H. sapiens* dispersed across the Asian continent and overlapped with Neandertals in western Asia. At this time, these archaic humans exhibited utilitarian tool use and culture essentially identical to that of the Neandertals with whom they shared the Middle Eastern environment.

By about 40,000 years ago, modern *H. sapiens* appeared in full possession of a richer culture. They entered Europe, joining Neandertals already present there. By 30,000 years ago, Neandertals were extinct. These first *H. sapiens* to colonize Europe, sometimes referred to as Cro-Magnon humans, were not only anatomically modern but culturally advanced as well. Their culture was richer and more varied than that of Neandertals, and they introduced technological changes in tool use and techniques of working with stone and bone. Cro-Magnon burials were clearly evident and were more elaborate; the deceased were accompanied ceremoniously with gifts and possessions fashioned by human artisans. Most likely, religious beliefs and rituals became more elaborate as well. However, Cro-Magnon humans are best known for their "cave paintings" discovered at sites in France and Spain (figure 15.5). This cave art was not graffiti. The hands of the artists were too certain, practiced, and skilled to attribute these paintings to any casual or bored Paleolithic individuals merely passing the time. Nor were the paintings displayed in the open for general public viewing. Instead, they were painted deep underground in dark, discrete corners of caves; away from general living quarters; in mysterious, secret, shrinelike lacunae; or tucked behind narrow ledges, recesses, or rock fissures. Small mammal contemporaries of the Cro-Magnon peoples such as rabbits, foxes, and wolves are not often depicted; human figures appear only a few times. Instead, the usual subjects of these paintings are large game animals, such as bison and deer. The significance of subject selectivity is not known, but choice was obviously made. Perhaps, like paintings of some primitive tribes today, these were important totemic or symbolic figures.

Besides paintings, these peoples produced carved figurines and small statues—nonutilitarian objects. Together these objects and the paintings are consistent with the view that Cro-Magnon peoples, perhaps like the Neandertals they replaced, practiced a form of cult or religious worship.

Modern races of *Homo sapiens* differentiated in late Cro-Magnon times, taking hold on the Eurasian and African continents by about 30,000 years ago. Crossing a land bridge of winter ice over the Bering Strait, humans reached North America and spread rapidly along the marine coastal areas, into the central plains, and into the subtropical and tropical regions of Central and South America, reaching the southern tip of South America. Humans reached North America in several pulses—some perhaps as early as the much-debated date of 40,000 years ago, others more recently.

HOMINID EVOLUTION— INNOVATIONS AND INSIGHTS

Hominid evolution is characterized by extensive diversification. Not so long ago, a simple display of our history could be represented with a few branches. Currently, more branches are added; more species are now

FIGURE 15.5 Cave Art Reindeer art from a cave in France. Cro-Magnon culture.

recognized (figure 15.1). Partly, this is the result of new fossil finds. *Ardipithecus ramidus* and *Australopithecus anamensis* are new finds of the 1990s, earning a place early in the phylogeny. Partly it is because distinct, new species are now recognized where before they might have been lumped together with a better-established fossil species. The result is that our hominid history becomes bushier, more species are recognized, branches are added, and gaps are filled. This will likely continue as our own rich ancestry becomes even better delineated.

Mosaic Evolution

Modern humans possess an assortment of characteristics. We can make a partial list: grasping hands, bipedal posture, increased brain size, reduced body hair, speech. The list could be lengthened, but for the moment let's just look at these features. Notice (figure 15.6) that, except perhaps for speech (language), these

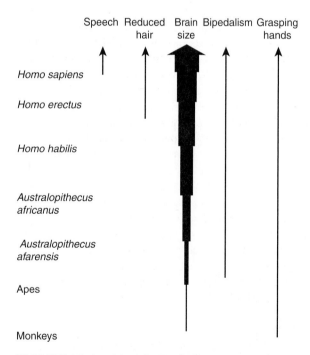

FIGURE 15.6   Mosaic Evolution   Characteristics of modern *Homo sapiens* include speech, reduced hair, enlarged brain size, bipedal locomotion, and grasping hands. Yet each of these adaptive features arose at different points in our history, not during one single evolutionary burst. Grasping hands arose long ago in monkeys; speech arose recently; other traits arose at other steps.

characters evolved before modern humans emerged. Grasping hands debuted very early, back within the earliest primates; so did the tendency for the brain to become relatively larger. Bipedal posture entered hominid history very early as well. Although only a guess, reduced body hair likely occurred first among *H. erectus*. Speech may have been a relatively new invention, perhaps emerging as recently as 40,000 years ago.

Modern humans are an assortment of characters, gathered at various times along the course of our long hominid history. We are endowed with many derived features, making our physical makeup complex. But our complexity did not evolve at one evolutionary moment in one, big evolutionary binge. Each of these features evolved separately, at different times; but collectively they add up to the physically complex species we are today. We are the consequence of a **mosaic evolution**, a montage or assortment of features entering the hominid radiation at separate evolutionary moments. Each feature evolved for the immediate adaptive advantages it conferred and continued on as it provided an important service in our evolutionary survival.

Human Variation

We might notice other features that have changed during hominid evolution as well. One change has been the transition from a drawn-out snout to a flattened face (figure 15.7). Another is the presence of a projecting chin. As recently as Neandertals, the chin was recessed, like in the earlier *H. erectus* (figure 15.3). In modern humans, the chin projects beyond the tooth row. Apparently this change results not from the forward extension of the lower jaw, but from the backward shift of the tooth rows. The advantages of this toothrow shift are not clear, but the result is to crowd the back teeth, especially the teeth you and your dentist know as the "wisdom" teeth. For some, this is a modern dental problem and the wisdom teeth must be surgically removed. For other people, the wisdom teeth are already reduced or absent, reflecting what is apparently a recent modification within *H. sapiens*.

Variation characterizes a species. Recall the use of natural variation by domestic breeders of plants and animals (chapter 3). And recall how, within the same species, local variations evolve to suit local conditions (chapter 9). Our own modern species is no different. Across the human species there is variation and variety. Consider the following.

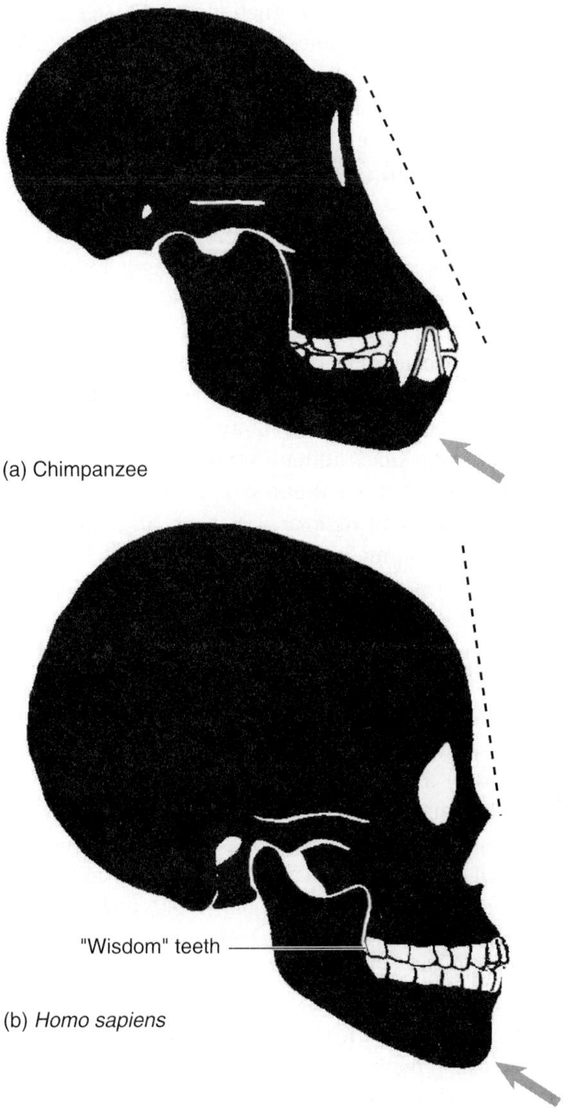

(a) Chimpanzee

(b) *Homo sapiens*

"Wisdom" teeth

FIGURE 15.7   Skulls of Chimpanzee and Human   (a) Chimpanzee skull. (b) Skull of modern human *Homo sapiens*. Differences in skull, teeth, and lower jaw are evident. Note the differences in the plane of the face (dashed line) and in the chin (arrow).

Noses

We noted earlier (chapter 13) the important function of nasal turbinates in an endothermic animal. These thin bones, situated in the nasal passage within the skull, hold mucous membranes that warm and moisten the entering air and collect heat and water upon exhalation, thereby preserving both. During prolonged exertion, such as during active hunting, larger amounts of air must be moved and the nasal chamber receives greater demands to condition the air. For the deep-set nasal chamber to continue its function, the first compartment must be enlarged to prepare the entering air even before it reaches the turbinates set in the skull. In humans, this compartment is what we call the *nose*. It preconditions the entering air, thereby reducing the trauma to the turbinates, and consequently increases the overall effectiveness of the nasal passageway in warming and moistening entering air. The nose rests on the bony skull. For the nose to enlarge, the skull must be modified—or, as has apparently happened in human evolution, the nose is lengthened outward, away from the face, producing a protruding structure in some humans.

Amongst modern humans, the size of the projecting nose correlates with climate to which the particular human population is suited. The humidity of inhaled air is the most important factor, and cold air is necessarily dry. In humans challenged by cold winter climates, the nasal chamber is called upon to compensate. This is accomplished by narrowing these deep chambers, bringing the warming and moistening surfaces closer to the core of the airflow. This narrows the external nose itself, which is then increased in height to retain the size of the overall passageway and pass the same volume of air. Increasing the forward projection of the nose also helps produce a large, warming air chamber. The result is a narrow, tall, projecting nose, characteristic of human populations evolving in, for instance, northern European regions (figure 15.8).

Neandertals were the first hominids to live along the advanced edges of glacial ice sheets. Narrow, projecting noses may have debuted within them as ice age adaptations. Northern European populations of modern humans adapted to cold, dry climates similarly come equipped with such specialized, jutting nasal chambers. In modern humans from subtropical or tropical regions, noses are less protruding.

Body Types

Human body shapes and sizes also vary as a consequence of different adaptive demands. For warm-blooded mammals living in cold climates, ears and limbs are short (Allen's rule) and bodies hefty (Bergman's rule) compared to mammals in warm climates. These changes, compared to their warm-climate relatives, reduce the amount of exposed surface area

FIGURE 15.8   Noses During a visit to China, a street vendor used scissors to cut this silhouette of the author from black paper. The vendor charged extra for the technological adaptation (glasses) and biological adaptation (large northern European nose).

and thereby reduce the loss of heat to the cold environment. In general, with some exceptions, the same rules of physiology apply to humans adapted to cold and to warm climates (figure 15.9). In general, cold-climate people are stocky, lateral in build (Inuit, Eskimos), or large (northern Europeans). Conversely, warm-adapted humans, wherein heat loss to cool the body is important, tend to be tall, lineal in build (some African peoples), or small (Pygmies).

Skin Color

All humans have the same skin structure, which is intrinsically yellowish in color and tinted with the red of underlying blood vessels. To this, shades of brown to black are imparted by melanin, a pigment secreted into the outer layer of skin (the epidermis) by special cells called *melanocytes*. The number of skin melanocytes is approximately the same in all humans, but the abundance and distribution of secreted pigment vary. The primary function of melanin is to absorb ultraviolet (UV) radiation in sunlight, striking a balance between too much or too little UV radiation. Too much UV penetration of the skin can cause considerable long-term damage directly to deep cells and can lead to cancer. UV radiation can also destroy folate, a form of vitamin B, circulating in the blood. Decreased levels of folate correlate with birth defects and infertility. On the other

hand, some UV light must penetrate the skin to catalyze the conversion of dehydrocholesterol in the blood into useful vitamin D. Normal bone growth and maintenance depend upon the presence of vitamin D. In northern climates, the midwinter sun stays close to the horizon—its light slants through the long atmosphere, and much of the useful UV radiation is filtered out. Reduction of skin melanin, producing light-skinned humans, allows reduced levels of UV light to still be effective in producing healthy levels of vitamin D. Humans living in tropical regions receive high levels of UV light, because the sun does not seasonally dip so low on the horizon; in these people, melanin is secreted in high levels, and their skin color is adaptively dark.

Certainly, final human skin color results from modification of the extremes of light or dark skin. Whether born with light or dark skin, all humans tan when exposed to high levels of sunlight. This is an immediate, physiological adjustment to seasonal variations in light levels wherein melanocytes respond by secreting temporarily higher quantities of melanin. Humans who live at very high latitudes (e.g., Inuit, Eskimos) obtain vitamin D from their diets, especially from fish and seals, compensating for the absence of UV radiation in midwinter. Natural melanin levels in the skin are also reduced in peoples living in regions where cloud cover or dense vegetation reduces UV radiation.

Variation within the human species is related to adaptive adjustments to the selective pressures of challenging environments. Noses, body types, and skin colors are adaptive responses within our single human species to varying demands of the local environment. We need no religious, political, or sociological theory to understand these differences or to interpret them.

PHYSICAL AND BEHAVIORAL FEATURES—REAL AND IMAGINED

Variation within modern humans is accessible to experimental study because modern humans are available to study. But as we step back in time, we meet only fossils that speak to us of our ancestry and of our hominid history. Soft parts and behaviors usually leave no traces, provide no evidence, give no clues, bequeath no fossils. Largely because they are unreferenced to the hard bony evidence of the past, conjecture about soft parts and social behaviors of earlier hominids invite untamed and often wild speculation. There is nothing wrong with this, so long as we recognize that anthropologists are extrapolating beyond the data; making best guesses;

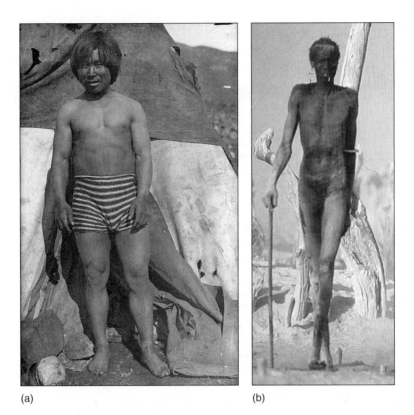

(a)   (b)

FIGURE 15.9   Human Body Types   Humans adapted to cold climates are often lateral in build; those adapted to warm climates tend to have a linear build. (a) Inuit from the Arctic (under five feet tall). (b) Nilote of near equatorial Africa (over six feet tall).

stretching the evidence. This is good fun, but let's remember that such speculations are only loosely linked to hard evidence. Consequently, they are susceptible to frequent revision.

Hairless Bodies

Body types are reasonably implied from fossil skeletons, but noses do not preserve, nor does skin color. Our bodies are hairless, compared to those of our primate cousins. When were hominids dressed down, shedding the thick coat of primate hair? A good guess would be in *H. erectus* (and/or *H. ergaster*). From *habilis*, which preceded it, *erectus* stepped up the emphasis on hunting. *H. habilis* walked on short hindlegs and, like the australopithecines before it, slowly foraged along the edges of savannahs, collected tubers and fruits, and pounced on unwary, small game animals. In contrast, *H. erectus* was long-legged and probably trekked long distances in search of game, adding the tracking of large game animals to its foraging repertoire. Pursuit of game over many days produced heat stress. This fa-

vored two changes—reduction in body hair and increased cooling capacity. Like shedding a winter coat, loss of covering body hair reduced the insulating furry coat, exposed skin directly to cooling air, and permitted more effective, direct loss of heat through the skin. Increase in sweat glands now served evaporative cooling. Fur loss coupled with increased evaporative cooling (sweating) suited *erectus* to such an active hunting existence. Reptiles do not sweat, but many mammals do. Racehorses sweat; dogs do not. Dogs hang a wet tongue from their mouths to accomplish the same thing, but they do not sweat through the skin. Compared to most other animals and even compared to other primates, humans are relatively sweaty animals. This is a feature that likely entered our history with *erectus*.

Language

And what of spoken language? Speech. Many animals vocalize, producing socially significant sounds, but only humans possess speech. Animal vocalizations are stereotyped calls evoking particular responses. There is little room for improvisation. The sound system is closed. But human speech is open, ready to produce new word structures that can be decoded by the listener. Human language has syntax and semantics. For example, consider the following sentence:

Know Texas comes we boy the who from.

Make any sense? Probably not. The words are right, but the syntax is missing. Add the syntax and we get:

We know the boy who comes from Texas.

The same words, just placed into a proper syntax, and now it makes sense to us. Our language is not simple stimulus (sounds) and response (understanding). It has both semantics (meanings of words) and syntax (grammatical structure).

We possess a capacity for language beyond that of even our nearest relatives, the pongids. Look at the two conversations shown in figure 15.10a,b. One is of a 21-month-old human child; the other is of a chimpanzee (the chimp's symbolic "speech" is translated into words for comparison). Cover the labels (a and b). Can you confidently tell which is which? Six months later, the human child produces complex sentences (figure 15.10c); the chimp is still rattling off a few simple words. Upon graduation from high school, the average human knows as many as 60,000 words, picked up with little effort. The chimp, after years of intense training and effort, moves little beyond what you see here. I do not mean to insult the impressive abilities of apes in other achievements. Instead, the point is that human speech is a substantial leap beyond the vocal repertoire of chimpanzees—much more than a few grunts or gestures strung together.

Speech has been a very important advance in human capabilities. Two areas of the brain—Broca's area and Wernicke's area—are crucial. In humans, these areas are enlarged. Broca's area is connected to the voice apparatus of the larynx, tongue, and throat. Wernicke's area organizes speech into the symbols, semantics, and syntax involved in the comprehension of language. Because these are expanded areas in the brain, their accommodation within the skull should be reflected in bulges in the bony skull. A search for such bulges in hominids as far back as *H. erectus* has produced mixed messages. Some anthropologists find evidence of such bulges in brain endocasts, molds of the skull interior occupied by the brain; others are not convinced. Evidence of a voice box, enlarged to produce speech sounds, is evident in Neandertals but not in earlier hominids. In-

APE

Toothbrush there, me toothbrush.
Sleep toothbrush.
(Sees picture of tomato.) Red me eat.
Berry, give me, eat berry.
Come…there.
Give eat there, Mary, me eat.
Give me berry.
Afraid, hug,
Mary afraid, hug.
Play.
Pull, jump.
Tired. Sleep. Brush teeth. Hug.

(a)

HUMAN (21 months)

Rock? (Rocks chair.)
Rock.
Chair.
Chair.
Chair.
House?
Chair.
Get up. (Asked if wants to get up.)
Get.
Please? (Asked if wants juice.)
Please. (Given juice.)
Thank-you.
Thank-you.
Apple. (It is apple juice.)
Fan.
Fan.

(b)

HUMAN (six months later)

I want to put the squeaky shoes some more, Daddy.
Let's get a piece of rock and make it go ding.
Where'd the ball go? Where's the ball?
There's Geoffrey. There's ya cookie monster.
    There's the nother cookie monster.
I saw Robert, and saw Kevin, and saw Luanna.
I did go in the kitchen throw it, Dad.
Didja sit down tray a give me a little pudding?
Was a good job I throw a diaper a rubbish.
I want to play catch. I'm gonna throw it to you.

(c)

FIGURE 15.10   Speech and Symbolic Language   (a) Chimpanzee, adult. The language of a signing chimp is here translated into the English words. (b) Human, 21 months of age. The spoken words are shown. (c) Human, 6 months later than (b). The actions or prompts of the interrogators are not included.   (From Bickerton 1990.)

directly we might note that Neandertals produced tools of some sophistication requiring skill to manufacture. It would be surprising, indeed, if Neandertals produced complex tools, passed these from generation to generation, but did not talk about them.

The adaptive advantage of human speech is not easily identified. Certainly we would be encumbered without it today, but why did it evolve in the first place? Hunting for big game was central to primitive human societies. During the hunt, vocal commands to coordinate the closure on prey and direct the attack could be useful. But during a socially coordinated hunt, members are dispersed beyond easy earshot. Other social predators, such as lions and wolves, hunt in a coordinated fashion without speech to guide them. Planning the hunt could entail communication, requiring speech, but unpredictable changes in prey evasiveness would quickly invalidate the best-laid plans.

In primitive societies of today, speech is used mostly in camp—by hunters as they relax after the hunt and by the nonhunting members at camp. Speech conveys information from one person to another. A few grunts and groans will not do. Speech transfers history; tales of the successes and failures of yesterday are passed to the young generation so that their future may not include the same mistakes and they may enjoy the survival benefits of past successes.

Speech distinguishes modern humans. No other living species has quite the same vocal richness, not even other living primates. We do not know when speech evolved within hominids. At 40,000 years ago, *Homo sapiens* lived in a rich culture and produced symbolic objects. Reasonably, like ourselves, they possessed speech. Archaic humans back to at least 130,000 years ago were anatomically like us, but by comparison they were culturally impoverished. In Neandertals and even *H. erectus*, accomplished adults taught eager young the skills of tool manufacture, generation to generation. Could such skills be taught by mimicking and not by word of mouth? At the moment, there is no consensus as to how far back in our history we can trace speech. But certainly speech is a hominid invention, not a fully formed behavior passed to us from chimpanzee cousins.

As some linguists once put it, bees dance, birds sing, and chimpanzees grunt, but human language is qualitatively different. Specifically, animal languages are basically *closed* systems of communication. Recall the vervet monkeys of chapter 12. They produce three alarm calls—one for snake, one for leopard, one for eagle. But they do not mix these calls into new significant sentences. The system is closed in its variety of information conveyed. On the other hand, human language is an *open* system of communication. No other species has the extraordinary capacity to combine words and sound units into an unlimited variety of meaningful sentences. For example, fill in the following blanks appropriately and speak the sentence out loud:

"I [fill in your full name], am reading this book chapter by the author Kardong on this ___ day of the month ___ in the year ___."

Okay? You have just spoken for the first time a sentence never before spoken, in this exact form, by anyone in all of human history. Yet even though it is unique, it is perfectly understandable to you and to colleagues, as English-speaking humans. A small example, certainly, but consider the great richness of human spoken culture, codified in writing. The rich and original literature of human culture testifies to the creative capacity of our open system of language to produce an almost unlimited variety of meaningful communication. Forty thousand years ago, perhaps earlier, this new innovation we call speech debuted in hominid history. Although the time and nature of its first onset in human society are still debated, the importance of human speech is not questioned, and its distinctiveness is a major reason for our current success as a species.

Religion

Religion is a state of mind, an attitude, an approach to life. It does not fossilize. We know of Protestants, Catholics, Jews, Muslims, Hindus, and others. These are recent religions, several thousand years old at most. But when did religion become part of our history? When did we become religious primates?

If religion itself does not directly fossilize, evidence of some of its rituals does. Burials serve to remove unpleasant, rotting corpses of the dead from areas occupied by the living. Burials also remove carcasses that otherwise would be easy foods, attracting the attention of formidable scavengers such as hyenas, lions, bears, and wolves. Grave goods, however, deflect no scavenger. They betray the emotional attachments of the people who bade farewell to the departed. Personal wealth such as necklaces and knives might be included. Ochre, a collected pigment, paints the sites of some buried remains. Burials that include such grave goods, personal property, and ochre are of *Homo sapiens* from 40,000 years ago. Before that, the evidence is ambiguous. *H. sapiens* dates to 160,000 years ago, but grave goods are not present that early, either because adding such personal effects to the grave had not yet become customary or because the grave goods did not survive. Neandertal graves include tantalizing evidence. From some sites, pollen grains suggest buried bodies were graced with newly picked flowers; ochre is added. But some anthropologists counter this interpretation;

flower pollen could have just as well drifted in on spring breezes. Practical tools of animal butchering are present, but these may be accidental. It is difficult to settle these ambiguities.

H. erectus sites contain no certain hints of religious activity. An *erectus* site in China preserves evidence of cannibalism, with the possibility it was a ritual act rather than predatory. A skull at an African site shows evidence of scalping. This, too, is ambivalent evidence. Religion and ritual are conspicuous by their absence.

At the least, religion was present early in human evolution, 40,000 years ago, maybe earlier, but probably not as early as *erectus*. What biological advantage might religion bring to the survival of early humans? Religion contained the codified information of successful generations that went before. Within religion was the historical memory of successes and failures of the past. Toxic foods that were tempting and tasty ended up on taboo lists, lending the authority of religion and its gods to their sanction, thereby ensuring avoidance and survival of the practitioners. Along with language, religion passed to each generation the secrets of survival— where resources were to be found, where dangers lurked, what to do in times of unseasonable climate. Religion, within its mythologies, contained the survival kit of adaptive practices sifted and sorted over decades of trial and error. Handed down from past generations to the new generation, and practiced faithfully, religious dogmas and mythologies carried forward the experience of the past to guide behaviors of the present in challenging environments.

WANDERLUST

Out of Africa

The first of the hominids to move out of Africa was *H. erectus*. Earlier forms of *erectus*, or its immediate ancestor *H. ergaster* (see earlier discussion), place it in Africa at 1.9 million years (figure 15.4). But certainly by 800,000 to 500,000 years ago, maybe earlier, it had reached Indonesia and China. The second hominid arising in Africa (600,000 years ago) but migrating out to occupy parts of Europe (500,000 to 200,000 years ago) was *H. heidelbergensis*. It most likely was the immediate ancestor to Neandertals, which populated parts of Europe and the Middle East, and to *Homo sapiens*. *Homo sapiens* first arose in Africa 160,000 years ago, but soon dispersed from there. By 100,000 years ago, archaic forms of *H. sapiens* had reached the Middle East, where they met and overlapped with resident Ne-

andertals. By 40,000 years ago they reached Western Europe. Earlier, they spread into Southeast Asia, where they may have met lingering populations of *H. erectus*, which soon became extinct. By 60,000 years ago, *H. sapiens* had colonized Indonesia and Australia. Taking advantage of winter ice, *H. sapiens* reached North America by 20,000 years ago and very quickly spread southward into the rest of the Americas.

Arrival of *Homo sapiens*

Two recent competing theories examine the relationship between *H. erectus* and *H. sapiens* (figure 15.11).

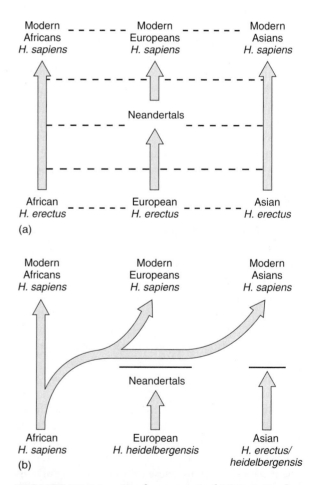

FIGURE 15.11   Replacement of *H. erectus* by *H. sapiens* (a) Multiregional Theory proposes that after *H. erectus* dispersed to various geographic regions, it continued to evolve in place, producing modern humans. (b) The Out-of-Africa Theory proposes that after *H. erectus* dispersed to various geographic regions, *H. sapiens* arose in Africa, then also dispersed to other geographic regions, displacing *H. erectus* as it spread.

The **Multiregional Theory** (in place) envisions *H. erectus*—dispersed across Africa, Asia, and Europe—evolving more or less in place directly into *H. sapiens*. The **Out-of-Africa Theory** (replacement) envisions *H. sapiens* originating in Africa, then dispersing to Asia and Europe, replacing lingering populations of *H. erectus* and Neandertals with no interbreeding between them. Because this theory envisions no intermingling or breeding species, it predicts a discontinuity between *H. sapiens* and earlier species.

DNA evidence so far supports the Out-of-Africa theory. Anthropologists extracted genetic material from well-preserved bones of European Stone Age *Homo sapiens* (Cro-Magnons) dating from 23,000 to 25,000 years ago and from Neandertals, and compared this genetic material to today's humans. The result? Cro-Magnons were similar to today's humans, but Neandertals differed sharply from both Cro-Magnons and modern humans. If early humans and Neandertals had intermingled (Multiregional Theory), their genetic structure should be similar. It is not, as the Out-of-Africa Theory predicts.

To the Americas

Humans entered North America about 20,000 years ago, when much of Earth's water was locked up in extensive glacial sheets of ice. The lowered sea levels exposed a land bridge, called Beringia, that allowed humans to migrate from Siberia to Alaska (figure 15.12). Although this was still a period of glacial advance, these migrating peoples, already adapted to cold climates, took advantage of an ice-free interior of Alaska, dispersing into warmer latitudes within North America, then into Central and South America, reaching these distant locations within about 7,000 years of first crossing Beringia.

We should be careful about how we think of this colonization of the New World. It was not a "gold rush" stampede, with great floods of people streaming along major highways through the Alaskan interior, and then racing to points farther south. Instead, small, independent groups likely made their way across Beringia, settled, and prospered for several generations; then some descendants moved on, looking for fresh areas of game,

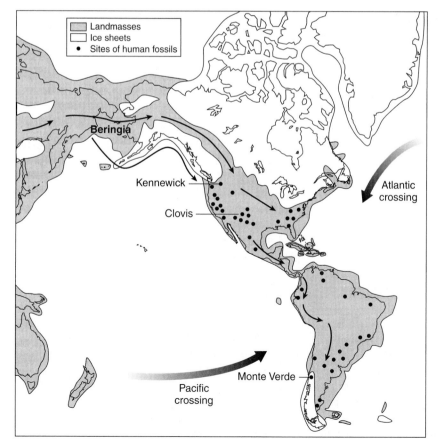

FIGURE 15.12 Human Colonization of the New World The most likely routes were through Beringia via the coastline or inland via a land corridor that opened in the glacial ice sheets. Other, although more speculative, routes include a possible Pacific crossing or an Atlantic crossing. (After Campbell and Loy.)

which would lie always to the south. Colonization progressed in waves or multiple surges of peoples from Asia into Alaska. Over the ensuing 7,000 years, they would come to populate the New World, but not in large numbers nor in heavy concentrations, and certainly not yet in cities. These were tribal peoples—hunter-gatherers.

Recall that the arrival of humans was accompanied by a demise of large mammalian fauna (see chapter 13). These arriving peoples were accomplished hunters. In 1932, near Clovis, New Mexico (USA), the remains of an extinct mammoth were found. Within its ribcage, finely crafted stone spear points were discovered. Tied to a wooden shaft, these could have been hurled at the mammoth, becoming embedded and inflicting multiple wounds. Eventually weakened by loss of blood, the animal would have fallen, providing a meat-meal for the relentless hunters. These distinctive spear points are called *Clovis points*, after the site, and were subsequently discovered at other sites from about 10,000 to 12,000 years ago across North America. The people who made these points, the Clovis peoples, were paleo-Indian peoples, the likely ancestors to the aboriginal Indian peoples met by Europeans when they arrived a few hundred years ago.

The herds of large herbivores in the New World experienced predation from accustomed enemies such as bear (grizzly and an even larger species), a species of lion, cougar, and an assortment of wolf species. But the arrival of these accomplished human hunters, with skills practiced and honed in the Old World, surprised many of these herbivores. Within a few thousand years of human arrival, the ground sloths, camels, mastodons, and mammoths were gone, and with them went some of the large predators that had been dependent on them. Today the standard story is that the New World was founded by these Clovis peoples and their likely descendants, Amerindians ("Native Americans"). In some ways, genetic evidence supports such a view. Nearly all current Amerindian groups carry DNA most similar to the DNA found today in indigenous peoples of Siberia, Mongolia, and Tibet, the mostly likely ancient source of Clovis peoples who crossed Beringia. But there are details that do not fit easily within this common view.

By Sea

Monte Verde is an archaeological site in Chile (figure 15.12). Humans set up camp there some 12,500 years ago. Yet Monte Verde is 10,000 miles from Beringia, a long trek for anyone, especially if the first crossing started at 20,000 years ago. This led some anthropologists to suggest that these peoples arrived by boat, not on foot, hugging the coast and taking advantage of the marine coastal community for food—sea mammals, salmon, shellfish. Even during the time of greatest glacial advance in North America, there were thawed pockets of coastline where food could be found and settlements established. Unfortunately, any such shoreline villages, and hence any surviving archaeological evidence, would have been flooded when the Earth entered its current interglacial phase, glaciers melted, and sea levels subsequently rose about 300 to 400 feet, covering these sites.

These seafaring peoples might have been accomplished mariners. Certainly other peoples elsewhere during the Paleolithic were skilled at coastal navigation. Lowered sea level expanded and connected islands from Southeast Asia throughout Indonesia, and helped peoples eventually reach Australia 50,000 to 60,000 years ago. Nevertheless, much of that route would still have been treacherous open ocean, requiring dependable craft and practiced seafaring skill to complete the journey. Perhaps peoples traveling north out of Asia and Siberia possessed similar navigation skills that served them well as they traveled the coastline into the New World.

Pre-Clovis

If these dates are correct—first crossings at 20,000 years ago, reaching the southern region of South America less than 7,000 years later—then the dispersals were indeed very rapid. Even by boat, this is too rapid for some anthropologists, who also point to an older date, 32,000 years ago for a site in Brazil, as evidence of an earlier entry of peoples into the New World across Beringia. DNA results suggest to some anthropologists an entry date of around 42,000 years ago. If there was an earlier wave of peoples to the New World, the fate of these peoples is not known. Perhaps when Clovis peoples arrived, they interbred with them or warred with them but eventually displaced them.

Other Peoples

Where skulls are available, the story has become even more complicated. Human skulls are distinctive. Bony features are highly heritable, and overall traits are characteristic of a particular people or ethnic group. Based on multiple measures of the skull, a characteristic **craniometric profile** can be produced. Using skulls of modern Amerindians from current tribes in the Great Plains, Great Basin, and southwestern United States, a standard craniometric profile database has been compiled. Not surprisingly, when compared to other modern-day groups, Amerindian skulls are similar to some *northern* Asian skulls, and some prehistoric sites yielded skulls similar to both. One is Buhl Woman (10,700 years ago) from Idaho; another is the Wizards Beach Man (9,200 years ago) from Nevada. What is surprising is that at many other sites, these prehistoric Americans do not have craniometric profiles that resemble Amerindians today. In fact, the closest match for them is with *southern* Asians, generally from the same colonization that populated Indonesia to Australia; and there are even some similarities to indigenous Europeans!

Recently this puzzle has deepened. Along the banks of the Columbia River in Washington State (USA), the skull and partial skeletal remains of a human eroded out of the riverbank. This is Kennewick Man, initially misidentified as a nineteenth-century European trapper. An ancient spear point embedded in his hip, and especially subsequent radiocarbon dating, testified instead that this was a 9,000-year-old human. Surprisingly, his cranial features were distinctively Caucasoid, not Amerindian.

These affinities of ethnic identity to southern Asian (craniometric data) and Caucasoid (Kennewick Man) groups have provoked some rowdy speculation. One new conjecture is that southern Asian peoples, upon reaching Australia, did not stop, but continued via a *Pacific crossing* to South America. Such a Pacific crossing over long stretches of open ocean placed them directly into South America, thereby accounting for the ethnic character of some prehistoric humans there and the early ages of the sites. Another conjecture proposes an *Atlantic crossing*, with seafarers from Europe following the edges of glaciers past Greenland into the eastern shores of North America. Humans may have reached the New World through a variety of routes, at different times, and from different ethnic source groups.

OVERVIEW

We have come a long way since our hominid ancestors departed from ape ancestors almost 7 million years ago. When on the ground, living apes (orangutans, gorillas, chimpanzees) move about on all fours—in quadrupedal walking or its derivative, knuckle-walking. Such apes are at home in the trees, but their quadrupedal walking is not economical on the ground. These apes occupy dense forests where slow and careful foraging over less than a mile a day returns enough food to sustain them. In contrast, hominids move about comfortably using two-footed gaits—bipedal walking. In open landscapes, bipedal walking is more efficient and economical than the quadrupedal gait of apes. Much of early hominid evolution took place not in dense forests, but in or next to open woodland and grass savannahs. If these early hominids foraged similarly to modern-day hunter-gatherers occupying such open landscapes, then a trek in search of resources may have taken them over 6 to 8 miles daily to return enough food to sustain them. Efficient bipedal walking brought adaptive benefits.

Correlated with these changes from ape to hominid was a change in climate and environment. Beginning about 5 million years ago, the African continent began to dry under a changing global climate. Forests gave way to grasslands, and resources became dispersed and irregularly clumped. Reaching such clumped and patchy resources required an overland hike across open landscapes. Quadrupedal apes stayed with the surviving dense forests; early bipedal hominids became far-ranging foragers who exploited the enlarging savannahs.

Two different foraging strategies arose subsequently in hominid evolution. The early australopithecine ability to process tough, low-quality plant materials received greater emphasis in the later robust hominids. *A. boisei* and *A. robustus* possessed conspicuous adaptations for grinding up fibrous plant material. These included reinforced skulls to bear the chewing forces and powerful jaw muscles to work the massive lower jaws that moved large upper and lower molar teeth against each other like grinding stones. Some meat may have spiced up their diets, as it does today amongst chimpanzees. But robust hominids primarily foraged for large quantities of low-quality fibrous plants—basically they were vegetarians.

The other foraging strategy emerged in gracile hominids, whose diets contained less plant material and more animal foods. Meat packs up to twice the energy of fruits and up to ten times the energy of leafy foliage. Hunter-gatherers of today derive 40% to 60% of their dietary energy from meat and other animal products. They are not strict carnivores like wolves and lions, but omnivores enjoying a mixed diet of both meat and plant foods. But the addition of meat to the diet is critical because it brings an energy-rich food to

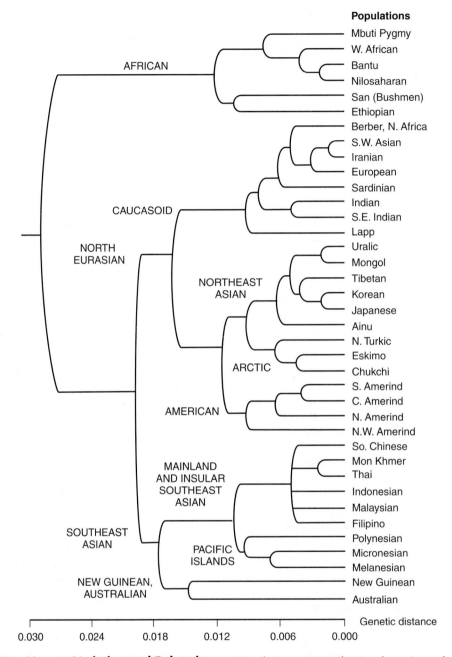

FIGURE 15.13   Human Variation and Relatedness   Based on genetic similarities, the various ethnic groups of humans can be compared for their closest human relationships and nested together accordingly.

the menu, thereby reducing the foraging effort for plants. For early hominids, the expanding African grasslands supported expanding herds of grazing animals, an environmental opportunity. Exploiting this were the gracile hominids whose diet now included soft but high-quality meat. Reflecting this, their skulls were more delicate in design (i.e., *gracile*), molar teeth were less like grinding mortars, and jaws and jaw muscles were less massive. Evidence of this shift to an omnivorous lifestyle is implied in the tools used by *H. habilis*—skinning and meat-scraping implements. But *H. erectus* was the first hominid to develop hunting-and-gathering as the centerpiece of its social economy. At *erectus* fossil sites, animal bones are now found in abundance, along with evidence of large animals butchered with fashioned tools. A hand ax and other tool improvements also debut. Generally, herbivores require less territory than carnivores because plants are concentrated but prey are dispersed over a larger area. Consequently, with its emphasis on hunting, the territory *H. erectus* searched necessarily became larger and its home ranges expanded. Just such an impetus may account for its spread out of Africa and colonization of parts of Europe and Asia.

Later, gracile hominids—Neandertals and modern *Homo sapiens*—continued and even expanded this hunting behavior. By this time, perhaps earlier, hominids were relatively hairless. With a hairless or nearly hairless body, skin color became important. African *erectus* was likely dark-skinned; European *erectus* was probably light-skinned, as were the later Neandertals who lived in northern Europe. Archaic *H. sapiens* was present at least 160,000 years ago, and modern humans by about 40,000 years ago. Speech and religion were most likely present in modern humans, but debate continues as to whether they were present even earlier.

Neandertals, or the most immediate ancestors, *H. heidelbergensis*, arose in Africa but soon dispersed out of Africa as *erectus* had done much earlier. Neandertals survived in subarctic conditions along the edges of great glaciers occupying northern Europe and Asia, where they were efficient and effective predators. Archaic *Homo sapiens* arose in Africa, and it too soon spread to Eurasia, displacing the earlier hominids, either by outcompeting them or by interbreeding with them. Humans crossed to the Americas relatively recently but the dates are not clearly settled, and the ancestral groups likely produced multiple waves of colonists.

Certainly, ideas about colonization of the New World are undergoing some rethinking. In the next few years, what is now untamed but intriguing speculation is likely to find evidence to help resolve, one way or the other, the conflicting features of dates (pre-Clovis) and cranial profiles (varying ethnicities). But at least three current ideas are likely to stand: First, colonization of the Americas occurred in multiple waves. Second, at least one wave came via Beringia, beginning about 20,000 years ago; perhaps by land, perhaps by sea. Third, the Old World ancestry of modern-day Amerindians traces to peoples in northern Asia (Siberia, northern China, Japan, Korea). What remains unsettled at the moment is whether there were earlier immigrations, before 20,000 years ago. Along with debate on dates, there is debate over the routes. The result has been to produce great diversity within the modern human species (figure 15.13).

Selected References

Balter, M. 2001. In search of the first Europeans. *Science* 291:1722–25.

Bickerton, D. 1990. *Language and species.* Chicago: University of Chicago Press.

Caramelli, D., et al. 2003. Evidence for a genetic discontinuity between Neandertals and 24,000-year-old anatomically modern Europeans. *Proc. Natl. Acad. Sci.* 100:6593–597.

Hauser, M. D., N. Chomsky, and W. T. Fitch. 2002. The faculty of language: What is it, who has it, and how did it evolve? *Science* 298:1569–79.

Jablonski, N. G., and G. Chaplin. 2002. Skin deep. *Sci. Amer.* 287 (October): 74–81.

Leonard, W. R. 2002. Food for thought. *Sci. Amer.* 286 (December): 106–15.

Nemecek, S. 2000. Who were the first Americans? *Sci. Amer.* 283 (September): 80–87.

Tattersall, I. 2000. Once we were not alone. *Sci. Amer.* 282:56–62.

Wong, K. 2000. Who were the Neandertals? *Sci. Amer.* 282:98–107.

Learning Online

Testing and Study Resources

Visit the textbook's website at **www.mhhe.com/evolution** (click on the book's title) to take advantage of practice quizzing, study/writing tips, timely news articles, and additional URLs for research on the topics in this chapter.

16

"The enemies of dynamism [creative progress] are forever arguing that we must destroy dynamic processes in order to save them. The open society, they insist, is its own greatest enemy and should therefore be closed."

Virginia Postrel, *The Future and Its Enemies*

Evolutionary Biology:
Today and Beyond

NATURE RED IN TOOTH AND CLAW

When we passed from the twentieth to the twenty-first century, no one turned off natural selection. Where human populations badly outdistance their resources, hunger lurks and large-scale famine threatens. Imported charity from elsewhere in the world may be the only difference between life and death for peoples on the brink of starvation. Disease still strikes, especially in less-developed regions of the world, taking out those weakened by malnutrition, poor sanitation, and low bacterial resistance. These tragic examples remind us that brutal Malthusian consequences still stalk peoples who lack the resources to sustain their unchecked numbers. Contentious religious and cultural wars deepen the tragedy and impede the arrival of enlightened political and economic systems that might lift the specter of death hanging over such peoples. But let's consider the science involved.

You may have heard it said that in a few hundred years our wisdom teeth will be gone, or our little toe will disappear, or some other useless part of our human anatomy will drop off; or an overused part will inflate, giving us, for example, huge brains. These are extrapolations of past trends into future consequences. If natural selection is the process, then the only way body parts can be lost is if those humans with undesirable traits (little toes, wisdom teeth) are brutally dispatched or prevented from having children, thereby preventing them from propagating these traits.

In a crude way, this harsh culling has been practiced by human societies for thousands of years. In the ancient Greek and Roman worlds, many infants born with deformities were put to death; in some Asian and African societies, defective infants were "exposed," meaning they were placed alone and unattended in the bush to meet whatever fate visited, which meant, of course, death. On the other side, socially pleasing traits could be promoted if individuals with such desirable traits are paired up and encouraged to spawn offspring. In fact, just such a movement arose in the early twentieth century. This view, called **eugenics**, discouraged breeding of humans with socially undesirable characteristics and encouraged breeding of those with desirable traits. The argument continued that successful medicines and public health programs relaxed the "natural" culling of persons with defects, allowing deficient individuals to breed and "flood" the human gene pool with undesirable traits. Consequently, proponents of eugenics often saw urgency in enacting a government-imposed program of selective breeding of the best to serve social ends. Many compassionate and thoughtful intellectuals of the day were drawn to this idea, as were many politicians who sought to galvanize public support for their bid to control public policies. What started as a scientific concern became a coercive social tool, confiscated by politicians waiting for just such scientific cover to justify their racial prejudices.

The tragic results are well known. By the mid twentieth century, the Nazis implemented a program to promote a "pure" Aryan race and exterminate "inferior" peoples. The Holocaust was the result. But such prejudices are already present in societies, not planted by science, as we have seen in recent genocides in areas ranging from Southeast Asia (Cambodia), to Africa (Uganda, Rwanda), and to Eastern Europe (Bosnia-Herzegovina, Croatia, Serbia). Science can be misused. But we should also consider science on its own merits. If we do so, then what do we make of eugenics?

On its own merits, eugenics fails. The reason is simple to understand. If you eliminate an individual with "bad" characters, you also eliminate an individual with "good" characters. Suppose we, as a society, decided to reduce the incidence of diabetes, and used the crude techniques of eugenics to accomplish this. If through harsh government legislation we banned the breeding of individuals carrying diabetes, we would reduce the future incidence of the disease in future generations. But the problem is that we would also reduce the incidence of favorable characters these same diabetic individuals carry. Many persons with great athletic talents who excel in their sport are also diabetics, as are many successful scientists, scholars, and business leaders. Along with detrimental diabetes, we would also be reducing favorable traits of brawn and brain. Eugenics throws out the proverbial baby with the bathwater.

So practiced, eugenics, which mimics natural selection by eliminating the whole individual, also eliminates the good traits carried by that whole individual. But because eugenics is applied by society, it relinquishes to society, and to its prejudices, the power to decide what is a "bad" trait and what is a "good" trait. In dealing with the elimination of a bad trait or the promotion of a good trait, the destiny of the whole individual is manipulated one way or the other, to sterilization or to procreation. To affect the trait, the entire lifestyle of the person is deliberately manipulated.

Collectively, each individual marches in lockstep to public policy. The Third Reich all over again.

Enter, Genetic Technology

Today, our options are different. For some detrimental traits that arise from defective genes, genetic engineering provides the tools to repair selectively and directly the malfunctioning section of DNA. Repair the genome, treat the defect, save the individual. This is **gene therapy**, the alternative to eugenics (figure 16.1). At the moment, we are not as close as we might like to routinely repairing defective parts of the genome, but the genetic technology is promising—and, some would add, disturbing. Are we to take over from the agency of natural selection the responsibility for our own evolutionary destinies?

Consider the following: Severe cases of Gaucher's disease arise from defects in a gene that produces a critical enzyme important in the disposal of a lipid (fat), glucocerbroside, produced during the normal recycling of old, battered blood cells. When the gene is defective, its synthesized enzyme malfunctions, the lipid abnormally accumulates and fouls critical organs, and the afflicted infant dies, but not right away.

Let's take an actual example. A Florida couple, professionals by trade, settled into their new marriage and soon produced their first child, a son. Sadly, and unknown to them, both parents were carriers of the gene-based Gaucher's disease. At the baby's six-month checkup, his abnormal behavior alerted an attentive pediatrician to the problem. Tests followed, which confirmed that the infant had a severe case of the disease. Worse yet, he had less than a year to live. Some mild cases of the disease yield to **replacement therapy**, wherein the defective enzyme is replaced by administered supplements of the normal enzyme. But not in this case. The gene therapy to replace the defective gene was not yet available. During the ensuing year, swollen organs and sudden throat spasms left the baby gasping for air. Nearly round the clock, day after day, the parents traded off administering mouth-to-mouth resuscitation two to three times a day. At the end of his year of agonizing torture and pain, the small child died, leaving the heartbroken parents emotionally and physically drained.

Your Choice

Some medical and ethical decisions, if viewed from a distance, are easy to delay, easy to preach about, and easy to leisurely ponder. But suppose this were your

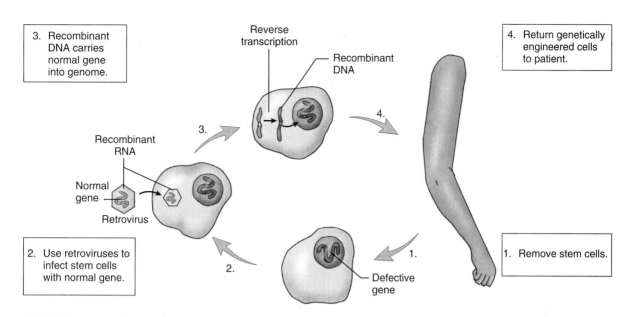

3. Recombinant DNA carries normal gene into genome.

Reverse transcription

Recombinant DNA

4. Return genetically engineered cells to patient.

Recombinant RNA

Normal gene

Retrovirus

2. Use retroviruses to infect stem cells with normal gene.

Defective gene

1. Remove stem cells.

FIGURE 16.1   Gene Therapy   Gene therapy is a method of inserting genetic material into a human to treat disease. For example, defects in the protective immune system often result from defective genes. (1) Stem cells, formative cells, are removed from the bone marrow of the afflicted infant. (2) The normal gene is carried into the defective stem cells by a benign virus, here a retrovirus, which has the ability to (3) indirectly insert the normal gene into the stem cell DNA. (4) The repaired stem cell, genetically engineered, is returned to the patient, where it produces new immune system cells that now function normally, restoring the immune system to its role in thwarting infections.

child. If gene therapy were available, would you choose to intercede? Would you allow medical intervention to repair the gene nature had broken? These will be our decisions to make in the future, because this is part of our future opportunities and ethical decisions. My guess is that few parents would turn away from their child and refuse effective gene therapy. However, society, distant from the heartbreak, might take another view and dictate parental decisions forbidding you from exercising your choice. In some ways these decisions are already upon us, even though the well-crafted, effective, and widespread gene therapies have yet to arrive.

Available today are *prenatal tests* that can determine if the fetus carries certain genetic diseases. Based on fetal age and anatomy, and guided by ultrasound imaging, a physician inserts a thin needle or tube into the pregnant woman's uterus to obtain a sample of the amniotic fluid surrounding the fetus (amniocentesis) or even sample the fetal part of the placenta (chorionic villi sampling, CVS). With the ability to administer a variety of tests for genetic defects, in a process called **genetic screening**, physicians can determine if the fetus carries genetic disorders (figure 16.2). A newer screening test can be done before the embryo implants

on the uterine wall, permitting an earlier decision about the fate of the fetus. This is PGD—preimplantation genetic diagnosis. The parents undergo *in vitro* fertilization; eggs are collected from the mother and sperm from the father, and an embryo is produced in a dish. The fertilized egg begins dividing. At only the eight-cell stage, one cell is removed and genetically screened. If the embryo is found to be disease-free, then the remaining cells are implanted in the mother's uterus for normal subsequent development. If the embryo is defective, the parents have the choice of discarding it.

Here is the rub. Genetic screening does not repair the defective genome. It only identifies the genetic disease, if any. Genetic screening is just a test; information is its result. What to do with this information is the ethical choice at hand. Some parents choose gene therapy; but for most genetic diseases, gene therapy is not available today. Other parents choose to continue the pregnancy. Replacement therapy is their hope. And still other parents choose to terminate the defective embryo/fetus and try again, hoping that in the next roll of the genetic dice, the tragic combination of deficient genes will not sort into the next embryo. For many people, this last choice—aborting a defective embryo/fetus—is unethical

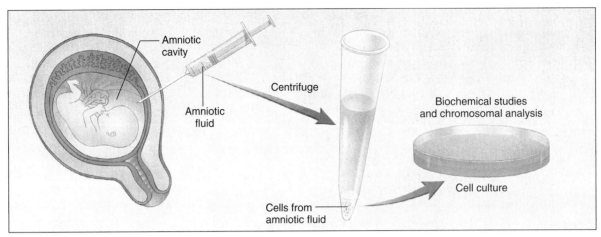

(a) Amniocentesis

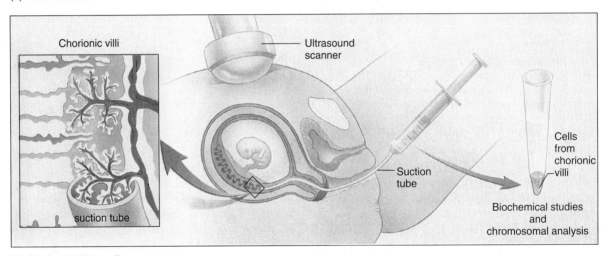

(b) Chorionic villi sampling

FIGURE 16.2   Genetic Screening of Fetus in Uterus   (a) In amniocentesis, cells are sloughed from the embryo into the amniotic fluid to be collected by syringe. (b) Chorionic villi sampling (CVS) is used in even earlier embryos. The separate but intertwined blood vessels of mother and fetus can be collected through an inserted suction tube guided by ultrasound images. By both methods, cells from the fetus or its placenta can be analyzed for possible genetic defects to inform parents.

on right-to-life grounds or because it is "dehumanizing." Hence, enter society and eugenics. Some arguments against abortion are cast as arguments against eugenics, charging that abortion eliminates an entire "individual" in order to avoid transmission of an undesirable trait.

The arrival of reliable and safe gene therapies may skirt this objection, at least for some. Several religious scholars argue that God created human beings as his co-creators, responsible for making the world a better place. As partners in a divine purpose, we are granted medical technologies to deploy. Yet secular issues persist.

Enhancement Therapy

Medical technology that improves genome performance instead of just treating or repairing deficiencies is *enhancement therapy*. But the distinction between repair and enhancement is not always clear. For example, some people carry a deficient mutation that prevents the proper processing of LDL, the "bad" form of cholesterol. The cholesterol collects at critical locations in major arteries and, in conjunction with blood vessel cells, forms a swollen growth that may hemorrhage, producing clots downstream, blocking blood flow to heart muscles, and causing a sudden, deadly heart attack. In severe cases, the onset of the gene-based disease may be relatively rapid, and afflicted persons may die in their early thirties. Chemical therapies are currently available wherein cholesterol-lowering drugs reduce the detrimental effects of the gene deficiencies. Gene therapies are on their way to repair defective parts of the genome. Both chemical and gene therapies fall within the category of straightforward treatments for heart disease in such afflicted individuals. But what of individuals without this severe mutation, who employ these same therapies anyway to keep their cholesterol low so they can engage in an epicurean lifestyle that includes rich foods high in cholesterol? Is this treatment? or preventive medicine? or simple enhancement?

Today, elite athletes spend long hours in the weight room, building muscle condition and stamina. Unethical enhancement or just good training practice? Most of us, I think, would call this dedicated conditioning, and encourage the athlete—no problem. But where do we draw the line? We might raise the same ethical question about dietary supplements many athletes take to gain important minerals and vitamins. Steroid supplements are banned in most sports, but not always for their "enhancement" effects but because of their adverse side affects (muscle and heart problems). At one time (perhaps still), endurance athletes engaged in "blood doping." Several months before a major event, quantities of their blood were removed and safely stored. Their bodies quickly responded, replacing the volume of blood removed, regaining normal blood volume within about a week. Then shortly before the endurance event (e.g., a marathon), the blood removed earlier was returned to their bodies, thereby increasing, at least temporarily, the number of circulating red blood cells available to carry oxygen to their muscles, thus increasing their stamina and athletic outcomes. The practice is banned.

Erythropoietin (EPO) is a natural body hormone. A reduced level of oxygen in the blood passing through the kidneys stimulates cells there to produce EPO. In turn, EPO promotes production of new red blood cells. In 1985, the gene for EPO was cloned in the laboratory, leading to a replacement therapy using what is called rhu-EPO. Currently, this agent is used to treat anemia (low blood cell levels) associated with kidney failure, HIV infection, and occasionally following cancer chemotherapy. Straightforward medical uses. No problem. But EPO has been used outside of clinical applications as well.

Injection of EPO into athletes preparing for competition mimics blood doping, by artificially, but substantially, increasing the number of oxygen-carrying blood cells. The practice is banned in most sports, and cheaters are caught by detection of abnormally high levels of EPO in their bodies. But such testing is not always reliable, and covert enhancement therapies are not always detectable.

Testosterone, a "male" hormone, has many effects, one of which is to enhance muscle stamina and performance. Some female athletes have surreptitiously used injections of testosterone at low levels to enhance their performance against other female competitors. The practice is banned in most sports, but it is difficult to detect from blood tests alone because the tests cannot always determine why a female athlete has high testosterone levels—she might simply have naturally high levels of the hormone, or she might be using birth control pills that chemically mimic testosterone and confuse such tests.

In one way or another, humans have been pushing and prodding their performance, going beyond what nature endowed, sometimes by more intense conditioning, sometimes by nutritional enrichment, and sometimes by the use of chemical stimulants. If we anticipate the consequences of coming gene therapy, then we can see the ethical issues ahead.

Better Genes

Soon, we may be able to increase intelligence or height or strength or physical appearance or rid ourselves of those pesky wisdom teeth by replacing average genes with supergenes. Parents can have designer kids. Nothing new about this hope. As we saw with eugenics, this mission is an ancient quest. A deformed, or sick, or "wrong"-sex child might be "exposed" (left to die) while the parents mate again to give the genetic dice another roll. Today, the difference is in the technology, not in the quest or the hope. Today (or soon), aborting embryos and "exposing" infants will not be necessary, to breed desired offspring. Assembling members of your F_1 progeny with desirable traits will be possible by choosing your children's genes ahead of time. For many, such a future is repugnant, a betrayal of human nature. To others, this is just inevitable, a chance to relieve suffering from disease and to enjoy a better future. In fact, such choices might reside not with the parents but with the child itself when she or he becomes an adult.

Consider the still very tentative work with artificial chromosomes. Like natural chromosomes, artificial chromosomes could be designed with pairs of nucleotides, but prerecorded with genes that may be useful. Seeded into body cells, these individually designed chromosomes could code for a missing enzyme. You have Gaucher's disease? No problem. Medical technicians will be able to design chromosomes with correct copies of the defective gene and seed cells in your liver. For individuals with sickle-cell anemia, artificial chromosomes might carry normal genes into their blood-forming cells to code for normal blood cells.

More Than Genes

As evolutionary biologists, we noticed earlier that an individual (phenotype) is more than just the sum of his or her genes. A child's nutrition, behavioral support, social opportunities, and general environmental health all affect the character of the adult it might become. Genes, even the best of the engineered genes, define only the general boundaries of adult possibilities; alone, even supergenes do not guarantee super outcomes. Nature and nurture are both important for phenotypic outcomes. As human engineering takes over nature's role in future outcomes, we need to remember that nurture too is part of this future.

Heritability of Choices

If gene therapy could change the DNA in gametes, then those changes would be fixed in the reproductive cells and passed in the **germ line**—the offspring of the next generation, and the next, and so on. Our choice to change the germ line now affects unborn generations down the line in the future. We might just set in motion consequences that cannot be anticipated. Because of such fears, research on gene therapy has been restricted to research on somatic (body) cells (figure 16.3). This way, at least changes in the genome are limited to the current generation and will not carry into future generations.

Evolution in Our Hands

Our world does not want for problems. The previous discussion is not intended particularly to burden us with more. It is intended to prepare us to meet the problems, but not necessarily in detail. A shift, underway for a century or more, is still in progress. It is a revolution, totally changing our biological world, as profoundly different as any other time in our evolutionary history—more so than when we first used tools, more so than when we acquired speech. It is simply this: We are changing from a time when our fate was determined by

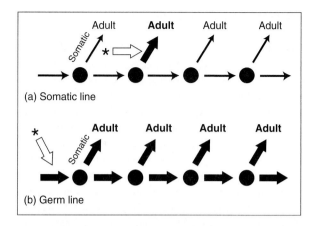

FIGURE 16.3   Somatic and Germ-Line Genetic Therapy   Traits pass in the germ line from generation to generation. Each fertilized egg (solid dots) grows the somatic body of an adult, setting aside gametes to produce the next generation, the germ line. (a) Somatic gene therapy. Changing the genome in the somatic cells of one individual changes the adult, but does not genetically propagate into future generations. (b) Germ-line therapy. Changing the gametes directly results in changes to adults, and these changes are also passed from generation to generation.

nature to a time when our fate will be determined by us. Literally, our fate is in our hands, guided by reason and enlightenment, not by blind accident. That has never been true before, at least not on this scale. It is true of no other species.

PEOPLE, PATHOGENS, AND PLAGUES

As we evolve and survive under our own direction, organisms with which we contend evolve as well, guided still by natural selection. This is especially true when dealing with pathogenic organisms. Wars and famine have threatened human survival for thousands of years. But most terrifying of all, because it has been least understood and most unpredictable, has been disease, in particular indiscriminant disease that explodes into virulent epidemics sweeping through communities, decimating cities, and leaving large-scale death in its wake. This is the **plague**. It has killed thousands in its path and brought down great political powers. Occasionally, a plague arrives when it is least expected, when all seems to be going well. Consider the case of the ancient Greek city of Athens in the fifth century B.C.E., about 2,500 years ago.

A Plague on Your City

In 490 B.C.E., a Greek army, made up mostly of soldiers from Athens, stood between Athens and the leading invasion force of the Persian army. The messenger Pheidippides was sent to nearby Sparta for help, covering the 140 miles in two days. He found the Spartans in the midst of preparations for a religious festival they would not abandon, but they promised to join the effort later (when finally they did arrive, the fighting was over). Pheidippides jogged back with the disappointing news, and rejoined the Athenians. After bitter and heroic fighting, the Athenians miraculously defeated that Persian force on the plains of Marathon. Joyously, Pheidippides dashed off again, this time to Athens about 22 miles away, proclaimed the victory to anxious citizens, and then expired, as legend has it, of exhaustion. The Athenians had defeated a force three times their size. Their preeminence on the Greek peninsula was established, and threats from Persia abated. But that is what started the problem with their neighbors on the peninsula.

The Peloponnesian peninsula became too small for both the prominent Athenians and the formidable Spartans. War was inevitable. In 431 B.C.E., the Athenians drew up in their seaport city behind impregnable barri-

Consider This—

The Marathon—Stretching It

In 1896, the modern Olympic Games were inaugurated in Greece. To commemorate the run of Pheidippides in 490 B.C.E., a marathon race was included coursing over 24.85 miles (40,000 meters) and ending in the Olympic stadium in Athens. The winner was a Greek postal worker who averaged over 7 minutes per mile. The marathon became the traditional final Olympic event. At the 1908 Olympic Games in London, the distance was stretched to reach the stadium, and then an additional 385 yards was tacked on to the end so it, and the Games, finished directly in front of King Edward VII's royal box, for a new total of 26.2 miles (42,166 meters). The 26.2-mile distance was established as the official marathon distance in 1924 and remains so to this day.

ers from a landward invasion, and with the most daunting navy of the ancient world enjoyed protection from a seaward attack. Time and might were on their side as they garnered allies and, by sea, struck Spartan outposts at will. A decisive blow on their Spartan enemies seemed certain and imminent. Then, without warning, in 430 B.C.E., disaster struck a crowded Athens—not war, but pestilence.

The plague that struck had arisen in Africa (Ethiopia), moved to Egypt, then entered Athens via marine commerce supplying the city. The disease might have been smallpox, measles, or typhus, but the best guess is that it was a severe form of scarlet fever. Reliable estimates of the carnage within the city are hard to come by. It is easier to count the dead than the living. A third and perhaps as much as two-thirds of the cooped-up Athenians died. Festering and diseased corpses piled up in the streets. The blow the Spartans could not deliver in war was delivered by a pathogen in the plague. Athens' moment for military victory passed. The war lingered on for another three decades, until a debilitated Athens fell.

Plagues, like the one that struck Athens, overwhelmed cities of the ancient world in periodic outbreaks. Careful chroniclers in China recorded 244 plagues and pestilences sweeping the country between 37 C.E. and 1718, an average of one every six or seven years. In the sixth century, Emperor Justinian, leader of the Byzantine Empire, set about to reestablish the Roman Empire by wresting the surrounding lands from the Persians. In the midst of doing so, his empire was struck by a plague moving out of Egypt. Striking Constantinople, the capital, its deadly virulence terrorized the citizens for more than three months. Many attributed it to God, or hallucinated demons visiting this scourge upon them. Stricken victims exhibited buboes, swellings of the lymph nodes in the armpits and groin that would redden and sometimes burst. Delirium and restlessness led for most to unconsciousness and then to death. This was the *bubonic plague*. It eventually swept all the way to Ireland, leaving devastation and a weakened Europe in its wake.

Its subsequent history is less well known, perhaps because later it was isolated in distant, remote pockets with few to record its local effects. But by the fourteenth century, the bubonic plague returned to Europe with a vengeance, where it was known as the "Black Death" (figure 16.4). Likely following trade routes out of central Asia, it hit major cities. By 1347, it emerged along the Mediterranean coast and spread inland. One in four Europeans died. It spread to China, Russia, and Scandinavia, then to Greenland via supply ships.

FIGURE 16.4   The Black Death In one of its outbreaks, the bubonic plague ("Black Death") spread throughout Europe in the fourteenth century, moving generally from east to west, and south to north. For five years it terrorized Europe, where one in four people died.

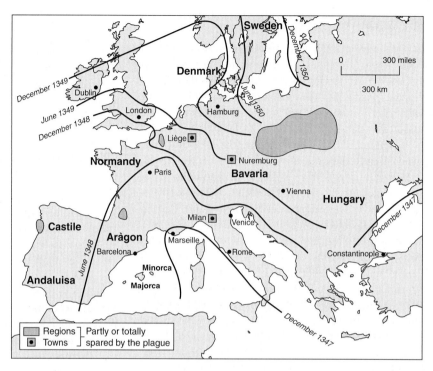

The Greenland colonies, founded by the Vikings in about 936 C.E., flickered out.

The plague returned again in the seventeenth century. Samuel Pepys, a citizen of London who survived it, recorded its impact on London in his journals. The Great Plague of London in 1665 claimed thousands; at its peak, 10,000 Londoners perished in a single week. But before this, death certificates record three notable outbreaks in 1630, 1636, and 1647, with cases also reported in years between. Surprisingly, after 1665, deaths from the plague dropped precipitously, to just under 2,000 deaths in the year 1666, then over the next few years faded away from London and from Europe in general. A grim description of the plague survives in a common, cheery nursery rhyme:

> Ring a ring of roses [buboes]
> A pocket full of posies [flowers to cover the odor]
> A choo, a choo [sneezes]
> All fall down. [death]

From Gods to Germs

We easily forget the comfortable insights we take for granted. People in the ancient world attributed pestilences to demons, gods, and fiendish enemies as punishment for their sins or other transgressions. Today, the cause of pestilence is obvious. A *germ* is just a lay person's catchall term for a pathogen. **Pathogens** are the agents that cause infectious diseases. But such an insight was long in coming. Not until the nineteenth century were pathogens isolated and their involvement in disease confirmed. In a species such as humans with a 40,000-year (perhaps 160,000-year) history, a hundred years ago is like yesterday.

Alexandre Yersin demonstrated in the late nineteenth century what a few had guessed: The bubonic plague involved a bacterium. In due scientific honor, that bacterium bears his name and its history, *Yersinia pestis*. Of Swiss background, Yersin trained as a medical doctor in Paris at the Pasteur Institute. Seized by adventure, he left this all behind to travel in Southeast Asia. Working as a mapmaker plotting remote areas for the French government, he heard of local outbreaks of the plague. In 1894, an outbreak occurred in the Chinese city of Canton, killing up to 100,000 people, and spread to Hong Kong, where Yersin traveled to study it. Robert Koch, working in Germany, had set down the rules to verify the causative agent of a disease. Yersin followed them. First, he collected fluid from the ruptured buboes of victims of the disease. Second, he isolated the bacterium using techniques he had learned in Paris and formed a pure strain of the suspected pathogen. Third, he inoculated samples of this pure strain back into a healthy host. Fourth, the host became sick, showing characteristic signs of the bubonic plague, confirming the initial identification. Even at such an early stage in formative medical research, the use of humans in steps three and four was then, and still is, clearly unethical. Yersin used animals, rats in particular. When they exhibited signs of the disease, Yersin had found and confirmed the agent of the disease.

As work on plagues by many expanded, suspicions were confirmed about other culprits behind this disease, the other factors contributing to deadly effects. The pathogen finds a friendly home in rats. Fleas carry the disease between rats, and between rats and humans. Before infected rats die of the disease, they act as living reservoirs for the pathogen. Fleas bite through their skin to drink a blood meal and inadvertently draw in some of the pathogen as well. When they leap and then land on an uninfected rat, fleas again drink a blood meal but in breaking the skin unintentionally inoculate the rat with the pathogen they carry, and so on through the rat population. By such means, fleas also carry the pathogen to and inoculate humans.

Epidemics

Plagues are devastating, but they come and go. They occur in some areas, but not others. A virulent pathogen itself may evolve, become mild, but then later revert suddenly to a deadly form. A pathogen spreads through a population taking advantage of available transport, usually not under its own power. If we are to understand this, we should recognize that pathogens are evolving organisms. They are adapted to us and to our culture—that is, to their environment. They evolve as our culture evolves and our habits change. By itself, a pathogen does not cause an epidemic to happen. One way to think of this is to consider a triad of factors that contribute to a plague.

These three factors, taken together, are the "cause" of infectious diseases (figure 16.5). Separately, any one factor is not enough to cause an epidemic. The *host* is the organism invaded by the pathogen. The *agent* is the immediate cause of the problem, the pathogen. The *environment* is the external surroundings. Together, the interaction of these three factors determines whether an epidemic occurs. If the host is a human, a strictly animal pathogen may not be effective; the resulting infection might be mild, and the epidemic non-existent. But if the host is an animal, then the animal-specific pathogen may be better adapted, the infection could be severe, and

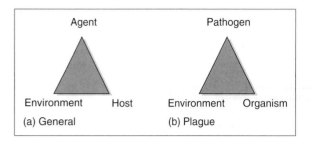

FIGURE 16.5   Epidemics   (a) Diseases spread in epidemics as a result of three interacting factors—an *agent* that infects a susceptible *host* promoted by favorable *environmental circumstances*. (b) Plagues, fierce epidemics, include a *pathogen* (agent), infecting an *organism* (host), where *environment* includes crowding, unsanitary conditions, and/or poor public health practices that promote spread of the disease.

in the animal population many might succumb to an epidemic. Managers of wildlife populations keep regular tabs on the health of animals under their jurisdiction. Wildlife can be as susceptible to plagues as are we. Occasionally, environment, host, and an agent harmless to humans will conspire to visit a devastating plague on susceptible herds of deer, elk, sheep, or musk oxen that fall prey in the hundreds to infectious diseases fatal to wildlife. For other pathogens, the reverse is true—human hosts are susceptible, but other animal hosts are not. Or a pathogen may be severe, but the host might be resistant and enjoy the protection of its immune system, which quickly disables the invading pathogen. The environment also can tip the balance. Epidemics are more likely if human conditions are crowded (e.g., Athens) or if rodents prevail (e.g., the Great London Plague). Commercial trade may bring contact with distant lands, returning economic goods and perhaps new pathogens. On the other hand, epidemics are less likely if sewers are covered, rats and their fleas are eradicated, and crowded conditions abate.

Viruses

The most common pathogens among the microorganisms are bacteria, which we met earlier (chapter 4), and viruses, which we have not. Viruses are obligate molecular parasites, infecting bacteria and eukaryotic cells. Viruses commandeer the replication equipment of the cell they infect, place it under the control of their genome, and use it to make more viruses. A virus consists of a *genome*, packaged in a protein casing, the *capsid* (figure 16.6). The genome may consist of a short section

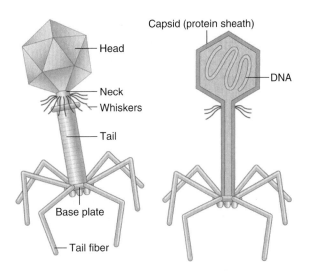

Capsid (protein sheath)

Head

Neck

Whiskers

Tail

DNA

Base plate

Tail fiber

FIGURE 16.6 Viruses Viruses attack bacteria, plants, and animals. Viruses are simple parasites consisting of a genome (usually DNA, sometimes RNA) held in a protective capsid. Shown is a T4 virus.

of DNA; in some viruses the genome is actually RNA. Viruses, bursting from their infected eukaryotic cells, may carry away part of the plasma membrane, which forms an *envelope* around the newly minted virus.

When isolated from their host cell, viruses tend to crystallize, entering a latent state. This inert phase was once thought to be evidence of their evolutionary intermediacy between nonlife and life, preceding the origin of all other life. Today, we see that this is just a dormant stage. (If you think about it, a parasite must evolve *after* the host it infects.) Instead, viruses may be escaped sections of bacterial or eukaryotic cell genomes, reaching a semi-independent state, but still using a cell host in which to propagate.

Making Mischief

Viruses that infect bacteria are *bacteriophages*, or just *phages.* Their life history may include up to two cycles (figure 16.7). The *lytic cycle*, as the name suggests, ends when the bacterium lyses, or ruptures. It starts when the virus attaches to the surface of the bacterium and injects its genome, and this genome, free in the cytoplasm, confiscates the cell machinery to replicate and assemble multiple copies. The thousands of new copies of the virus burst from the host bacterium to find new bacteria to infect and, so, quickly proliferate. In the *lysogenic cycle*, bacteriophages do not kill the cell they infect, at least not right away. They inject their genome,

called a *prophage*, which then becomes integrated into the DNA of the bacterium, replicating along with the bacterium as it replicates.

Viruses that infect eukaryotic cells may exit peacefully, taking some of the plasma membrane with them, or destructively, causing deadly lysis of the cell. Although it has some idiosyncrasies, HIV (human immunodeficiency virus) is an example of such a virus, structurally resembling the flu or mumps virus. HIV is a *retrovirus*, so named because instead of the transcription DNA $\longrightarrow$ RNA, it reverses this flow of genetic information (DNA $\longleftarrow$ RNA), starting with RNA that makes DNA. Upon first infection of the cell, the RNA genome of HIV produces DNA, which enters the nucleus and becomes integrated, as a *provirus*, into the chromosomal DNA of the host cell. Here, the provirus acts, like any active section of DNA, to synthesize products, in this case more viral RNA and its associated proteins. These assembled new viruses leave the cell to infect other cells (figure 16.8).

By lysing cells, some viruses directly kill the host. HIV does not. Instead, HIV weakens the immune system of the host, a condition known as AIDS (acquired immune deficiency syndrome). At first, HIV enters and proliferates in *macrophages*, cells that assist the immune system. Later, HIV enters critical cells of the immune system, the white blood cells known as *T-lymphocytes*, wherein the virus multiplies, filling and eventually lysing these cells as they exit. As the disease progresses, the exploding numbers of the virus infect and gradually depopulate the body of T-lymphocytes, leaving the individual defenseless.

Individuals do not die directly from the disease; but indirectly, from a weakened condition produced by the virus. So weakened, the individual falls prey to passing opportunistic diseases or unchecked cancers that a healthy individual might otherwise ward off. A slow death ensues.

EVOLVING PLAGUES AND PATHOGENS

Our future is likely to be a future of plagues. The body counts will be high because our human population is at an all-time high and growing. Consider just a few recent examples.

The "killer flu" of 1918 traveled home with the soldiers of World War I. Worldwide, more than 21 million people died of this virulent flu, many more than the

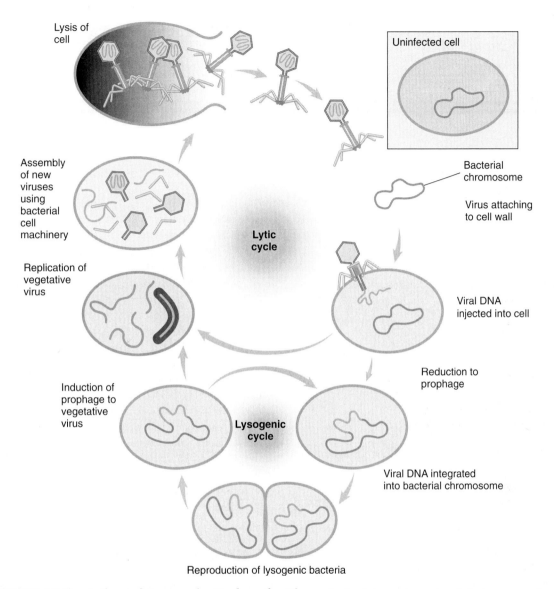

FIGURE 16.7   Lytic and Lysogenic Cycles of a Virus   In the lytic cycle, a virus infects, propagates, and bursts out. In the lysogenic cycle, a virus infects, remains in residence, and propagates.

numbers of soldiers who died in the war itself. The "Asian flu" of 1957 spread around the world. In the United States alone, 100,000 died, and many more were infected. The "Hong Kong flu" of 1968 infected 50 million Americans; 70,000 died. Eruptions of the *Ebola* virus in Africa can be particularly deadly. It spreads through contact. Within a few days of infection, symptoms may include high fever, stomach pain, bloody diarrhea, and vomit reddened by intestinal hemorrhage; death usually soon follows. Local mortality rates can

reach up to 90%. Imagine the consequences if this *Ebola* virus broke out of Africa and became a worldwide epidemic.

The Origin of Diseases

New human diseases usually originate in tropical areas where seasons are less pronounced, sparing parasites from the need to survive freezing or drying weather. Tropical humidity and warm temperatures promote

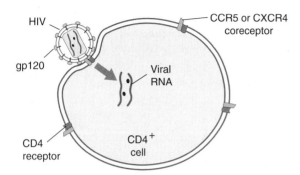

The gp120 glycoprotein on the surface of HIV attaches to CD4 and one of two coreceptors on the surface of a CD4+ cell. The viral contents enter the cell by endocytosis.

(a)

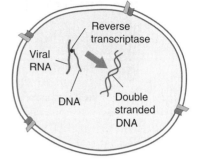

Reverse transcriptase catalyzes the synthesis of a DNA copy of the viral RNA. The host cell then synthesizes a complementary strand of DNA.

(b)

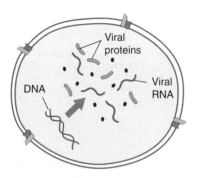

The double-stranded DNA directs the synthesis of both HIV RNA and HIV proteins.

(c)

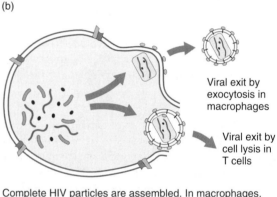

Complete HIV particles are assembled. In macrophages, HIV buds out of the cell by exocytosis. In T cells, however, HIV ruptures the cell, releasing free HIV back into the bloodstream.

(d)

FIGURE 16.8   The HIV Infection Cycle   (a) Surface molecules (gp 120 glycoproteins) on the HIV attach to surface molecules (CD4) on the white blood cell, permitting the HIV to dock and insert its RNA into the cell. (b) In co-operation with the co-opted host cell replication machinery, HIV RNA produces, with the aid of a promoter molecule (reverse transcriptase), a DNA copy of itself to which the deceived cell produces a complementary copy, yielding a double-stranded DNA copy of the HIV. (c) This double-stranded viral DNA synthesizes copies of HIV. (d) After successful replication, the new HIV copies exit the cell, traveling in macrophages or in the blood to infect other white blood cells.

proliferation of disease-causing microorganisms. The great diversity of tropical species provides great opportunity to find a *disease vector*—an organism, such as an insect, that itself already has a life history that brings it into contact with the pathogen's host. Animals are another source of human diseases.

In addition to living in a tropical area, peoples in southern Asia live in close proximity to domestic animals (pigs, chickens, ducks), which are also hosts to influenza viruses. The Hong Kong flu arose through genetic recombinations between duck and human viruses, in turn producing a more virulent virus eventually passed back to humans. In Africa, HIV arose orig-

inally from a related form in chimpanzees, SIV (simian immunodeficiency virus).

Evolution happens. It happens today among the pathogens that afflict us. And it happens quickly in such rapidly proliferating microorganisms. Humans at special health risk (e.g., the elderly) should get a flu shot each year because the pathogens from last year have evolved and new ones threaten this year. Mutations of viral and bacterial genomes, together with genetic recombination, produce novel gene combinations. Some of these new genomes notch up the virulence; others permit a pathogen to jump from animal to human. People and pathogens are in an evolutionary race. Still.

CO-EVOLUTION OF PEOPLE AND PATHOGENS

Humans

Humans, like other animals, and for that matter like plants, protists, and even bacteria, are challenged by a veritable zoo of pathogens. Plants produce antimicrobial chemicals that inhibit infection or kill pathogens. Invertebrate animals have cell surfaces dotted with marker proteins. Prowling amoeboid cells attack and engulf any invading cells that lack such identifying labels. Vertebrates, humans included, possess an immune system based on white blood cells. Pathogens and other foreign objects, such as dust and pollen, entering the body are collectively *antigens*. Because these are foreign (nonself) to the individual (self), they are treated as a threat. This job of interdiction falls to the immune system, which produces *antibodies* to neutralize these invaders.

Immune System Protection

The immune response is twofold, mostly involving white blood cells. First there is *recognition* of the antigens. Early in embryonic development, the immune system "learns" the chemical signature of its own normal cells. In a sense, it is learning "self." Lesson learned, the immune system becomes involved in self-surveillance. Abnormal cells that lose control of their orderly mitotic division go on a dividing rampage, producing cancers. If detected, the immune system can respond before the cancer overwhelms the body, destroying these out-of-control cells. Underactivity of the immune system may permit these cancers to escape detection and grow; an overactive immune system might mistake some of the body's own tissues for antigens and attack them, producing an autoimmune disease. Rheumatoid arthritis is an example of an attack on self, focused at the joints.

Antibodies belong to a class of macromolecules called immunoglobulins, which have a base with two arms forming a *Y*-shaped molecule. Although only a few hundred genes code for antibodies, the reshuffling of the arms of the antibody during its manufacture produces upward of 1 million different antibody combinations, each with the ability to recognize a specific foreign antigen. This manufacture usually occurs early in life. Each antibody is stored in its own white blood cell, a lymphocyte dispersed to lymph nodes throughout the body, waiting to come into contact with the particular antigen it can recognize. Antibodies reside on the outer surface of the lymphocyte, meeting, in turn, the surface characteristics of the antigen. When making contact with the particular antigen it is programmed to recognize, the lymphocyte is stimulated to proliferate, thereby making more copies of itself and its antibody type.

This proliferation of the lymphocyte sets into motion the second response of the immune system: *neutralizing* the antigen. This is carried out in two ways. Upon exposure to the antigen, the stimulated lymphocytes produce antibodies that leave the cells and enter the blood circulation, reaching the antigen, surrounding it, and immobilizing it. This is the *humoral* (fluid) *immune response*. Other cells, such as macrophages, arrive to break down the immobilized antigen. The other method of neutralization is the *cellular immune response*. Responding white blood cells move to the side of the antigen, surround it, and break it down directly. If the white blood cells are overwhelmed by the antigen, casualties occur, and dead white blood cells collect. This is the "pus" often associated with an infection or with a splinter puncturing your finger. Tissue grafts and heart transplants meet this cellular immune response, wherein the immune system sees these transplanted tissues as "foreign." Reacting normally, white blood cells crowd aggressively around the foreign tissue and can eventually cause rejection. Following such transplant surgeries, the immune system of the recipient is often deliberately suppressed until the foreign tissue is accommodated and welcomed by its new body.

Environmental Protection

Social circumstances change the environment and thereby change how a host and infectious agent interact. Dietary taboos and superstitions direct people away from foods that may carry unseen parasites. In some parts of the world, pigs are raised as export trade goods to win allies in war and as a dietary staple to supplement food supplies. However, in other parts of the world, pigs naturally harbor and pass debilitating parasites to humans that eat lightly cooked meat. The trichina worm, *Trichinella*, is such a parasite (figure 16.9). Historically, it is found in the Middle East, and there religious taboos against eating pigs grew up among the ancient Egyptians, Muslims, early Christians, and Hebrews. The taboos did not credit the disease itself for the proscribed sanctions, nor could they. Microscopes and effective medicines were inventions of later centuries. Not until the twentieth century was the parasitic agent discovered, its life cycle revealed, and enlightened intervention adopted. Pigs were no longer fed gruel contaminated with the parasites, and people were taught to cook pig meat thoroughly to kill any embedded parasites. The cycle was broken and

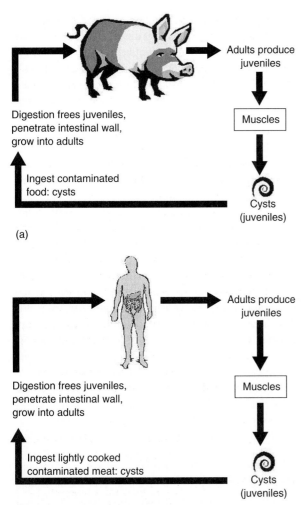

(a)

(b)

FIGURE 16.9   Trichinosis (a) Domestic omnivores, such as pigs, act as hosts that pick up the parasite in contaminated food. Digestion frees the juvenile parasites (cysts), which then penetrate the intestinal wall and mature to adults. Adult parasites produce juveniles, which migrate to muscles of the host where they roll up into cysts. (b) Humans (and carnivores such as bear and cougar) eat the meat, which contains the cysts. If the meat is not properly cooked, the juvenile parasites survive, are freed by digestion, and infect the human.

pigs, properly cooked, were now medically safe to eat. The medical strategy for addressing the disease of trichinosis centered less on directly killing the *trichina* parasite (agent) and more on modification of food processing and preparation (environment).

Medical strategies of treatment may target any of the three factors that contribute to a disease (figure 16.5):

Consider This—

Friendly Fever

With a severe cold come headaches, runny noses, coughs, and usually a fever—elevated body temperature. A fever is not a malfunction of the body's thermostat, disrupted by the infection. Instead, a fever is the body's measured response to the infection, a way to slow it. If a rat with a 2-degree fever is placed in a hot room, its cooling mechanisms activate to lower its temperature down again to 2 degrees above normal. The reverse happens too. Cool an infected rat, and its heat-conservation mechanisms raise its temperature again to 2 degrees above normal. Even cold-blooded lizards, when infected by a pathogen, produce a fever, but not by boosting their internal thermostat (they are ectotherms, after all), but instead by seeking out a hot spot to bask. A high fever can be serious and can even produce delirium, so medicines to bring it down are reasonably administered. But, generally, fever is the body's desperate but adaptive response to a threatening infection.

Environment. Crowded social conditions favor the spread of a pathogen; thin population densities reduce it. Quarantine of infected individuals effectively reduces social contact between the healthy and the sick. Closing open sewers, ridding streets of rats (and their fleas) and other animal vectors, and basic hygiene all help reduce environmental disease risks to a healthy human.

Host. Vaccinations boost the preparation of an individual's immune system, improving host resistance to infection.

Agent. Sometimes the medical strategy is to directly attack the agent, the pathogen itself. In the next section we look at this approach more closely.

Pathogens

As humans are challenged by pathogens, pathogens are challenged by human defenses. Human and pathogen are co-evolutionary contenders. Pathogens cause illness by destroying host cells or by releasing toxins. They have evolved and continue to evolve and adapt as host and environment change. Consider some of their innovative ways.

Our respiratory tract is the tubular system carrying oxygen-rich air to the lungs and expelling oxygen-depleted air from the lungs. But also carried within this tidal flow of air, in and out, are pathogens riding on the wind. Each inhalation carries them deep within the body. As a defense, the respiratory tract is coated with thin, sticky mucus. Passing pathogens are collected in the mucus, where they meet antibodies. The mucus also collects dust and airborne debris. When this channel gets clogged or congested, you cough or sneeze, basic reflexes that clear the breathing passageway. Some invading pathogens take advantage of these reflexes by promoting elevated levels of mucus secretion. Your nose runs, your lungs may congest, and you cough. And cough. And cough. The infecting pathogens are carried out on a hard cough and shower down on anyone around you. Thanks to you, and the exaggerated cough the pathogens provoke, the aerosol pathogens enjoy an airborne ride to a new host, and the disease spreads.

Malaria spreads not by such airborne transfer, but via an animal vector, the *Anopheles* mosquito. The agent is a protist, *Plasmodium*, living a complex life cycle between mosquito and human (or other mammal) (figure 16.10). When a mosquito vector bites a human, the *Plasmodium* in its sporozoite stage is transferred and soon enters first the liver, then the red blood cells. Proliferation eventually ruptures these red blood cells, reducing their numbers, and producing anemia (abnormally low red blood cell levels). Symptoms include fever and flu-like illness, including chills, headaches, and muscle fatigue. Untreated, it may progress to kidney failure, seizures, coma, and death. Into the nineteenth century, malaria occurred as far north as the Eastern United States and into Europe. The draining of swamps and use of insecticides (mosquito controls) led to its retreat into primarily tropical areas, especially sub-Saharan Africa and India, where it remains today a devastating disease. Each year perhaps a quarter of a billion people are infected and a million die, ranking malaria with AIDS and tuberculosis as currently one of the most serious of human diseases.

At low levels, malaria parasites may last for years in an infected individual. But how do they avoid detection

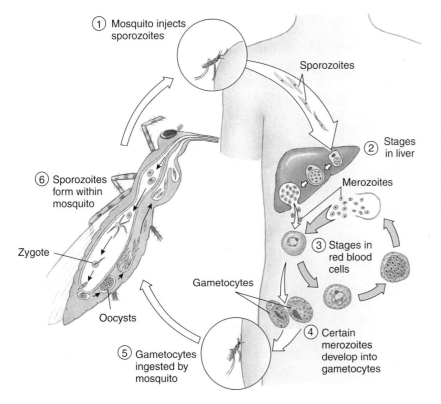

① Mosquito injects sporozoites

Sporozoites

② Stages in liver

Merozoites

⑥ Sporozoites form within mosquito

Zygote

③ Stages in red blood cells

Gametocytes

Oocysts

⑤ Gametocytes ingested by mosquito

④ Certain merozoites develop into gametocytes

FIGURE 16.10 Malaria
The causative agent of malaria is *Plasmodium*, carried from host to host by a species of mosquito. When the carrier mosquito bites a human, the sporozoite stage of the parasite is inadvertently delivered to the blood of the host, where it travels in turn to the liver, and then eventually to the red blood cells. When another mosquito drinks of the host's blood, it picks up and carries away the reproductive stages of the parasites (4, 5).

and destruction by the host's immune system? One way pathogens can evade the host's immune response is by hiding inside body cells, where they are hard to detect. Another adaptation is their behavior of frequently shedding their identifying surface protein coat. Just as antibodies begin to recognize the particular protein signature of the parasite and prepare for action, the pathogen changes, shedding this protein coat for a new one, and in a sense the immune system must start again in its work of identifying and responding to the organism.

Some medical specialists have suggested, cautiously, that causing fatigue in the infected human might be another adaptive strategy of the *Plasmodium*. Forced to lie down and rest, the distracted patient becomes an easier target for the marauding mosquito vectors; more insects alight without being noticed or swatted, drink their blood meal seeded with the parasite, and set out for other humans. Parasites that induce this fatigue symptom eventually enjoy the advantage of higher transfer rates to new hosts.

A change in the virulence of a pathogen during its epidemic spread is another adaptation to changing environmental conditions. Some pathogens strike with what seems a vengeance, overwhelming bodily defenses. They rapidly bring on the full fury of the disease, as it surges through a region, ending usually in a quick death to the thousands stricken. Oddly, as the disease lingers, its virulence often subsides. The pathogen is doing us no favors. It is just adapting to the changing environmental conditions it has helped to create.

The beginnings of agriculture 8,000 to 10,000 years ago brought together in small villages formerly dispersed peoples and their domesticated animals. Trade crisscrossed continents. Larger villages and eventually cities grew up at strategic locations, further concentrating people and their livestock. Such environmental conditions readied the social stage for plagues of biblical proportions. Virulent pathogens entered and quickly killed their current host. With conditions crowded, they passed easily by contact to another nearby host, in food, or as aerosols. Large numbers of people died; many fled to the countryside. As death and diaspora depopulated the cities, the deteriorating conditions were increasingly unfavorable to the spread, survival, and propagation of the virulent pathogen. Fewer people meant fewer contacts, so direct transfer from one host to another declined.

A plague may debut with a virulence that decreases long-term advantages. A deadly virulence reduces the numbers of available hosts. A quick death to its current host leaves the pathogen little opportunity to build up its numbers and move on. Consequently, a nasty pathogen is often replaced by a more benign variant, infecting but not quickly killing the host, and the overall virulence of the epidemic declines.

Other conditions may work against virulent pathogens as well. For example, some pathogens spread from host to host through contaminated water supplies. Such waterborne pathogens enter the water supply, are ingested by a human where they grow and thrive in the gastrointestinal tract, and then are expelled in the feces, eventually returning to and again contaminating the community water supply. But if the water supply is cleaned up, this avenue for reaching new hosts is denied.

This co-evolutionary coupling of human and pathogen is not a killing and carnage contest. As the afflicted party, that certainly is our personal perspective. But from an evolutionary perspective, it is a persistence and propagation contest. Under the right initial environmental conditions (crowding, contamination), a virulent pathogen enjoys the advantages of rapid propagation. But as the environment changes, it loses in competition with less-deadly variants that replace it.

Like all other organisms, pathogens evolve too. Because they propagate rapidly, their evolution can be rapid. New mutations produce new varieties that may find ready and waiting a hospitable environment in which they thrive. That seems to have happened with the toxic shock syndrome (TSS) that reached public notice in the early 1980s. Then, its predominant target was menstruating women who used tampons.

Recall that disease is "caused" by the fortuitous conjunction of three factors—agent, host, environment. Together, these factors conspired to produce TSS. At its worst, TSS sends the victim into shock, and death may soon follow. The agent is the bacterium *Staphylococcus aureus*. It waits on unwashed hands, in soil—almost anywhere within a contaminated environment. A "staph" infection is a troublesome illness that usually gains access through cuts and open sores. This pathogen produces toxins that circulate in the blood, with the levels of the toxin increasing as the pathogen multiplies. Upon first exposure, an individual experiences, at most, flu-like symptoms. But then her or his immune system becomes confused and begins to destroy the body's own protective white blood cells. Upon a second or third exposure to the staph pathogen, the compromised immune system further self-destructs, leaving the body with no cellular response to challenge the toxin that produces TSS.

Looking back on it, *S. aureus* (agent) apparently arrived in the United States in the mid 1970s, but in the form of a new and especially virulent mutant, unknown before then. Once taking hold in the body, it grew up to 10,000 times faster than normal staph, spewing out large, ever-increasing quantities of toxin that flooded the blood of the host. And it was resistant to common antibiotics. It may have journeyed in with travelers from elsewhere, or arisen in place from genomic mutation/recombination. In the United States, it met a large population of women in the midst of a social change (environment). Larger numbers of human females were becoming accustomed to the convenience of inserted tampons, replacing the external diaperlike pads used earlier for menstruation.

The female uterus is an optimist. Each month the uterine wall builds up a nutrient-rich lining anticipating the arrival of a fertilized ovum; implanted, the ovum develops a placenta with participation of the uterus to support the growing fetus. However, if implantation does not occur, the blood-rich uterine wall is shed via the vagina, exiting the body as short-term menstrual bleeding. As more and more women pursued industrial and professional careers, they sought ways to deal with the bleeding so they could get comfortably through long workdays. Tampons were the answer. Superabsorbent tampons were even better. Or so it seemed.

Although tampons provided the convenience many women demanded, they became a nutrient-rich incubator for the new staph pathogen, especially if left in for an extended time. Invasion of the staph during one menstrual cycle produced flu-like symptoms, from which a woman recovered, leaving the bacteria in her vagina, but this first exposure started the autoimmune disease. With the next menstrual flow the bacteria burst out, aided by the favorable incubation medium provided by the blood-enriched tampon. Now with the immune system in self-destruct mode, the body had no answer, and the uncontested toxin produced toxic shock syndrome. People of Scandinavian and German extraction (host) have a high incidence of TSS, presumably due to special genetic susceptibility.

This new, severe strain of the staph bacterium is not gender-specific, just opportunistic. Boys and men are also vulnerable to it—not through tampon use, of course, but this staph pathogen can enter via skin cuts, deep wounds, and exposure during surgery. More care and attention to tampon use has reduced the number of cases of related TSS. Use of tampons with the lowest absorbency that still meet menstrual needs reduces the culture medium supporting the bacteria; changing tampons frequently (wearing each tampon for less than 8 hours) reduces bacterial proliferation to low levels. Without this bacterium, tampon use does not cause TSS.

EMERGING PLAGUES

It started well. Sewer systems were closed and effluent washed away from city centers; drinking water was protected; public health agencies monitored disease outbreaks and responded quickly. By the early twentieth century, humans were fighting back. By the middle of the century, there was reason for even great optimism. Medical technology had found new ways to combat infections.

Medical Technology

Immunization is a major medical weapon against diseases, especially against viral infections. It takes advantage of the body's own natural immune defense. Harmless variants of a disease-causing pathogen, the inactivated virus itself, or derivative parts are collected to make a *vaccine*. The injected vaccine provokes the immune system to mount a response. Should the pathogenic virus now enter, the mobilized antibodies are already present at high levels to intercept and neutralize it.

The technique is actually a few centuries old. Edward Jenner, an English physician, first developed this strategy late in the eighteenth century (figure 16.11). Smallpox was then a fearsome scourge, killing many and leaving the survivors pockmarked and often weakened. Jenner noticed that in farm country, the milkmaids who had contracted cowpox, a mild disease picked up from infected cows, were resistant to smallpox. So close are cowpox and smallpox, that an immune system exposed to one can resist the other. In an experiment, Jenner scratched a young farm boy with a needle dipped in the cowpox sore of a milkmaid. Later exposure to smallpox confirmed that the boy was resistant to the more severe disease.

Producing new vaccines for new viruses remains one of the most important means of interdicting viral diseases. Unfortunately, once a virus gains a foothold, modern medicine can do little to eradicate the viral infection. The common cold and flu are viral infections.

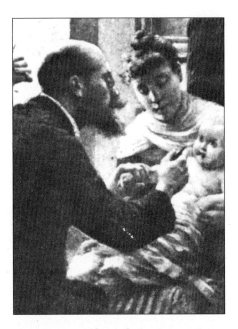

FIGURE 16.11   Edward Jenner This painting shows Jenner in the 1790s inoculating an infant with the less harmful cowpox, which thereby confers protection from the related, but much more severe, smallpox virus.

Once an individual is stricken, it is usually just a matter of toughing it out, dripping and coughing and sneezing, hoping for sympathy, and waiting for the immune system to gear up and clear out the disease.

Antibiotics work directly against bacteria. Antibiotics are chemicals that inhibit bacterial enzymes or actually destroy the bacteria. Viruses have few such enzymes to disrupt and no cell wall to maintain, so antibiotics are useless against them. Discovered in 1929 by Alexander Fleming, and produced in commercial quantities during the Second World War, penicillin was known as a "wonder drug." Fleming chose to study the bacterium *Staphylococcus aureus*, which he grew in agar plates. About to throw out a couple of these plates contaminated with a fungus, he noticed that around the fungus, the plate was staph-free. He investigated and discovered that the fungus produced a secretion, later named "penicillin" after the fungus, that protected the fungus from the bacteria.

The Magic Bullet

Antibiotics were magic bullets. Now physicians could administer the antibiotic to persons already infected,

selectively disrupt the action of susceptible bacteria, and help cure the disease. More antibiotics were to follow. The "age of antibiotics" had arrived. But what was new to enlightened human medicine was old to ancient pathogens. They had faced such chemical warfare before, mainly from plants and fungi. Now these pathogens are adapting to the challenges thrown at them by human-concocted antibiotics. They are developing resistance. From an evolutionary perspective, each new antibiotic is a selective agent. Bacteria with the ability to resist its destructive effects survive and reproduce. In the microscopic world of bacteria, it is survival of the fittest. From the survivors, resistant strains quickly emerge.

Revenge of the Germs

The writer Laurie Garrett has called this emergence of resistant bacterial strains the "revenge of the germs." The promising success of antibiotics in the treatment of otherwise hopeless diseases was a miraculous achievement. An ultimate weapon against pathogenic bacteria at last seemed at hand. For a time, the basics of Darwinian evolution were forgotten; natural selection was ignored. Antibiotics were used widely and, in retrospect, carelessly. The pathogens evolved resistance, then surged back in a virulent "revenge." People and pathogens are in a co-evolutionary relationship. The advance of one (humans and their antibiotics) selects for adaptive adjustment of the other (pathogen resistance). Remember the Red Queen (chapter 13)?

In 1952, *Staphylococcus* was 100 percent susceptible to penicillin; by 1982 fewer than 10% of staph clinical cases could be cured with the drug. The bacterial agent *Mycobacterium leprae* produces the biblical scourge leprosy. The antibiotic of choice was dapsone, until 1977 when a dapsone-resistant strain surfaced in Ethiopia. Within 10 years, bacterial strains all over the world had appeared that were invulnerable to the drug. Gonorrhea is a sexually transmitted disease (STD). By the 1970s it had acquired widespread resistance to penicillin; by the 1980s it was resistant to spectinomycin. Next up were the drugs tetracycline and cefoxitin; by about 1985 gonorrhea was resistant to tetracycline, and by 1992 strains resistant to cefoxitin emerged. Several strains of *Streptococcus pneumoniae* are involved in upper respiratory infections, pneumonia, and meningitis. By the 1980s they were essentially resistant to penicillin. One medical strategy of treatment is to hit the pathogen with several antibiotics simultaneously. Unfortunately, bacteria evolve multiple resistances. One strain of strep isolated in South Africa is resistant now to thirteen different antibiotics.

Bacteria also develop resistances to disinfectants. Antibacterial chemicals are now added routinely to hand soaps, detergents, cotton swabs, and band-aids. Low levels of these antibacterial chemicals select for the survival of the hardy. Cholera is a water- and food-borne disease causing severe diarrhea. During the height of a cholera epidemic in Lima, Peru, city officials dumped additional chlorine into the water supply, successfully breaking the back of this infectious disease. In the United States, chlorine is added routinely to our water supplies for similar reasons of public safety. Unfortunately, some well-meaning but naïve citizens feared they would get cancer from chlorine. Never mind that actual chemical carcinogens fall within the quite different polychlorinated biphenyls (PCBs) and dioxins. Public complaints and political organizations such as Greenpeace cast a blanket condemnation of all chlorine compounds, and local municipalities wavered. With chlorine levels low, resistant bacteria evolved, and one strain in particular has been found that can survive in household bleach.

Plasmids

The evolution of such remarkable resistances in pathogens testifies to their genetic flexibility, and to their long-standing evolutionary battle with environmental chemicals. Bacterial resistances are genetically based adaptations to chemical assault. Some adaptations incapacitate the antibiotic; others safely sequester it from harming the cell; others prevent its entry; and some deny it access to recognition sites on the cell surface, permitting the bacterium to go unnoticed.

More remarkable still are *plasmids*—small mobile circles of DNA that travel like wandering nomads from bacterium to bacterium. As they move, they pick up sections of host DNA, which they may subsequently insert into the DNA of the next bacterium they visit. When they pick up a section of the genome responsible for antibiotic resistance, they can deliver this to other strains of bacteria, not just to the same strain. For example, thanks to plasmids, the resistance established in strains of bacteria in animals has jumped to strains of bacteria that infect humans.

Antibiotics Everywhere

In most countries today, the first line of defense against bacterial infections is the administration of antibiotics. Lots of disease, lots of antibiotics. Antibiotics do not work against viruses, only against bacteria. But in some countries, health care is delivered by paramedics who lack the training necessary to distinguish between diseases of viral origin and diseases of bacterial origin. To be on the safe side, they often give antibiotics automatically. In some countries, antibiotics are unregulated and sold over the counter. Antibiotics might be given as a prophylactic (preventative) to persons headed into surgery who are asymptomatic (no signs of illness). Of the 25 to 30 million Americans receiving surgery each year, most will be given, as a matter of routine, preoperative antibiotics—"just in case."

To the millions of humans receiving antibiotics each year we can add even larger numbers of animals. Early on it was discovered that adding antibiotics to livestock food improved their health and increased their growth. Today, as a standard prophylactic, cattle, chickens, pigs, sheep, and other asymptomatic domestic stock routinely are given antibiotics to improve their husbandry.

Running Out of Bullets

Antibiotics change the competitive arena in which pathogens contend. Assailed by drugs, the resistant strains survive. Now, when infecting a human host, resistant pathogens come equipped with the chemical arsenal to defeat the assaults of current medical technology. Pathogens prosper as humans succumb—for a moment, pathogens have the upper hand. Then, new antibiotic bullets are developed to take out the resistant strains, and for a moment humans have the upper hand. Co-evolution continues, back and forth, between adversaries—people and pathogens.

Rebounding resistant pathogens are just one threat. To this must be added the emergence of entirely new strains hopping from animal to human, or simply mutating from a benign to a virulent form. HIV and *Ebola* viruses are just the most recent pathogens to appear. They will not be the last.

Disease is caused by three factors, so we have three areas in which to intervene. Certainly, favorably changing the social practices or stabilizing the shaky economic infrastructure of a country may improve the *environment* and promote sound public health. But often new or resistant diseases appear in distant and dusty parts of the world, where "intervention" by Western medicine is not always welcome and where sometimes its complicated technologies cannot be implemented safely. New strategies of immunization protect the *host*. A growing arsenal of new antibiotics can target the *agent*. But for these things to occur, we must realize that the effort is ours, and in our hands.

OVERVIEW—EVOLUTION TODAY AND TOMORROW

For some, our future is disturbing—a thing to fear, a place to avoid. New technologies are "arrogant," acting "against nature," "playing God." Such ravings do not prepare us usefully for our destinies. Ancient mythologies will not help today because we live now, for better or worse, in a world unreferenced to those mythologies. Could we go back? Probably not, but even if we could, would we want to? How far back? Would we return to times of plagues and pestilence, to times when most newborn children died in the first year of life, to times when few lived past the age of 30?

In some ways, ethical issues of today began thousands of years ago. By about 40,000 years ago, humans were basically of a modern mold. In their basic size, shape, and anatomy, humans back then were indistinguishable from ourselves today. Humans back then were equipped with speech, and certainly were engrossed in religion and in a rich array of cultural artifacts. Their basic anatomical and mental equipment were essentially the same as we are born with today. The difference is what we have made of it from then to now. To hunting and gathering, we have added farming and animal domestication; with farming, villages soon followed; and with villages, cities came next; and so on. In each of these changes, we take a small part of our destinies into our own hands. To this we can add medical technology, perhaps the most remarkable human innovation of all. But with this technology comes the responsibility for an enlightened ethic that supports its thoughtful use. As we take our fate from the hands of nature and put it into our own hands, we must see this as inevitable, not arrogance. We must be mindful of the need for an accompanying system of ethics. Medical technology becomes dangerous only when our ethical standards fail; medical technology serves us best when ethical guidance is enlightened.

Remember that evolution is not just a thing of the past, something that happened to the dinosaurs and not to us. Drugs aimed at disease pathogens or agricultural pests set up an ancient and familiar battle. Bacteria and bugs that can withstand the chemical assault survive; those that cannot, perish. From the hardy survivors come the resistant generations of the future. Some use the metaphor of an "arms race" to describe this co-evolutionary battle. When two hostile countries contend, the advance of weapons in the one country prompts the response of the other to meet and exceed that threat; in turn, the first country now placed at a disadvantage must reply with ever-new weapons, and so on and so forth.

In the biological world, our arms race with pathogens and pests is not grim. We simply must recognize that we are engaged in such a challenge and not falter in our response, nor be intimidated by those suggesting that we are "playing God."

The culling process once left to nature is now more and more in our hands. With gene therapy we can correct the deficiencies that nature botched. With this comes the obligation to develop a reasoned and ethical moral system that will be our guide and point of reference. This ethical system will certainly borrow from the successful guiding moral codes of the past, but incorporate the insights that enlightenment has brought.

Finally, we should now realize that we are products of the evolutionary process, stretching back over 6 million years of hominid history, coming in turn out of 4 billion years of preceding biological sifting, and sorting, and culling. We appreciate that childhood experience shapes the adult who will emerge; good or bad, the child becomes the adult. Similarly, our evolutionary experience has built the individual that we are. Indelibly stamped on our human character is the imprint of our evolutionary history. Perhaps the greatest contribution that evolutionary biology brings is the insight into who we are and what we can become.

Selected References

Achtman, M., K. Zurth, G. Morelli, G. Torrea, A. Guiyoule, and E. Carniel. 1999. *Yersinia pestis*, the cause of plague, is a recently emerged clone of *Yersinia pseudotuberculosis*. *Proc. Nat. Acad. Sci.* 96:14043–48.

Alexander, B. 2002. The remastered race. *Wired* 10:68–74.

Ewald, P. 1994. *Evolution of infectious disease.* Oxford: Oxford University Press.

Garrett, L. 1994. *The coming plague.* New York: Penguin Books.

Nesse, R. M., and G. C. Williams. 1994. *Why we get sick: The new science of Darwinian medicine.* New York: Random House.

Palumbi, S. R. 2001. Humans as the world's greatest evolutionary force. *Science* 293:1786–90.

Wills, C. 1996. *Plagues: Their origin, history, and future.* London: HarperCollins.

Learning Online

Testing and Study Resources

Visit the textbook's website at **www.mhhe.com/ evolution** (click on the book's title) to take advantage of practice quizzing, study/writing tips, timely news articles, and additional URLs for research on the topics in this chapter.

Appendix 1

Cell Division—A Review

CELL DIVISION IN PROKARYOTES

In bacteria, cell division is a simple process of binary fission, in which the cell divides into two nearly equal halves. Each cell has one chromosome—a single, circular, double-stranded DNA molecule. Just prior to division, this chromosome replicates. Each chromosome becomes attached to the cell membrane, and the cell membrane elongates to carry the chromosomes apart. When about twice its original size, the cell membrane constricts, forming a new cell wall and two new cells, each with its own chromosome (see figure A1.1).

CELL DIVISION IN EUKARYOTES

In eukaryotes, there are two types of cell division, each with different consequences. One type is *mitosis*, which occurs in somatic cells and gives rise to more cells to replenish those lost to injury or to increase cell numbers during growth. It is also the basis of asexual reproduction. The other type of cell division is *meiosis*. It occurs in germ cells (sex cells) and gives rise to gametes (eggs or sperm) that participate in reproduction. Steps of the two processes of cell division are intricate and significantly differ from one another. Thus, the general consequences of both should not be lost in the details of the divisional processes. Specifically, mitosis produces more somatic cells for growth and replacement of injured tissues; meiosis produces gametes for reproduction.

The two types of cell division also differ in their effects on chromosome number. Both start with diploid cells, but one produces more diploid cells (mitosis); the other produces haploid cells (meiosis). For each species, there is a standard number of chromosomes carried in each cell. For instance, a mosquito has 6 chromosomes per cell; corn, 20; a human, 46; a dog, 78; a goldfish, 94. This characteristic chromosome number per cell is a species' *diploid* number, represented as $2n$. Mitosis produces somatic cells with a diploid number of chromosomes preserved. However, meiosis produces gametes with half the normal chromosome number, the *haploid* number, represented as n. At fertilization, two gametes fuse and the diploid number is restored. Without the pre-

ceding process of meiosis, each fertilization would double the chromosome number per cell, $2n + 2n$ giving $4n$; then, in the next generation, $4n + 4n$ giving $8n$; and so on. Meiosis produces gametes with half the chromosome number of somatic cells. When haploid gametes fuse during reproduction, the chromosome number is restored, not doubled. Thus, meiosis produces haploid gametes; mitosis produces diploid somatic cells. Let's examine each type of cell division more closely.

Mitosis

Compared to prokaryotes, eukaryotes contain about 1,000 times more DNA per cell; the DNA of eukaryotes is linear, not circular, and is associated with proteins to form the chromosome; each cell usually contains several chromosomes. The *cell cycle* consists of two parts: the relatively short mitosis (cell division) and the longer interphase, a time of replication and preparation for division (figure A1.2). Entering mitosis, each chromosome consists of two identical parts, the *chromatids*, joined at the *centromere*.

During *interphase*, the chromosomes are thought to be loosely coiled; thus, little of the genetic material can be seen under the light microscope. However, as the cell enters mitosis, the chromosomes begin to condense and so become visible. This first and longest of the next four phases of mitosis is *prophase*. During this phase, chromosomes continue to condense, individual chromosomes (by now, pairs of chromatids) become visible, centrioles migrate to what will be opposite sides or poles of the cell, the nuclear envelope surrounding the chromosomes disappears, and the contractile spindle fibers make their initial appearance and establish contact with the chromosomes at specific sites, the kinetochores. Next, the cell enters *metaphase*. Now chromosomes (still chromatid pairs) become situated in the middle of the cell and the spindle fibers become completely formed. The spindle fibers stretch from centriole to centriole with chromosomes halfway between. Next, during *anaphase*, chromatids move to opposite poles of the cell, following or perhaps being drawn by the spindle fibers. With chromosomes parted, the cell enters *telophase*; the spindle fibers disperse, the nuclear

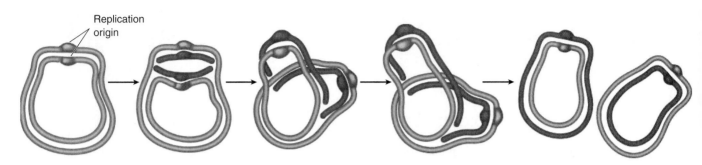

FIGURE A1.1   DNA Replication in a Bacterium   The circular bacterial DNA begins replication at a single site, the replication origin. Replication proceeds out in both directions, until copies of each strand of DNA are produced.

envelope reforms around each set of chromosomes, and the chromosomes uncoil. Constriction of the cytoplasm at the middle of the cell, *cytokinesis*, completes the separation of the single cell into the two cells.

Each of the daughter cells now enters interphase. The early part of interphase is a time of rapid growth in cell size and replication of organelles in the cytoplasm. During the middle of interphase, each chromosome undergoes duplication. However, the chromatid pair that results will not become visible under the microscope until later when the cell ends interphase and enters mitosis.

Meiosis

Meiosis occurs only in eukaryotes. It consists of two successive divisions of the genetic material. The result is the formation of haploid cells, the gametes (figure A1.3). At fertilization, gametes (n) from two parents (the egg and sperm) fuse to restore the diploid number ($2n$). This cell produced from fusion of gametes is the *zygote*. It now undergoes rapid division (mitotic divisions) to grow and differentiate into the adult individual. This means that every diploid cell from the zygote to adult contains chromosome pairs, one from each parent. These pairs of chromosomes are known as *homologous chromosomes*, or simply as *homologues*. The two homologous chromosomes resemble each other in size, shape, and the genes they carry.

No special pairing of homologues occurs during mitosis. But during the first part of meiosis there is pairing. As a result, during the first phases of meiosis, the two members of each homologous pair separate to opposite poles of the cell and the chromosome number is halved. During the second part of meiosis, the identical chromatids of each single chromosome separate. Cytokinesis occurs, and meiosis draws to an end. Reducing meiosis to its simplest, the two successive divisions involve first separation of homologues, then separation of identical chromatids. Specifically, the steps are as follows:

1. A cell entering meiosis begins in *prophase I*. Early in this phase, homologous chromosomes pair up, or *synapse*. This is a very precise and tight zippering-together of homologues. Each chromosome underwent duplication during interphase, so by now each consists of two identical chromatids. Thus, a synapsed homologous pair totals four chromatids—two of each paired chromosome. From this chromatid foursome, the complex derives its name, a *tetrad*. While synapsed, matched segments of chromatids may exchange places, an event known as *crossing-over*. Crossing-over provides a way of mixing and exchanging genetic material. Crossing-over occurs only among homologues, and only while they are synapsed in prophase I. Near the end, the spindle fibers complete their organization and the nuclear envelope disappears.

2. Next, in *metaphase I*, the paired homologous chromosomes line up along the center of the cell. Spindle fibers connect to the waiting tetrads of aligned pairs.

3. During *anaphase I*, the two homologues in each pair separate to opposite poles of the cell.

4. The cell now enters *telophase I*, which varies in length depending upon species. There is also some variation among species as to whether a nuclear envelope reappears now and whether cytokinesis occurs. Regardless, all cells eventually enter the second part of meiosis, beginning with prophase II.

5. During *prophase II*, a nuclear envelope, if present, disappears, and chromosomes, if uncoiled, condense again. The spindle fibers once again appear. Because homologous chromosomes have parted ways, no synapsing occurs.

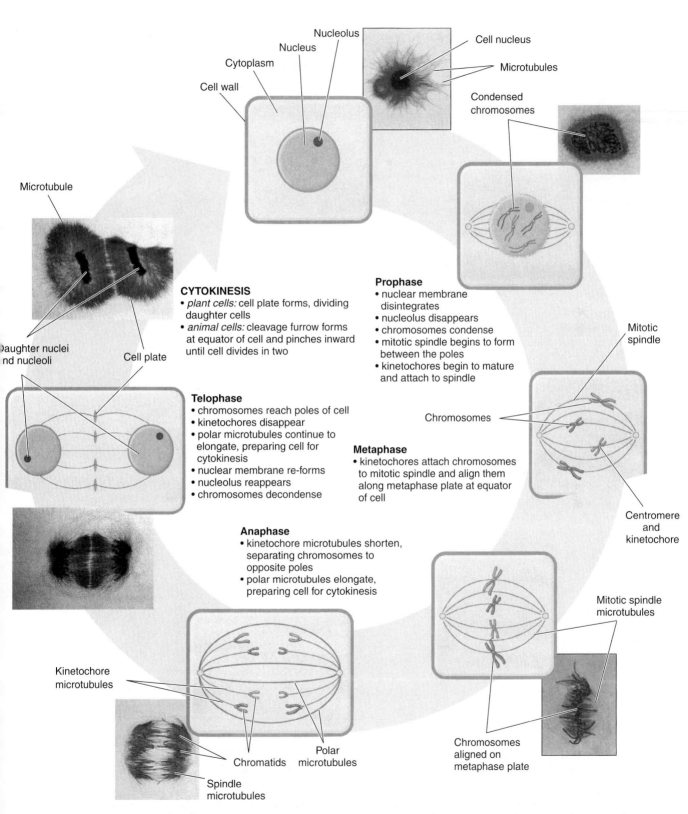

Cytoplasm

Nucleus

Nucleolus

Cell wall

Nucleus

Cell nucleus

Microtubules

Condensed chromosomes

Microtubule

Daughter nuclei and nucleoli

Cell plate

CYTOKINESIS
- *plant cells:* cell plate forms, dividing daughter cells
- *animal cells:* cleavage furrow forms at equator of cell and pinches inward until cell divides in two

Prophase
- nuclear membrane disintegrates
- nucleolus disappears
- chromosomes condense
- mitotic spindle begins to form between the poles
- kinetochores begin to mature and attach to spindle

Mitotic spindle

Telophase
- chromosomes reach poles of cell
- kinetochores disappear
- polar microtubules continue to elongate, preparing cell for cytokinesis
- nuclear membrane re-forms
- nucleolus reappears
- chromosomes decondense

Chromosomes

Metaphase
- kinetochores attach chromosomes to mitotic spindle and align them along metaphase plate at equator of cell

Centromere and kinetochore

Anaphase
- kinetochore microtubules shorten, separating chromosomes to opposite poles
- polar microtubules elongate, preparing cell for cytokinesis

Mitotic spindle microtubules

Kinetochore microtubules

Chromatids

Polar microtubules

Spindle microtubules

Chromosomes aligned on metaphase plate

FIGURE A1.2   Mitosis   The mitotic division of the genome occurs in four stages—prophase, metaphase, anaphase, and telophase. Cytokinesis—division of the cell into two daughter cells—completes the process of cell division.

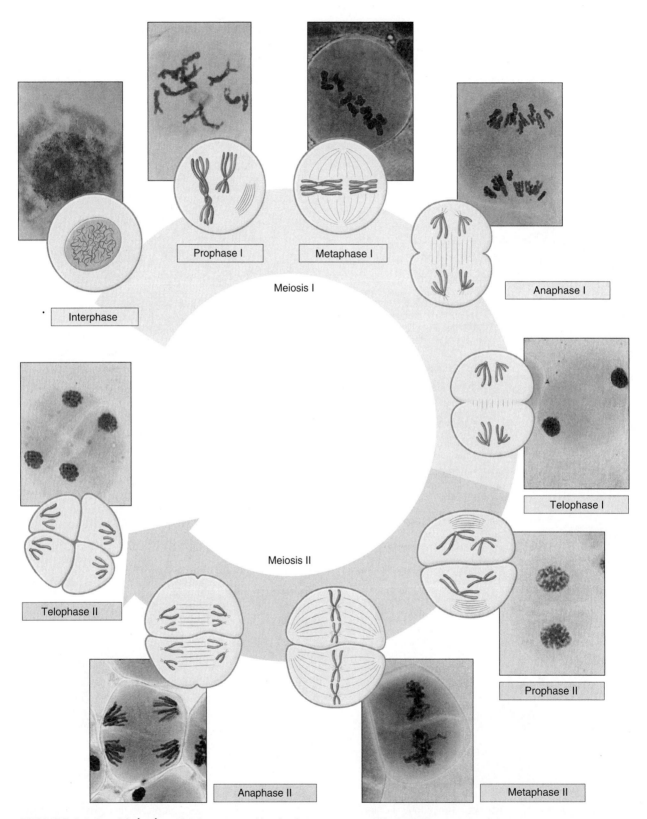

FIGURE A1.3   Meiosis   The formation of haploid gametes proceeds through two sets of stages.

6. Next, during *metaphase II*, individual chromosomes line up individually along the middle of the cell; spindle fibers become attached.

7. During *anaphase II*, the identical chromatids of each chromosome are drawn to opposite poles.

8. Finally, during *telophase II*, spindle fibers disappear, and the nuclear envelope reforms. Cytokinesis proceeds to finally delineate the four separate cells (gametes) of these two successive divisions of the genetic material.

9. *Interphase,* non-dividing interval.

Appendix 2

Taxonomy

SYSTEMATIZING LIFE

There is no going back. Life is a one-way street from the past toward the present. The course of evolution from ancestors to descendants, known as **phylogeny**, can be summarized in graphic schemes termed *phylogenetic trees*, or **dendrograms**, which depict treelike, branched connections between groups. Dendrograms are one-way streets following the events of the past toward the present. Ideally the representation is a faithful depiction of the relationships between groups. But the choice of dendrogram is based on intellectual bent and practical outcome. Dendrograms summarize evolution's course. This brevity gives them their attractiveness. All have risks, all flirt with oversimplification, and all take shortcuts to make a point.

The evolution of life is a continuous and connected process from one moment to the next. New species may evolve gradually or suddenly, but there is no point of discontinuity, no break in the lineage. If a break occurs in the evolving lineage, the consequence is extinction, a finality that cannot be redeemed. When taxonomists study current living species, they examine an evolutionary cross section of time in that they view only the most recent but continuing species with a long diverging history behind them. The apparent discreteness of species or groups at the current moment is partly due to their previous divergence. When followed back into their past, the connectedness of species can be determined. A dendrogram showing lineages in three dimensions (figure A2.1) emphasizes this continuity. If reduced to a two-dimensional, branching dendrogram, the relationships stand out better but imply an instant distinctiveness of species at branch points. The sudden branches are a taxonomic convention but might not faithfully represent the gradual separation and divergence of species and new groups.

Most dendrograms intend to make a point and are simplified accordingly. For example, the evolution of vertebrates is depicted in figure A2.2a to make a point about steps along the way. Although this representation is considerably simplified, it is a convenient summary; but if taken literally, the dendrogram is quite implausible.

The first four species are living, so they are unlikely ancestral species in the steps. A more plausible representation of their evolution is shown in figure A2.2b. Species at each division point lived millions of years ago and are certainly extinct by now. Only distantly related, derivative ancestors survive to the present and are used to represent steps in the origin of vertebrates.

Living vertebrates derive from a succession of distant ancestors and differ considerably from them. Modern vertebrates carry forward the collective results of these changes upon changes—thousands of them. Taken together, these collective changes produce the modern groups as we meet them today. To reconstruct this history, we may examine particular characters, using them to track the history of these changes. Formally, the earlier (or ancestral) state of a character is its primitive condition; its later (or descendant) state after transformation is its derived condition. A **taxon** is simply a named group of organisms. A taxon may be a natural taxon, one that accurately depicts a group that exists in nature resulting from evolutionary events. Or a taxon may be an artificial taxon, one that does not correspond to an actual unit of evolution. A *sister group* is the taxon most closely related to the group we are

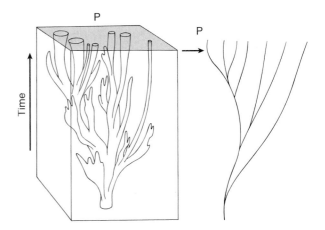

FIGURE A2.1 Evolution and Phylogenetic Trees The actual course of evolutionary events is depicted on the left; the representation of these changes is summarized in a phylogenetic tree on the right.

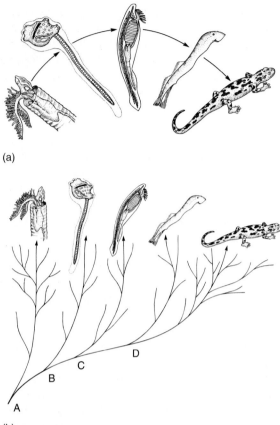

(a)

(b)

FIGURE A2.2   Expressing Evolution   The same five species are presented in both of these phylogenetic trees, but there is a trade-off between simplicity and complexity of their relationships. (a) All are living species, so they are not likely the immediate ancestors of each succeeding group, as this simple scheme implies. (b) Actual ancestors (from A to D, respectively) lived millions of years ago and are now extinct. The five species living today are representative of the lineage, but likely are specialized in their own ways.

studying. Using transformed characters as our guide, we inspect the pattern of vertebrate evolution and assign names for taxa accordingly, but we may do so with different goals in mind.

If a group of organisms carries a large number of distinctive derived characteristics, we might wish to recognize this by suggesting that the group has reached a new stage, step, or *grade* in its organization. In a traditional sense, a grade was meant to be an expression of the degree of change or level of adaptation reached by an evolving group. In the past, groups have been treated

as grades in some taxonomic schemes. For example, the fused and distinctive shell of turtles might be seen as a drastic reorganization of the skeleton requiring taxonomic recognition. This could be done by elevating turtles to a distinctive taxonomic rank co-equal with birds. In this sense of grade, evolving groups collect such a large number of derived characteristics that they pass an imagined threshold that earns them a high taxonomic rank. By such a view, mammals could be considered a taxonomic grade; so could birds at equal level. Although sometimes useful as a way of recognizing the degree of anatomical divergence between groups, grades can be misleading. The group Reptilia traditionally includes members with scales and a shelled (cleidoic) egg. But such a grade does not represent a single evolving group. Instead, the reptilian grade has been reached independently, within the line to modern reptiles but also early within the line to mammals. Conversely, some current groups that do not seem to look alike are survivors of a common lineage. For example, crocodiles and birds are survivors of a common lineage that places them more closely related to each other than either is to modern reptiles. Therefore, we may prefer to recognize groups based on their genealogy rather than on a subjective judgment of the degree of change alone.

If members of a group of organisms share a unique, common ancestor, we can recognize this by naming the lineage itself. A *clade* is a lineage—all organisms in a lineage plus the ancestor they have in common. *Traditional taxonomy* places together organisms with similar or homologous characteristics. The newer *phylogenetic taxonomy* places together organisms belonging to the same clade, and hence this method is also called *cladistics*. Within cladistics, the taxon name refers to the clade—to the genealogy itself—not necessarily to characters per se. Clades are recognized without concern for the amount of anatomical variation within the taxon. Consequently, some clades might include members very homogeneous in their basic morphology (e.g., birds, snakes, frogs) or quite heterogeneous (e.g., ray-finned fishes). Genealogy, not within-group variation, is the basis for recognizing a clade. The dendrogram depicting this genealogy is a *cladogram*, a hypothesis about the lineages and their evolutionary relationships. The advantage of cladograms is the clarity and ease with which they may be critiqued. A practical disadvantage is the swiftness by which a cladogram may be replaced with a newer cladogram, leaving us with abandoned taxon names replaced with newer names for more recent hypotheses of relationship. Character transformations play a central part in producing cladograms. In particular, derived characteristics are most important.

Relationships between groups are recognized on the basis of derived characteristics. The more derived characteristics shared by two groups, the more likely it is they are closely related. The assortment of taxa we are interested in examining is our *ingroup*; the *outgroup* is close to but not part of this assortment and is used as a reference. In particular, the outgroup helps us make decisions about which character state represents the derived condition. The sister group is the first outgroup we might consult, because it is the most closely related. But we might also successively make comparisons to more distantly related, second or third outgroups. Often, at this point, fossils may play an important reference role so that we can better decide primitive and derived states of a character. Once the degree of shared, derived characteristics is determined, we can represent associations in a Venn diagram (figure A2.3a). Because evolution proceeds by descent with modification, as Darwin helped establish, we expect those groups most closely related to be part of a common lineage. Therefore, from this diagram, we produce our hypothesis of genealogy, the cladogram (figure A2.3b), based on the characters we have examined. The layers of brackets above the cladogram represent the levels of inclusiveness of our groups within clades. As we name each clade, we would be producing our classification of this ingroup. In our cladogram, we could mark the sites at which particular character transformations occur. We could thereby use the cladogram to summarize important points of character transformation in the evolution of the groups and identify the distinctive derived characters that are associated with each clade.

Cladistics demands that we staunchly follow the practice of naming clades that recognize genealogy (figure A2.4). A clade is *monophyletic* in that it includes an ancestor and all its descendants—but only its descendants. Groups formed on the basis of nonhomologous characters are *polyphyletic*. If we combined birds and mammals together because we mistook their endothermic physiology (warm-bloodedness) to be the result of common descent, we would be forming an artificial, polyphyletic group. Groups that include a common ancestor and some, but not all, of its descendants are *paraphyletic*. This can happen with some traditional definitions of Reptilia. Modern reptiles and birds derive from a common ancestor. If birds were left out of the clade that represented this common lineage, then what remained would be a paraphyletic group. When paraphyletic groups are used for convenience, the names are usually placed in quotation marks to signal the unnatural composition of the group. Both polyphyletic and paraphyletic taxa are artificial taxa. They do not reflect the actual, complete course of evolution within a common lineage. Further, within cladistics we discover a second meaning for the term *grade*, as a synonym for a *paraphyletic group*.

By producing explicit and uncluttered hypotheses of relationship, cladograms have become part of the modern language of evolutionary analysis. But the starkness of these straightened cladograms should not obscure the bushiness of the evolutionary pattern they represent. If, for reasons of convenience or incompleteness, fossils are excluded, then the resulting cladogram, based only on living taxa, can be rather barren (figure A2.5a). This does not suggest that modern birds evolved from crocodiles (or crocodiles from birds), only that among recent taxa birds are more closely related to crocodiles than they are to any other group. Adding only a few of the fossil taxa (figure A2.5b) should make it clear that the cladogram could be enlarged to better reflect the richness and actual diversity of evolution within these vertebrate groups.

READING PHYLOGENETIC TREES

Because there is no going back, taxonomic dendrograms are one-way routes from the past forward in time. Dendrograms are summaries of information about the course of vertebrate evolution. But dendrograms also contain, even if inadvertently, hidden expressions of intellectual preference and personal bias. Dendrograms are practical devices designed to illustrate a point. Sometimes this requires complex sketches, and other times just a few simple branches on a phylogenetic tree serve our purposes.

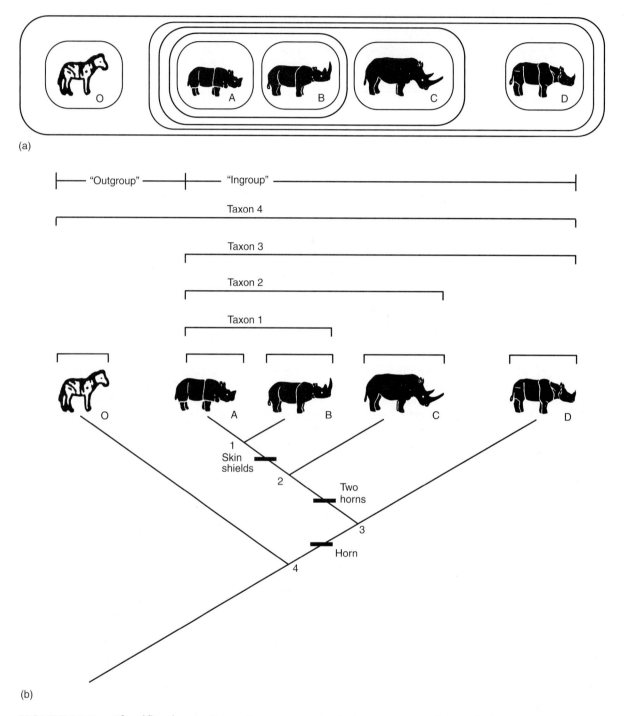

(a)

(b)

FIGURE A2.3   Classification   (a) Venn diagrams sort individuals into successive boxes of relatedness. Individuals of the same species are most closely related and put together in the smallest group—A, B, C, D, and O. If species A and B share more unique, derived features in common with each other than with any other species, then they are placed in a common group, and so on, expanding the diagram to include those more distantly related. (b) The genealogy of these species can be represented in the branching diagram, with the brackets representing Species A and B, together with their common ancestor, 1, at the branch, or divergence point. To make this dendrogram more useful, we could identify some of the many characters that change. For example, a horn first arises between divergence points 4 and 3; a second horn between 3 and 2; thick skin shields between 2 and 1.

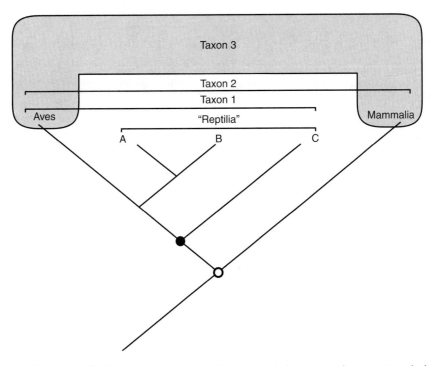

FIGURE A2.4   Taxonomic Concepts   Taxon 1 is a monophyletic group because it includes all natural members of the lineage—the ancestor (solid circle) and all of the descendant groups (A, B, C, plus Aves). Taxon 2 is also a monophyletic group, including the common ancestor (open circle) and all descendants. However, the "Reptilia" taxon is formally a paraphyletic group because it is somewhat artificial by not including Aves, one of the natural members of that lineage. Taxon 3 is a polyphyletic group, also artificial, because placing Aves and Mammalia together fails to include all other natural members between.

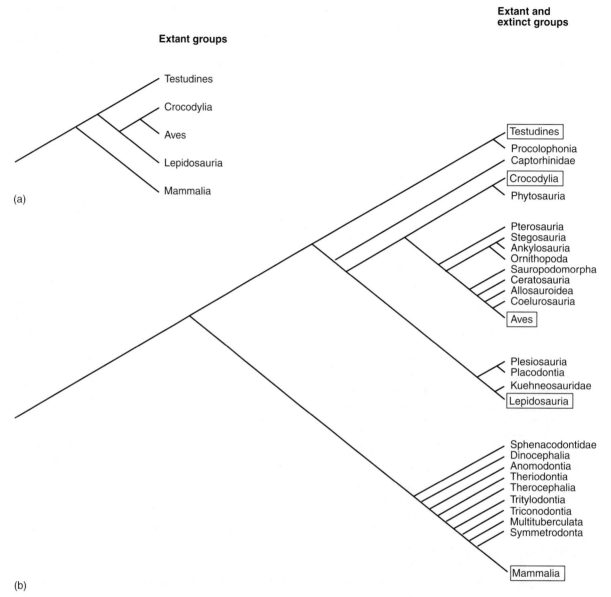

Extant groups

Extant and extinct groups

(a)

(b)

FIGURE A2.5   Extant and Extinct Groups   The richness of a dendrogram depends upon the number of species included. (a) Living (extant) groups are indicated. There is a close relationship between birds (Aves) and crocodiles (Crocodylia), which together are closely related next to lizards and snakes (Lepidosauria), and all these to turtles (Testudines). Mammalia is the outgroup. (b) Adding extinct groups illustrates the actual richness of the historical associations to the living groups (boxed).

Appendix 3

Molecular Clocks

The principle is very simple, known to all who own an attic, garage, or closet: The more time that passes, the more junk collects. The more time, the more junk. The same occurs in species. Two species just formed are very similar. But as time passes and they continue to diverge, more and more differences accumulate. The more time, the more differences. Consequently, the degree of difference between species generally reflects evolutionary divergence. We can use morphological differences—changes in bones, or shells, or floral displays, for example. Or we can use molecular differences—changes in the structure of important organic molecules.

MOLECULAR DATA

Mutations in DNA accumulate as speciation occurs and species diverge. Closely related species have fewer differences in base-pair sequences of DNA; distantly related species have more differences. Therefore, we can inspect differences in DNA as indicators of phylogenetic distance between species. Imagine that we have a group of species that share a particular gene. For each species, we can read the sequence of nucleotides in this gene and then compare the gene sequences. Closely related species should exhibit similar nucleotide sequences; distantly related species should exhibit more dissimilar sequences.

Because DNA codes for proteins, chains of distinct amino acids, we could also examine protein sequences between different species and use the same criterion to plot relationships of species—more closely related species have similar amino acids in the protein; more distantly related species have fewer amino acids in common. For example, the amino acid differences in hemoglobin, the oxygen-carrying molecule of blood, yield differences between humans and five other animals. The lamprey, a kind of primitive fish, departs in 125 amino acids from us, the macaque in 8, and other animals to varying degrees. These differences testify to different degrees of evolutionary relatedness, summarized in a molecular-based phylogenetic tree (figure A3.1).

MOLECULAR CLOCKS

If gene mutations, changes in nucleotide sequences, are neutral and not tied to adaptation, then they can accumulate without consequence to survival. Such neutral changes pile up at a fairly constant rate, or so it is assumed, and thus can be employed as a kind of molecular clock. Mitochondrial DNA (mtDNA) changes about ten times faster than nuclear DNA. Therefore, when clocking recent speciation events, mtDNA is preferred.

Nevertheless, molecular clocks must be calibrated. The rates at which neutral mutations accumulate, and hence the rates at which molecular clocks "tick" away, vary for different species and for different parts of the genome. Calibration is done by using a well-dated fossil within a molecular phylogenetic tree. The age of the fossil tells us when the fossil species diverged; the number of mutations to this divergence point tells us how many molecular changes accumulated during this time. Hence, the number of mutations per unit of time can be calculated, and this conversion can be used in the rest of the phylogeny to turn mutation numbers into years (figure A3.2).

ASSUMPTIONS AND LIMITATIONS

Molecular clocks assume that the accumulation of mutations through time occurs at a constant rate and that these amassing mutations are neutral, and therefore exempt from the effects of directional selection. (If selection reached these molecular changes, it might accelerate the rate at which beneficial mutations spread; or it might arrest the rate of detrimental mutations, confounding the molecular clock.) Using different parts of the genome may produce different results.

For the most part, data for molecular clocks comes from living species, as fresh, intact DNA or proteins must be used. In rare cases, DNA survives in ancient bones. Using very sophisticated laboratory procedures, mtDNA (360 base pairs) was successfully isolated from Neandertals discovered in 1856 in the Neander Valley of Germany. Neandertal and modern human

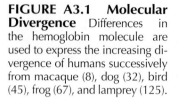

FIGURE A3.1   Molecular Divergence Differences in the hemoglobin molecule are used to express the increasing divergence of humans successively from macaque (8), dog (32), bird (45), frog (67), and lamprey (125).

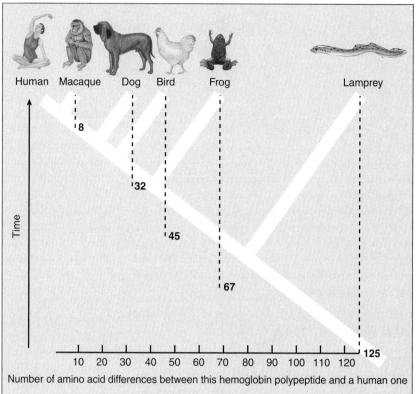

mtDNA were then compared, and a molecular clock calibrated. Looking only at these molecular data, Neandertals and modern humans diverged 317,000 to 853,000 years ago. If modern humans moving into Europe 40,000 years ago interbred with Neandertals, then present-day European descendants should show especially close genetic evidence of this interbreeding (see chapter 15). These molecular data do not support this, but instead argue that Neandertals did not contribute to the gene pool of modern *Homo sapiens*. However, a fossil from Portugal seems to be a morphological hybrid of Neandertal and modern human characteristics. Morphology and molecules are still contending partners in evolutionary studies.

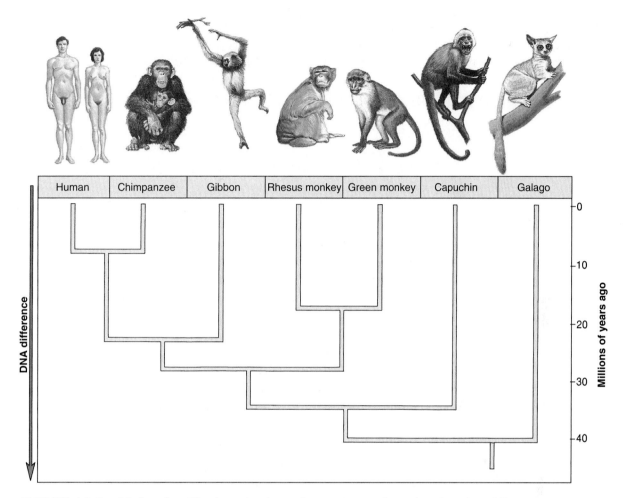

FIGURE A3.2   Molecular Clocks   This figure shows primate relationships based on differences in genomes. Lengths of the branches express the relative number of nucleotide differences between groups. Calibrated from the fossil record, these divergence points, branch points, can be expressed in millions of years.

Glossary

abiotic factor Not biological; usually refers to physical features. Cf. biotic.

acquired characteristics Somatic traits of an organism developed as a consequence of activities during the organism's lifetime; Lamarck's theory that characters arise from use or are lost from disuse.

adaptive radiation The diversification (increase of the number of species) of an evolving group of organisms as a result of directional selection pressures arising in different niches experienced.

advanced organism An organism arising later in time; derived. *See* primitive organism.

alarm call A distinctive vocalization uttered when a threat approaches.

allele One of two or more alternative expressions of a gene.

allopatric Apart; populations occupying different geographic locations. Cf. sympatric.

alternation of generations In plants, the characteristic life cycle wherein a diploid generation (sporophyte) alternates with a haploid generation (gametophyte).

altruism Unselfish behavior; biologically, a behavior reducing the fitness of the individual exhibiting the behavior but increasing the fitness of others. Cf. selfishness.

amnion An embryonic membrane forming a bag of fluid around the developing embryo.

amniotes A taxonomic group of vertebrates whose embryos possess an amnion—reptiles, birds, and mammals. Cf. anamniotes.

anaerobic Without oxygen.

analogy Similarity between two or more parts as a consequence of similar functions. Cf. homology, homoplasy.

anamniotes An informal taxonomic group of vertebrates whose embryos lack an amnion—fishes and amphibians.

ancestral organism An organism preceding later organisms that trace their history back to it. *See* primitive organism.

angiosperm A plant that produces flowers; a flowering plant.

animal vector The animal pollinator. Also used in medicine to describe the animal carrier of a pathogen.

annual plants Plants that live only one year or growing season. Cf. perennial plants.

aphotic zone The depth below which useful light does not penetrate an aquatic environment and is thus unavailable for photosynthesis. Cf. photic zone.

aposematic coloration Warning coloration.

artificial selection The culling out of individuals with heritable traits by humans to serve human purposes. Cf. natural selection.

asexual reproduction Formation of a new organism not involving fusion of haploid gametes; simple replication. Cf. sexual reproduction.

atavistic structure A throwback structure. The reappearance of a lost ancestral structure in a modern organism.

autotroph An organism that synthesizes compounds sufficient to sustain its own cellular metabolism; e.g., photosynthetic organisms. Cf. heterotroph.

background extinctions *See* uniform extinctions.

basal organism An organism arising earlier in time. *See* primitive organism.

Batesian mimicry Superficial similarity between two (or more) species, one of which (the model) is dangerous or unpalatable and the other of which is harmless (the mimic). Cf. Müllerian mimicry.

biased learning Restricted learning; behavior that is modified, in a limited way, usually during a short period of an organism's lifetime, by a limited array of environmental information; imprinting.

biogenetic law The generalization that ontogeny (embryology) recapitulates (repeats) phylogeny (evolutionary history); proposed by E. Haeckel. Cf. preservationism.

biogeographic realms A defining assortment of distinctive fauna and flora on land masses.

biological radiation The diversification, or increase in numbers of species, of an evolving group of organisms.

biological species A reproductive community wherein individuals potentially or actually interbreed among themselves, but not with others. *See* morphospecies.

biotic factor A biological element in the environment; usually predator, prey, or competitor.

bipedal Designed to walk upright, comfortably, on the two hindlegs. Cf. quadrupedal.

bottleneck effect A type of genetic drift wherein breeding members of a population represent a restricted and unrepresentative genetic sample of the whole population. Cf. founder effect, genetic drift.

brachiation Moving through trees by means of alternating arm swings.

camouflage Features of an organism's color, shape, and behavior that conceal it from easy detection. Cf. cryptic.

carbon fixation The direct incorporation of carbon, from carbon dioxide, into stable organic compounds, as in plants.

carnivore An animal-eating organism. Cf. herbivore.

catastrophic extinctions *See* mass extinctions.

cell wall A rigid boundary to the cells of plants, some protists, fungi, and many bacteria. Cf. plasma membrane.

center of origin The geographic area where the first members of a group first arise evolutionarily.

chloroplasts Organelles in plants that are dedicated to photosynthesis.

chromosome A gene-carrying structure composed of DNA and packing proteins.

cladogenesis The formation of new lineages through speciation; hence, the pattern of evolution characterized by branching. Cf. phyletic evolution.

cleidoic egg The leathery shelled egg (reptiles, monotremes) or brittle shelled egg (birds) that holds the vertebrate embryo wrapped in an amnion.

cline A geographic gradient in a character within a species.

clutch A group of eggs laid by a female; a brood of eggs.

codon A triplet sequence of messenger RNA nucleotides; each codon triplet specifies a particular amino acid to be placed in the forming protein.

coefficient of relationship A mathematical expression of the degree or fraction of shared, identical genes between two individuals.

co-evolution The reciprocal selection pressures between two or more species as a consequence of their ecological interaction.

commensalism A symbiotic relationship in which one partner benefits and the other is not harmed. Cf. mutualism.

competition The condition occurring when two organisms attempt to utilize the same limited resource; specifically, *intra*specific competition when contending organisms belong to the same species; *inter*specific competition if contending organisms belong to different species.

convergence The pattern of resemblance wherein distantly related species take on a similar appearance. Cf. parallelism.

craniometric profile The sum total of multiple skull measurements that anatomically characterize an ethnic or taxonomic group.

cross-fertilize Exchange of gametes between different organisms.

crossing-over The exchange of alleles between homologous chromosomes when they are synapsed during meiosis.

cryptic Hidden; camouflaged.

decomposers Organisms, specifically bacteria and fungi, that break down dead organisms to obtain nutrients themselves, but in so doing also recycle nutrients to the soil where they may again be gathered up and reincorporated into tissues of other living organisms. Cf. producers.

dendrogram A diagrammatic representation of evolutionary history, usually shown as a branching system of related organisms. *See also* phylogenetic tree.

derived organism Within a phylogenetic lineage, an organism that appears later, hence more recently, with no intended meaning as to whether it is superior or inferior to earlier organisms; hence an advanced organism. Cf. primitive organism.

deuterostomes A taxonomic group of animals characterized by, among other features, an embryonic blastopore that becomes the adult anus. Cf. protostomes.

diploid Full normal number of characteristic somatic cell chromosomes, represented "$2n$." Cf. haploid.

DNA Deoxyribonucleic acid; a double-stranded helix of nucleotides that constitute the hereditary material of the cell; the agent of inheritance and cell metabolism; transcribes RNA.

ecdysis To molt; to shed the outer layer of the body. *See* molt.

ectotherm An animal whose body heat is largely absorbed from the environment. Cf. endotherm.

endotherm An animal whose body heat is largely produced internally, through metabolic expenditure of energy molecules.

environment An organism's external world, where it resides, contends with threats and stresses imposed, and reproduces. Cf. internal environment.

eugenics The social view that humans with socially desirable traits should be paired up and encouraged to breed; and conversely, that humans with undesirable traits should be prevented from breeding, thereby improving overall the human species or a human race.

eukaryote An organism whose cells include a nucleus and cytoplasmic organelles, and that is capable of sexual reproduction. Cf. prokaryote.

extinction The loss, or disappearance, of a species. Cf. mass extinctions, uniform extinctions.

F_1 generation In an experiment of inheritance, the first progeny produced by an initial cross of the parental generation.

Subsequent crosses of each generation are represented as F_2, F_3, and so on. Cf. parental generation.

fauna The animals characteristic of a region. Cf. flora.

fitness The relative reproductive success of individuals, within a population, in leaving offspring in the next generation. At the genetic level, fitness is measured by the relative success of one genotype (or allele) compared to other genotypes (or alleles).

flora The plant life characteristic of a region. Cf. fauna.

founder effect A type of genetic drift wherein the restricted genetic variation of colonizing organisms limits subsequent evolutionary variety. Cf. bottleneck effect, genetic drift.

free-living Living on its own; nonparasitic. Cf. parasite.

fundamental niche The full range of available resources *potentially* available to an organism. Cf. realized niche.

gamete A male or female sex cell; sperm or egg.

gametophyte In plants, the phase in the life cycle where individuals are haploid, and gamete-producing. Cf. sporophyte, alternation of generations.

gene The unit of inheritance affecting the characteristics of a trait; a section of DNA that codes for a particular protein through an RNA intermediate.

gene flow The spread of alleles through a population or species as a result of successful interbreeding.

gene therapy Direct repair of defective genes. Cf. replacement therapy.

genetic drift A chance event wherein some alleles spread in a population as a result of random actions rather than as a consequence of natural selection.

genetic screening A variety of tests performed to determine whether a fetus carries any genetic disorders.

genome Genotype.

genotype The genetic makeup of an individual; genome.

germ cell A reproductive cell; sperm or egg. Cf. somatic cell.

germ line The generation-to-generation continuance of heritable features.

group selection A culling process acting to produce differential survival of groups of organisms; selection acting on groups. Cf. individual selection.

growth ring An annual growth pulse in the vascular cambium of a tree.

habitat Site of residence; the specific location in the environment where an organism resides. Cf. niche.

half-life The amount of time it takes for half the atoms of an unstable isotope to transform (decay) into a stable form.

haploid Half the normal number of somatic cell chromosomes, represented "$1n$." Cf. diploid.

herbivore A plant-eating organism. Cf. phytophage, carnivore.

heterotroph An organism that does not synthesize its own necessary compounds, but depends on finding these in the immediate environment. Cf. autotroph.

heterozygote An individual carrying two different alleles for a trait. *See* homozygote.

homeotic gene Master gene that brings under its control legions of secondary genes; *Hox* gene.

homology Similarity between two or more parts as a consequence of common ancestry. Cf. homoplasy, analogy.

homoplasy Similarity between two or more parts as a consequence of accident or incidental events. Cf. homology, analogy.

homozygote An individual carrying identical alleles for a trait. *See* heterozygote.

Hox **gene** A master control gene that brings under its control large numbers of secondary genes; a homeotic gene.

hyphae The thin, long filaments making up a fungus, which expand into the temporary reproductive fruiting body aboveground and below ground spread through the soil or substrate as the feeding network, the mycelium.

ice age A prolonged period when Earth's climate cycles between phases of glacial ice sheet spread (glacial phase) with temporary phases of glacial ice melt (interglacial phase).

inclusive fitness The total measure of evolutionary success based on genotype propagation directly (via offspring) and indirectly (shared by kin).

index fossil Based on independent and previous study, a fossil species uniquely indicative of a particular rock strata.

individual selection A culling process that acts directly on individuals within a population to produce differential survival of individual organisms and reproductive success; selection acting on individuals. Cf. group selection, species selection.

inheritance of acquired characteristics Lamarck's theory that metabolic changes can provoke changes in "characters" (characteristics) as a consequence of need (or disuse), and that these acquired characteristics are then passed on intact to future generations.

innate behavior Species-specific behavior that is expressed in a highly stereotyped and complete manner the first time it is deployed. Cf. learned behavior.

internal environment The structural, metabolic, and physiological features within an organism.

kin selection A type of individual selection wherein offspring and relatives benefit.

learned behavior Behaviors built up from the lifetime of successful and unsuccessful experiences of an organism. Cf. innate behavior, biased learning.

lichen A symbiotic association of fungus and green algae.

life cycle The full series of stages through which an individual passes during its lifetime, from egg through adult; typically, in animals, beginning with fertilized *eggs*, that grow into *larvae* or *juveniles* (sexually immature), that become *adults* (sexually mature stage). Cf. life history.

life history The life cycle of an organism, with attention to the adaptive consequences of each stage and suite of characteristics.

locus The particular site along a chromosome occupied by an allele.

macroevolution Evolutionary events examined at large scale, and usually over a relatively long period of time. Cf. microevolution.

mass extinctions The loss of large numbers of species, across various groups, during a relatively short period of geologic time; catastrophic extinctions. Cf. uniform extinctions.

meiosis Cell division resulting in haploid gametes. Cf. mitosis.

metabolic pathways Within a cell, the series of chemical steps that sequentially capture and transfer energy from molecules.

metamorphosis During the life cycle of an animal, the point at which there is a rapid and extensive anatomical transformation of one stage to another, e.g., amphibian larva (tadpole) to juvenile.

microevolution Evolutionary events examined closely, and usually over a relatively short period of time. Cf. macroevolution.

microhabitat The specific and particular location an organism occupies.

mimicry The superficial resemblance between two or more organisms as a result of a co-evolutionary relationship. Cf. Batesian mimicry, Müllerian mimicry.

mitochondrion An organelle in eukaryotic cells where energy, from organic fuels, is obtained and then stored in ATP.

mitosis Cell division resulting in more diploid cells. Cf. meiosis.

molt The shedding of an external skeleton in a growing animal; ecdysis.

morphospecies A group of individuals that are phenotypically similar. *See* biological species.

mosaic evolution The pattern of evolution within a lineage wherein characteristics appear at different times and change at different rates.

Müllerian mimicry Superficial similarity between two (or more) species, both of which are dangerous or unpalatable. Cf. Batesian mimicry.

mutagen An environmental agent, such as radiation or chemicals, that directly damages DNA to produce a mutation.

mutation A spontaneous error or mistake in DNA duplication, leading to a change in the genome.

mutualism A symbiotic relationship wherein both partners benefit.

natural selection The culling process by which individuals with beneficial traits survive and reproduce more frequently, on average, than individuals with less favorable traits.

naturalist In general, a person engaged in the study of nature—the physical and biological features of life; philosophically, persons drawn to a study of nature for its evidence of God's purposes and handiwork.

Neolithic The cultural period in human history characterized by the advent of farming; follows the Paleolithic. Cf. Paleolithic.

niche An ecological "profession"; the specific lifestyle of an organism; how an organism utilizes resources. Cf. habitat, fundamental niche, realized niche.

nitrogen fixation The conversion, by specialized bacteria, of unusable atmospheric nitrogen (N_2) into nitrogen compounds that can be used by other organisms, mostly plants.

nuclear envelope The membrane within eukaryotic cells that holds and bounds the chromosomes.

ovary A general term for a female structure involved in the production of eggs.

ovule In seed plants, a container holding the female gametophyte that produces eggs; in angiosperms, the ovule resides within a protective ovary. *See* ovary. Cf. pollen.

Paleolithic The cultural period in human history characterized by the manufacture of tools, up to the Neolithic. Cf. Neolithic.

parallelism The pattern of resemblance wherein closely related species take on a similar appearance. Cf. convergence.

parasite An organism that draws its sustenance from another, much larger living organism on or within which it lives. Cf. free-living.

parental generation In an experiment of inheritance, the adults that are used to begin the cross, noted as the **P** generation. Cf. F_1 generation.

pathogen An infectious agent that contributes to disease.

perennial plants Plants that live longer than a year. Cf. annual plants.

peripheral isolates Within a species, populations that are geographically distant from the main population.

phenotype The outward appearance of all traits, including anatomical and behavioral characteristics; the realized expression of the genotype and of environmental influences.

phenotypic plasticity Fine-tuning of the phenotype by simple physiological or behavioral alternation, and not by genetic changes.

photic zone The depth to which useful light penetrates an aquatic environment and is available for photosynthesis. Cf. aphotic zone.

photosynthesis The process, in plants and some microorganisms, whereby solar energy is captured and used to fuel the manufacture of organic molecules.

phyletic evolution The transformation through time of one species into another without frequent speciation along the way. Anagenesis. Cf. cladogenesis.

phylogenetic tree A diagrammatic representation of allied organisms, usually shown as a branching system of related organisms. *See also* dendrogram.

phylogeny A historical lineage of related organisms. *See also* phylogenetic tree, dendrogram.

phytophage A herbivore; usually applied to plant-eating insects.

phytoplankton A community of aquatic microorganisms (cyanobacteria, green algae) that are at the base of the food web and responsible for production of most of Earth's free oxygen.

placenta A specialized fetal-maternal organ through which the young, growing embryo receives nutritional and metabolic support.

plague Virulent epidemic.

planktonic An aquatic organism, usually small, living unattached and carried about passively by tides and currents. Cf. sessile.

plasma membrane The defining boundary layer of a cell, which presides over the selective entry and exit of compounds to and from the internal contents. Cf. cell wall.

plasmid A short segment of DNA that jumps between bacteria and becomes incorporated into the host DNA.

plate tectonics The energized movement of Earth's crustal plates.

polarity In embryonic development, the establishment of regional body identity in the early embryo—anterior and posterior, dorsal and ventral—hence, establishment of positional information.

pollen In seed plants, a tiny transport grain holding a much reduced male gametophyte that produces sperm. Cf. ovule.

pollination The successful transport of pollen, via wind, animals, or self-pollination, to the female part of a plant.

pollinator The animal involved in pollen transfer to the female part of the plant. *See* animal vector.

polygenes Many genes that govern the characteristics of one trait.

polymorphism Within the same species, individuals are found with two or more conspicuous and distinctive forms of the same trait, such as color.

polyploidy The duplication or addition of whole sets of chromosomes, doubling or more the chromosome number.

population Within a species, a local subset of organisms.

preadaptation Possessing the necessary form and function to meet the demands of a particular environment before the organism experiences that particular environment.

predation The lifestyle wherein an organism consumes another, its prey; usually restricted to the situation wherein one animal eats another.

preservationism The concept that the embryology of current organisms retains some of the same embryonic stages and their sequence of appearance as in ancestors. Cf. biogenetic law.

primitive organism Within a phylogenetic lineage, an organism that appears early, with no intended meaning as to whether it is inferior or superior to later organisms; hence, basal organism or ancestral organism. Cf. derived organism.

producers Organisms, specifically cyanobacteria, algae, and plants, that harvest light energy in photosynthesis to meet their needs in growth and proliferation.

prokaryote An organism whose cells do not include a nucleus and cytoplasmic organelles, and that reproduces by means of asexual reproduction. Cf. eukaryote.

protein A chain of amino acids; a polypeptide.

protostomes A taxonomic group of animals characterized by, among other features, an embryonic blastopore that becomes the adult mouth. Cf. deuterostomes.

punctuated equilibrium The observation that long geologic periods with little change (equilibrium) in a species are interrupted (punctuated) by sudden change; eventually recognized as a pattern (cladogenesis) that often invokes a macroevolutionary process (species selection). Cf. quantum evolution.

quadruped With four legs; designed to comfortably walk on all four legs. Cf. bipedal.

quantum evolution Accelerated microevolution producing short bursts of rapid, adaptive changes between major groups. Cf. punctuated equilibrium.

radiometric dating A method of dating rocks or fossils that takes advantage of the constant rate of transformation of an unstable isotope to its more stable form.

realized niche Out of all resources available, the suite of resources actually utilized by an organism. Cf. fundamental niche.

regulatory gene A gene that manages parts of the genetic program, especially as it controls structural genes. Cf. structural gene.

replacement therapy The medical use of drugs to substitute for or replace missing or defective enzymes, hormones, or other critical metabolic chemicals. Cf. gene therapy.

reproductive isolating mechanisms (RIMs) Phenotypic features—morphological, physiological, behavioral—that prevent interbreeding.

respiration (cells) Within cells, the sequential breakdown of energy molecules through a series of metabolic pathways using oxygen and giving off carbon dioxide as a by-product.

respiration (whole organisms) Breathing, by lungs, gills, or other surfaces. The uptake into the blood of oxygen from air (e.g., lungs) or water (e.g., gills) with the usual release back of carbon dioxide.

RNA Ribonucleic acid; the single-stranded sequence of nucleotides, transcribed from a DNA template, that presides over the assembly of proteins (messenger RNA) or helps stabilize the process (ribosomal RNA).

seed A small plant capsule composed of a protective seed coat around a plant embryo and its supply of nutrients.

selection pressure The intensity of environmental factors, biotic and abiotic, acting upon the survival and reproductive success of members of a population.

selective agent The particular biotic or abiotic element in the environment that causes elimination of unsuited individuals within a population.

self-fertilization The exchange of gametes between reproductive parts of the same organism.

selfishness Biologically, a behavior that increases the fitness of the individual who exhibits the behavior and decreases the fitness of others. Cf. altruism.

sessile Attached; firmly affixed to a substrate.

sexual dimorphism Distinctive anatomical differences between males and females of the same species as a consequence of different secondary sexual characteristics.

sexual reproduction Reproduction in which haploid gametes fuse to produce a diploid organism. Cf. asexual reproduction.

sexual selection A type of natural selection wherein individuals of the same sex compete with each other for success in attracting a mate.

somatic cell Body cell. Cf. germ cell.

speciation The process by which new species arise. *See* biological species, morphospecies.

species–area relationship In general, the more inhabitable area (habitat) available, the more species; conversely, loss of area results in loss of species.

species selection The hypothesized culling process that acts directly on species, causing their differential survival. Cf. individual selection.

spontaneous generation The view that new organisms arise directly from nonliving, inanimate substances rather than from living ancestors.

sporophyte In plants, the phase in the life cycle where individuals are diploid, and spore-producing. Cf. gametophyte, alternation of generations.

stratigraphy A geological method of relative dating of rock layers, and hence the fossils they contain, by placing rock layers in chronological sequence from oldest to youngest, bottom to top.

structural gene A gene that makes products involved in building the phenotype. Cf. regulatory gene.

symbiosis Living together; two or more species in a co-evolutionary relationship. Cf. mutualism, commensalism.

sympatric Together; populations occupying or overlapping in the same geographic locations. Cf. allopatric.

synthesize Assemble; put together. Usually used to describe cellular processes that manufacture larger molecules from smaller chemical building blocks.

taxon A named group of related organisms.

tetrapods A taxonomic group including all amphibians, reptiles, birds, and mammals. Cf. quadrupeds.

time and energy budget The accounting of how an organism takes in limited resources in limited time and, in turn, how it advantageously allocates these resources amongst basic survival functions.

transcription The cellular process by which the DNA template codes for, or transcribes, RNA. Represented: DNA $\longrightarrow$ RNA. Cf. translation.

translation The cellular process by which messenger RNA codes for proteins. Represented: RNA $\longrightarrow$ proteins. Cf. transcription.

uniform extinctions Loss of species within a taxonomic group over geologically long periods of time; background extinctions. Cf. mass extinctions.

vascular tissue In plants, the conducting system of tubular cells that conveys water and minerals from roots to shoots, and manufactured sugars from shoots to roots.

vegetative growth In plants, the growth, replacement of lost parts, or asexual production of new plants by cell proliferation, simple mitotic division.

vertebral column The "backbone"; the series of supportive bony or cartilaginous elements along the length of the vertebrate body.

vertebrates A taxonomic group of animals, most of which possess a vertebral column.

vestigial structure Rudimentary structure. An underdeveloped structure that was functional in some ancestor but no longer participates in its previous role and is consequently reduced in prominence. Cf. atavistic structure.

warning coloration Conspicuous, easily seen coloration that serves the biological role of publicizing a dangerous organism; aposematic coloration.

Credits

Dedication page (top): © The McGraw-Hill Companies Inc./Barry Barker, photographer; **(bottom):** © The McGraw-Hill Companies Inc./Richard Weiss, photographer

Brief Contents: U.S. Geological Survey

Contents: © The McGraw-Hill Companies Inc./Barry Barker, photographer

Chapter 1

Opener: © Christopher Ralling; **1.1:** Image #326697 by permission of American Museum of Natural History Library; **1.2:** Historical Pictures/Stock Montage, Inc.; **1.3a:** Stock Montage, Inc.; **1.3b:** Bibliotheque Centrale MNHN Paris; **1.4:** By courtesy of the National Portrait Gallery, London; **1.5 a:** Journal of Researches into the Geology and Natural History, 1890

Chapter 2

Opener: U.S. Geological Survey; **2.3a-e:** National Park Service; **2.7:** Courtesy Lowell Carhart; **2.8:** Courtesy Dr. Rupert Wild, Staatliches Museum für Naturkunde, Stuttgart; **2.9:** © John D. Cunningham/Visuals Unlimited; **2.10a, b:** Courtesy Dr. David Taylor, North West Museum of Natural History, Portland, Oregon

Chapter 3

Opener: Mendelianum Musei Moravie, Brno; **3.1:** Mendelianum Musei Moravie, Brno; **3.13:** Library of Congress

Chapter 4

Opener: A. Chesley Bonestell Space Art Chronology, "Formation of the Earth's Continents"; **4.1:** Courtesy J. William Schopf

Chapter 5

Opener: © Charlie Ott/Photo Researchers, Inc.

Chapter 6

Opener: © John Conrad/Corbis; **6.4a:** © Corbis/Vol. 62; **6.4b:** © Corbis/Vol. 96; **6.4c:** © Corbis/Vol. 71

Chapter 7

Opener: © Corbis/R-F Website ; **7.5a,b:** © Michael Tweedie/Photo Researchers, Inc.; **7.8a:** From Albert & Blakeslee "Corn and Man," Journal of Heredity, 1914, v. 5, p. 511; **7.10:** Diane R. Nelson; **7.11:** © Tom and Pat Leeson/Photo Researchers, Inc.; **7.13a:** © The McGraw-Hill Companies, Inc./Barry Barker, photographer; **7.13b:** © The McGraw-Hill Companies, Inc./Richard Weiss, photographer

Chapter 8

Opener: Corbis Website; **8.3a,b:** © Bill Longcore/Photo Researchers, Inc.

Chapter 9

9.6: (photo 1): © John Shaw/Tom Stack & Associates, Inc. **9.6:** (photo 2): © Rob & Ann Simpson/Visuals Unlimited; **9.6:** (photo 3): © Suzanne L. Collins & Joseph T. Collins/National Audubon Society Collection/Photo Researchers, Inc.; **9.6:** (photo 4): © Stephen P. Lynch

Chapter 10

Opener: © The McGraw-Hill Companies/Barry Barker, photographer; **10.1:** © The McGraw-Hill Companies/ Richard Weiss, photographer; **10.2:** © N&C Photography/Peter Arnold; **10.3a:** © Michael Fogden/DRK Photo; **10.3b:** Robert & Linda Mitchell; **10.4:** The McGraw-Hill Companies/ Carlyn Iverson, photographer; **10.6:** © Diane R. Nelson; **10.7:** Dr. James L. Castner; **10.8a:** Ken Kardong; **10.8b:** © The McGraw-Hill Companies/Barry Barker, photographer; **10.11 a,b:** Lincoln P. Brower; **10.12 a-d:** Courtesy F. Harvey Pough; **10.13:** © Corbis/Vol. 53

Chapter 11

Opener: © The McGraw-Hill Companies/Richard Weiss, photographer

Chapter 12

Opener: © The McGraw-Hill Companies/Barry Barker, photographer; **12.1a:** © S. Osolinski/OSF/Animals Animals/Earth Sciences; **12.4:** © Roger Wilmshurst/The National Audubon Society Collection/Photo Researchers, Inc.

Chapter 13

13.1 b: © Stephen Krasemann/Photo Researchers, Inc.; **13.13:** Ken Kardong

Chapter 14

Opener: © The Cleveland Museum of Natural History; **14.2:** Thomas D. McAvoy/Time Life Pictures/Getty Images; **14.9:** © The Cleveland Museum of Natural History

Chapter 15

Opener: © Bossu Regis/Sygma/Corbis; **15.5:** Image # 15038 by permission of the American Museum of Natural History Library; **15.8:** Ken Kardong; **15.9a:** Image # 231604 by permission of the American Museum of Natural History Library; **15.9b:** © Kazuyoshi Nomachi/Photo Researchers, Inc.

Chapter 16

Opener: © Bettmann/Corbis; **16.11:** © Visuals Unlimited

Appendix I

A 1.2: Dr. Andrew Bajer; **A 1.3:** Claire A Hasenkampf/Biological Photo Service

Glossary: U.S. Geological Survey

Credits: © The McGraw-Hill Companies Inc./Barry Barker, photographer

Index: © The McGraw-Hill Companies Inc./Barry Barker, photographer

Index

Page citations in **boldface** refer to boldface terms in the text. Italicized *f* refers to figures, *t* to tables, and *b* to boxed material.